Techniques in Inorganic Chemistry

Techniques in Inorganic Chemistry

Edited by

John P. Fackler, Jr. • Larry R. Falvello

CRC Press
Taylor & Francis Group
Boca Raton London New York

CRC Press is an imprint of the
Taylor & Francis Group, an **informa** business

CRC Press
Taylor & Francis Group
6000 Broken Sound Parkway NW, Suite 300
Boca Raton, FL 33487-2742

Printed in the United States of America on acid-free paper
10 9 8 7 6 5 4 3 2 1

International Standard Book Number: 978-1-4398-1514-4 (Hardback)

Library of Congress Cataloging-in-Publication Data

Techniques in inorganic chemistry / editors, John P. Fackler, Jr., Larry Falvello.
p. cm.
Includes bibliographical references and index.
ISBN 978-1-4398-1514-4 (hardcover : alk. paper)
1. Chemistry, Inorganic--Technique. I. Fackler, John P. II. Falvello, Larry. III. Title.

QD152.3.T434 2011
546.028--dc22 2010019526

Visit the Taylor & Francis Web site at
http://www.taylorandfrancis.com

and the CRC Press Web site at
http://www.crcpress.com

Contents

Preface

Comments on Inorganic Chemistry from time to time has solicited and received manuscripts that describe modern techniques used by practicing inorganic chemists. Since there is considerable current interest in inorganic chemistry, with applications ranging from materials to biology and medicine, inorganic chemists need to be made aware of the scope and limitations of the techniques that are available for the study of their chemistry.

It was appropriate for us to group these articles, somewhat revised from the original published works, into a book form that could be used as reference or textual material for students of inorganic chemistry. The original articles were published as early as 2003, but most have been updated by the authors.

Diffraction has become the dominant tool today in inorganic chemistry since modern x-ray diffractometers have made it possible to determine three-dimensional structures using very small crystals of materials. Neutron diffraction has not achieved this level of availability to date, but new neutron sources, such as the Spallation Neutron Source (SNS) at Oak Ridge National Laboratory, have recently come on line and will greatly expand the availability of neutron techniques to the wider research community. With currently available computer power, powder diffraction also has become a successful analytical tool for structures of increasing complexity. Hence, the article by Clegg, "Current Developments in Small-Molecule X-Ray Crystallography," is a good general article on x-ray diffraction. Senior author Maschiocchi's article, "X-Ray Powder Diffraction Characterization of Polymeric Metal Diazolates," while focused on a particular class of materials, describes how *ab initio* x-ray powder diffraction (XRPD) methods are used today to determine structure. Finally, an excellent article by Piccoli, Koetzle, and Schultz, "Single Crystal Neutron Diffraction for the Inorganic Chemist—A Practical Guide," brings the reader up to date on this powerful structural technique.

We have chosen to follow these diffraction papers with an excellent reporting of modern computational chemistry by Tsipis, "Adventures of Quantum Chemistry in the Realm of Inorganic Chemistry." Density functional calculations are becoming commonplace in the community, again a result of the availability of modern computers.

Finally, two important spectroscopic approaches to the understanding of the chemistry of inorganic materials are described. The first, by senior author Macchioni, "NMR Techniques for Investigating the Supramolecular Structure of Coordination Compounds in Solution," is an up-to-date presentation of the important new tools in solution NMR studies. The chapter by senior author Reber, "Pressure-Induced Change of *d-d* Luminescence Energies, Vibronic Structure, and Band Intensities in Transition Metal Complexes," presents a special approach capable of giving considerable electronic structural information.

It is hoped that a knowledge of this selection of topics, described by experts for a more general audience, will provide the practicing chemist with a solid background in the array of techniques available in the researcher's toolkit.

John P. Fackler, Jr.
Larry R. Falvello

Contributors

Gianluca Ciancaleoni
Dipartimento di Chimica
Università degli Studi di Perugia
Perugia, Italy

William Clegg
School of Chemistry
Newcastle University
Newcastle upon Tyne, United Kingdom

Kari A. Frantzen
Département de Chimie
Université de Montréal
Montréal, Canada

Simona Galli
Dipartimento di Scienze Chimiche e Ambientali and CNISM
Università dell'Insubria
Como, Italy

John K. Grey
Department of Chemistry
University of New Mexico
Albuquerque, New Mexico

Thomas F. Koetzle
Intense Pulsed Neutron Source
Argonne National Laboratory
Argonne, Illinois

Etienne Lanthier
Département de Chimie
Université de Montréal
Montréal, Canada

Alceo Macchioni
Dipartimento di Chimica
Università degli Studi di Perugia
Perugia, Italy

Norberto Masciocchi
Dipartimento di Scienze Chimiche e Ambientali and CNISM
Università dell'Insubria
Como, Italy

Paula M. B. Piccoli
Intense Pulsed Neutron Source
Argonne National Laboratory
Argonne, Illinois

Christian Reber
Département de Chimie
Université de Montréal
Montréal, Canada

Arthur J. Schultz
Intense Pulsed Neutron Source
Argonne National Laboratory
Argonne, Illinois

Angelo Sironi
Dipartimento di Chimica Strutturale e Stereochimica Inorganica
Università di Milano
Milano Italy
and
Istituto di Scienze e Tecnologie Molecolari del CNR
Milano, Italy

Constantinos A. Tsipis
Laboratory of Applied Quantum Chemistry
Faculty of Chemistry
Aristotle University of Thessaloniki
Thessaloniki, Greece

Cristiano Zuccaccia
Dipartimento di Chimica
Università degli Studi di Perugia
Perugia, Italy

Daniele Zuccaccia
Dipartimento di Chimica
Università degli Studi di Perugia
Perugia, Italy

1 Current Developments in Small-Molecule X-Ray Crystallography

William Clegg

CONTENTS

X-ray crystallography is one of the most important and powerful methods for investigation of the structures of chemical compounds. It has undergone revolutionary changes within the last two decades, vastly increasing its speed, extending its scope, and improving its already excellent reliability. This article identifies some of the major advances, discusses current attitudes toward crystallography and uses of it, and considers current challenges and future prospects.

INTRODUCTION

The term *small-molecule crystallography* is really a poor one to describe this subject, since many chemical substances studied by x-ray crystallography are nonmolecular (particularly with the burgeoning interest in supramolecular coordination networks, microporous materials, high-temperature superconductors, and other ionic compounds), and others are far from small (including multinuclear arrays, oligopeptides, and polyporphyrin arrays, for example), but it is well established in chemical vocabulary. It serves as a convenient distinction from macromolecular crystallography, which focuses on the biological realm of proteins, nucleic acids, and viruses, and from noncrystalline diffraction techniques used in the investigation of materials such as plastics, muscle, fibers, and other polymers. It plays a vital role in modern chemistry, organic as well as inorganic, and probably ranks alongside nuclear magnetic resonance (NMR) and mass spectrometry as the three most important structural techniques, being less widely applicable than these other two in its requirement for a crystalline solid sample, but far more powerful in the richness of its detailed results when successful. The impact of crystallography in chemistry is immediately obvious in a cursory scan of research journals (especially in inorganic chemistry), as well as in depictions of structures in undergraduate textbooks, posters and other promotions of chemistry, and the Internet.

X-ray crystallography has seen enormous development since the discovery of x-rays in 1895, the demonstration of x-ray diffraction by crystals in 1912, and the application of this in the first crystal structure determination in 1913, almost a century ago. It has been involved in research leading to the award of numerous Nobel Prizes, and its results adorn the front covers of many research journals. A generation or two ago it was commonly regarded as difficult, slow and expensive, and a last resort when more "sporting" techniques had been exhausted, particularly by organic chemists (this attitude has not completely died out).

Developments in the techniques of crystallography over about the last three decades have brought about dramatic improvements in speed, range of application, and reliability, to the extent that many chemists now see it as the structural method of first choice. This change has probably been most marked in inorganic chemistry (and in the pharmaceutical industry, for different reasons), where the need for detailed geometrical results rather than basic connectivity and qualitative stereochemistry arises from the great flexibility of bonding interactions, variability of coordination geometry, and unpredictability of reaction outcomes.

Improvements in x-ray crystallography have been particularly rapid and marked in the last 10 to 20 years, especially in the technological development of diffraction data collection procedures. Within my own research career of nearly 40 years, the almost complete displacement of photographic and serial four-circle diffractometer techniques by various kinds of area detectors, the transformation of low-temperature data collection from heroic pioneering to routine use, the increasing availability of high-intensity x-ray sources, the move from time-shared mainframe computers and minicomputers such as the DEC VAX and Data General Eclipse to desktop and laptop PCs of far greater power, the progress from punched cards, paper tape, and reel-to-reel magnetic tapes to floppy disks, CD, DVD, and DAT media and the now ubiquitous flash disk or memory stick, the explosion of the Internet, and the availability of convenient computer graphics have been revolutionary and breathtaking—except that memories seem to be short, and all these modern advantages are simply taken for granted. "What's a four-circle diffractometer?" ask most young graduate research students in a regular intensive crystallography course I help to teach,[1–3] and x-ray cameras are museum pieces. I wouldn't dream of calling the 1970s "the good old days," of course, and would not wish to inflict visual estimation of photographic intensities on anyone, but unfortunately the positive developments have also brought with them some unwelcome changes in practice and attitude, to be discussed later.

SOME RECENT DEVELOPMENTS

Undoubtedly the most dramatic change in x-ray crystallography over the last 15 years has been the widespread introduction of area detectors for data collection. Previously, diffractometers measured diffracted beams serially one at a time, so that the time required for a data collection depended on the size of the structure; a larger structure gives a denser diffraction pattern, the number of independent reflections up to a given maximum Bragg angle being directly proportional to the volume of the asymmetric unit of the crystal structure, and hence approximately proportional to the number of symmetry-independent atoms. The sampling of the diffraction pattern only at the expected positions of Bragg reflections by the small detector also meant that it was necessary to determine the unit cell geometry and the orientation of the crystal on the diffractometer before the full data collection could proceed.

An area detector is effectively an electronic version of photographic film: more convenient, cleaner and faster to use, and reusable. Although area detectors were widely used in macromolecular crystallography already, and there were some image plate and TV-based systems in chemical crystallography, serious uptake began only in 1994, when the first commercial charge-coupled device (CCD) diffractometer became available, consisting of goniometer, detector, control system, and software in a complete package. The first shipped system was still in operation in Newcastle in 2008 (it has now been replaced), but in the intervening years CCD-based area detectors have become more sensitive and less expensive, and they now dominate the market.

They bring several major advantages. The most obvious is their ability to record many reflections simultaneously, considerably speeding up the process of data collection; data for a larger structure need not take longer than for a smaller structure, since it simply gives a higher density of reflections. At the same time, many symmetry-equivalent data are usually collected, which would multiply the data collection time with a serial detector, and this has benefits for overall data quality and its assessment, as well as for effective absorption corrections and more reliable space group determination. Since an area detector records the whole diffraction pattern (all of reciprocal space) and not just the Bragg reflections (the reciprocal lattice points), it is not actually necessary to establish the unit cell and crystal orientation before collecting data, and the unit cell determination is usually based on far more observed reflections than with a serial diffractometer, so it is more reliable and less subject to potential difficulties arising from weak subsets of data in cases of pseudosymmetry, superlattices, and other problems. Dealing with multiple diffraction patterns from twins and other imperfect samples is made far more tractable with an area detector, and the speed and sensitivity of the devices have greatly extended the scope of application for weakly diffracting materials and poor quality crystals. Sample screening is rapid (essential at synchrotron sources). Last but not least, teaching and training in crystallography is made easier by the immediate and clear visual display of diffraction patterns on a screen instead of lists of numbers.

A more gradual and less dramatic change, but also important, has been the development of facilities for low-temperature data collection. About 25 years ago, low-temperature devices were unreliable and temperamental beasts, often homemade or modified, and their use was relatively scarce. The most common type of device generates a cold gas stream by evaporation of liquid nitrogen and aims to maintain a constant temperature at the sample by monitoring and feedback control. Formation of ice from atmospheric moisture must be prevented, usually by a concentric dry sheath of nitrogen or air around the coolant gas stream, possibly with the help of other factors, such as a dry enclosure and avoidance of drafts. Such devices have become much more reliable, and their acquisition and running costs are moderate, so they can now be used routinely and run continuously for months at a time. Temperatures down to about 80 K, just above the boiling point of nitrogen, can be readily achieved and maintained. More recently, helium-driven devices have become available and can reach even lower temperatures; obviously, the costs are considerably higher and the technical requirements are more demanding. Some newer devices also run on nitrogen, but extract it from the air, avoiding any need for a supply of liquid nitrogen.

The main advantage of low-temperature data collection is the significantly reduced thermal motion of atoms in the sample, leading to greater intensities, especially at higher Bragg angles, and a more precise structure. Thermally sensitive samples are obviously better handled this way, and flash cooling of oil-mounted crystals is a much more convenient way of dealing with air-sensitive compounds than sealing them in capillary tubes; even highly sensitive main group organometallics can be successfully studied without recourse to those cumbersome old methods. The use of low temperatures also helps in the modeling of structural disorder, by giving a clearer resolution of static disorder components and reducing dynamic disorder. It is probably the most cost-effective improvement that can be made in basic diffraction data collection, and ought to be standard practice. The only serious drawback is the occasional experience of an undesirable phase transition on cooling, especially if this involves a major structural change and results in destruction of the single-crystal specimen.

Low-temperature data collection is essential in some specialist areas of crystallography, such as in charge-density analysis and, obviously, in the study of temperature-dependent phenomena such as spin-crossover transitions and superconductivity. It has also made it possible to investigate the solid-state structures of compounds that are liquids or gases at normal room temperature, such as small-molecule derivatives of several main group elements.[4]

The basic laboratory x-ray tube has changed little in its fundamental design since the early days of crystallography, though modern tubes are more compact, safer, and better engineered than the originals. They remain highly inefficient devices, since that is the nature of the physical principle on which they operate, the conversion of electron kinetic energy into x-rays through impact on a metal target with resultant ejection of photoelectrons and secondary emission. Rotating-anode tubes spread the thermal load on the target and hence permit higher electron currents and generate more x-ray intensity, but the gain is usually no more than about an order of magnitude, as it is with other recent developments in electron-impact x-ray generation. Further intensity gains can be achieved by various focusing optics systems, whereby more of the emitted x-rays are captured and concentrated on the sample instead of going to waste in unwanted directions. The aim of all these attempts is to bring more x-ray photons per second on to the sample crystal, so that the diffraction intensities are greater, either improving the signal or reducing the time for a measurement.

A much greater increase in intensity, of several orders of magnitude in most cases, is achieved by setting up crystallographic diffraction equipment on a synchrotron storage ring. These, however, are major national and international facilities and the venture is a considerable undertaking. A number of technical difficulties have to be addressed, including the polarization properties of synchrotron radiation, which involves building diffractometers on their sides relative to the standard laboratory arrangement, and the variation in intensity from storage rings that are not operated in a constant top-up mode; each synchrotron source operates its own more or less bureaucratic system of user access, authorization, training, and safety. Nevertheless, the potential advantages of exploiting such sources are enormous and well worth the effort.[5] The dedicated small-molecule crystallography diffraction facility of Station 9.8 at the Synchrotron Radiation Source (SRS) at Daresbury Laboratory, UK, has been extremely successful since its construction more than ten years ago, and was oversubscribed by user applications through most of its operational lifetime (the SRS closed in July 2008).[6] Since 2001 we have operated a national service there (and subsequently moved to Diamond Light Source, its much more powerful successor) for UK academics eligible for support by the UK Engineering and Physical Sciences Research Council (EPSRC), who fund the service, so that the experimental work does not require cumbersome formal applications months in advance by individual users for occasional use, and so that it is carried out efficiently by trained experts. Up to about 12 data sets per day have been collected in this way. The samples, originating mostly from university chemistry departments across the UK, are not amenable to study anywhere else in the country; in many cases, crystals giving barely visible diffraction patterns with long exposures on a rotating-anode system with advanced mirror optics (probably the most powerful small-molecule crystallography facility in the country, at Southampton University) have produced excellent data within a couple of hours at the SRS,[7] and even greater enhancement and overall productivity is expected at Diamond beam-line I19 when it is fully operational (some of its facilities are still under development at the time of writing).

Small-molecule crystallography facilities at synchrotron sources elsewhere in the world have generally been less successful, mainly because they tend to be shared with other techniques (powder diffraction, spectroscopy, etc.) on the same beam-line, and so are not optimized for this particular use, though dedicated beam-lines are being developed at a number of centers, some of them modeled on the success of SRS Station 9.8.

While the high intensity is the most obvious advantage of synchrotron radiation for crystallography, there are also potential benefits in the ability to select the x-ray wavelength from a wide range in most cases. Very short wavelengths are useful for reducing effects such as absorption and extinction (usually at the expense of intensity), and are particularly important for charge-density work, and they can penetrate effectively through sample containers such as high-pressure cells and chambers for controlled atmospheres and *in situ* reaction studies. Alternatively, effects such as anomalous scattering can be maximized and exploited for the determination of absolute configuration of chiral structures.

There are few experimental methods available for the determination of absolute rather than relative configuration. With an appropriate combination of sample composition and quality and x-ray wavelength, establishing the absolute configuration is straightforward and highly reliable with crystallography. Without going into details of the theory, anomalous scattering is the small but significant phase shift that can occur when an atom scatters x-rays with a wavelength close to one of its absorption edges; the effect is negligible for atoms lighter than about Si or P with MoKα radiation, which is most widely used in chemical crystallography, but generally increases for heavier atoms and higher wavelengths. Anomalous scattering by a particular element can be maximized by wavelength tuning at a synchrotron source, though in many cases it is quite adequate for the determination of absolute configuration with standard laboratory facilities. In the presence of significant anomalous scattering, the diffraction pattern of a noncentrosymmetric crystal structure and that of its inverse are not identical, and one of these structures will fit the observed diffraction pattern better than the other. In the case of enantiomerically pure chiral materials, this corresponds to selecting the correct absolute configuration. Computational and experimental techniques have been developed for optimizing the procedure and assessing the statistical significance of the result.[8–10] It has obvious applications in current research emphases such as asymmetric catalysis and chiral solid-state networks.

Crystallography makes extensive use of computers, from the control of diffractometers and the processing of the measured data, through the solution and refinement of structures, to the graphical, numerical, and statistical representation and interpretation of results and their archiving and publication. The rapid development in personal computing power and resources has been very important for the success of new technologies in crystallography, especially with the huge quantity of data generated by area detectors. Powerful, affordable, and easily used computer graphics have played an important role in solving and refining complex structures and in understanding them once they are complete; such facilities were an expensive luxury a generation ago. Inexpensive magnetic and optical storage media are also vital if accumulated data and results are not to be lost.

By contrast, the real costs of crystallographic software are not tumbling. There is much time and effort involved in programming new ideas and theories, and under pressures of research assessment, quality assurance measurements, and economic difficulties in science subjects, academics are no longer so easily able to carry out such work as a personal interest and make their programs freely available to others. Software, especially in integrated packages, is increasingly being offered commercially rather than in the public domain, and much of the widely used software is showing its age as its originators retire from active research, with few younger researchers able and willing to take over (except in macromolecular crystallography). Commercial software seems to have a greater focus on presentation and automation than on flexibility and user control. These are unwelcome developments, and are being addressed to some extent by a British project with research council funding, aiming to produce open-access modern crystallographic software.

Crystallographers are fortunate in dealing with data and results that are ideally suited to systematic study, classification, and standardization in the form of electronic databases. The maintenance and development of these have been much aided by the adoption of some internationally agreed upon standard formats for crystallographic data and results, of which the Crystallographic Information File (CIF) has had the biggest impact in terms of information exchange, storage, retrieval, and publication.[11] The automatic computer validation of crystallographic results is readily achieved and is a major advantage in submitting them for publication in mainstream chemistry journals as well as specialist crystallographic ones, some of which now operate in an entirely electronic mode from submission, through validation and peer review, to proof checking and publication, achieving very short publication times and reducing both effort and the risk of errors.

Crystallographic databases have become not only the primary exhaustive repository of published crystal structures, but also research areas in their own right, through data mining, statistical analysis, and the discovery of structural patterns and trends. Their use has led to advances in our understanding of such topics as hydrogen bonding, conformational preferences, and coordination geometry.[12–14]

A welcome feature of small-molecule crystallography these days, though one that could be developed further, is a measure of cross-fertilization of ideas with related disciplines. As the subject has expanded its borders into larger and more complex structures through modern technological enabling, it has found more common ground with macromolecular crystallography than previously. Equipment used by the two communities is often very similar now, and it is interesting to note that designs for macromolecular and small-molecule diffraction at Diamond have much in common. Problems of low data resolution, extensive disorder of solvent and peripheral groups, and the mounting of very small crystals for synchrotron study in small-molecule crystallography have seen approaches and solutions imported from established macromolecular crystallography techniques. The successful investigation of smaller and smaller crystals has blurred the distinction between single-crystal and powder diffraction, and these communities have increasingly shared computational ideas and methods. Information from other solid-state techniques, such as NMR-derived interatomic distances, has been incorporated as restraints in difficult structure refinements. Furthermore, crystallography and theoretical calculations (molecular modeling and quantum mechanics) have made profitable use of each other's results, especially in charge-density studies and through idealized fragment geometries for fitting to electron density maps or as starting points derived from databases for theoretical calculations.

ATTITUDES, USES, AND ABUSES

The enormous advances made in crystallography in recent years have inevitably brought with them some problems. Among these, perhaps the most significant are some consequences of the speed and power of the technique, together with a growing "black box" mentality of automation and a frequent lack of understanding of some fundamental principles. To a large extent these problems are fueled by the promotion of commercial systems that can supposedly be used (and very often are, without trouble) by nonexperts, by funding restrictions, and by the growing use of monitoring and performance indicators, especially in academic research, with their impact on ambitions and working relationships. Also, to a large extent, crystallographers are the victims of their own success.

Modern diffractometer systems can generate good quality data sets from suitable crystals within hours, or even in minutes. In many cases, the determination of the unit cell and space group, and the solution and refinement of the structure, proceed smoothly and largely automatically. For a well-defined discrete molecular structure without disorder or other problems, and with no unusual features, the whole process is rapid and simple, and requires no special skill or training. This is push-button crystallography at its most effective. One eminent crystallographer infamously said a few years ago that crystallography is now so easy that monkeys can be trained to do it, and many chemists might agree. This may be true for straightforward structures, but a lack of real training and a black-box mentality, together with too much haste and a lack of attention to detail, lead all too often to poorly determined crystal structures in which small but significant problems are overlooked or not understood. This may produce structures that are qualitatively correct, but with errors in details of geometry through failure to recognize structural disorder or twinning, the consequences of anomalous scattering effects, poor treatment of encapsulated solvent molecules, incorrect symmetry, or misassigned atom types, among other errors. I have seen many such instances from eight years' experience (nearly 20,000 papers!) as joint section editor of *Acta Crystallographica, Section E: Crystal Structures Online*, a purely electronic crystallographic journal that specializes in rapid publication while maintaining high standards, and have come across numerous published results, ranging from poor to outrageous, in highly reputable international chemistry journals, where the work has clearly been done badly and the errors have not been spotted by the authors, referees, or editors. Some of these are subsequently addressed and corrected, particularly where space groups have been wrongly assigned, which may have disastrous consequences for molecular geometry when an inversion center has been overlooked.

I once had to reject a submitted paper that reported a complex of a highly unusual terminal one-coordinate copper atom, which was almost certainly a bromine atom (the atom bonded to it was also wrongly identified), and other crystallographic referees and editors have reported similar experiences. More recently, an organic structure that would have made the front cover of *Science* or *Nature*, had it been correct, was found, on careful investigation, almost certainly to have fluorine atoms rather than hydroxy groups, and a nitro group rather than a carboxylic acid function; the correct structure was, in fact, already known, published, and in the Cambridge Structural Database. The purely crystallographic report contained no analytical or spectroscopic characterization and the authors (who will remain anonymous) had to be persuaded that x-ray crystallography is a very poor technique for chemical elemental analysis. Some cases of mistaken identity have made it into publication and have been subsequently recognized, including nitramine ligands[15] that were actually rather less interesting acetates,[16] and the spectacular case of the amazing compound $[ClF_6]^+[CuF_4]^-$, correctly identified later as the rather mundane $[Cu(OH_2)_4]^{2+}[SiF_6]^{2-}$.[17] Fortunately, such errors rarely get all the way into publication, because of the inherent self-checking character of much of crystallography and the availability of programs for thorough testing of results. The best known of these is PLATON,[18] which usually generates an impressive string of warning messages if a structure is wrongly determined or of relatively low precision.

Another, quite different, kind of mistaken identity is the case, far from unknown, in which the crystal structure proves to be unexpected. Although this usually indicates that the reaction has proceeded in a different way from that planned or desired, it sometimes turns out to be a minor product (or residual starting material) in the sample. Failure to recognize this may lead to incorrect deductions about the synthesis. This is another reason why many chemical journals refuse to accept a crystal structure as the only evidence for the identity of a product. In order to demonstrate that the selected single crystal is representative of the bulk material, there

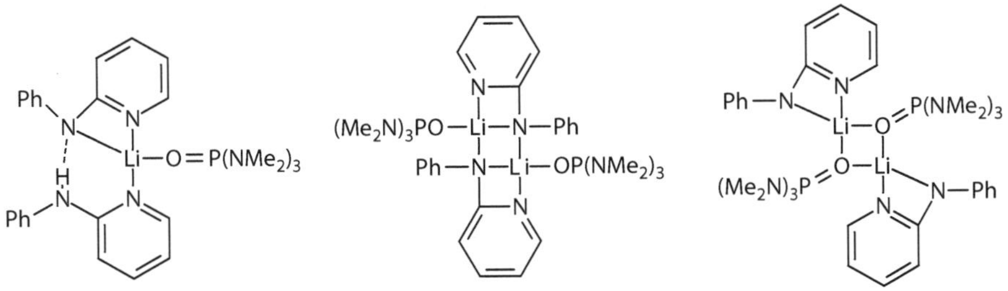

1 2 3

FIGURE 1.1 Three distinct compounds obtained in a single reaction.

should also be one or more of chemical analysis, appropriate spectroscopic results, and comparison with a powder diffraction pattern. A good illustration of this, from some 25 years ago, is one of the first results of my collaboration with Ron Snaith, then at Strathclyde University, in lithium chemistry. Crystals were obtained of a product from a 1:1:1 reaction of 2-anilinopyridine (PhNH-2-Py), ^{n}BuLi, and HMPA [$OP(NMe_2)_3$] and were mounted in Lindemann capillary tubes because of their air sensitivity (we were not routinely using low-temperature facilities at that stage). The structure of a selected crystal proved to be that of a 2:1:1 product, (PhNH-2-Py)(PhN-2-Py)Li(HMPA), **1**. Because of good spectroscopic, analytical, and cryoscopic evidence that the major product really did have a 1:1:1 stoichiometry, other crystals were examined and a different structure was found. Interestingly, the crystal structure was found to contain two different isomeric dimers, **2** and **3**, one with bridging (and simultaneously chelating) amido ligands and the other with bridging HMPA. The minor product **1**, which was actually present in only small quantities but formed the best crystals, could subsequently be prepared in high yield from a reaction with 2:1:1 stoichiometry of reagents.[19,20]

The advances in crystallography outlined earlier mean that we are able to tackle much more difficult structural problems now than a generation ago; the data collection and computing capabilities are available. However, this is precisely where expertise and time are needed and where automatic approaches fail. We have found with our extensive synchrotron work, tackling samples that come to us as a last resort, that they often demand many hours of painstaking work to resolve problems of pseudosymmetry, extensive disorder or twinning, and the slow emergence of structural details (especially in cases where there are several molecules in the asymmetric unit of the structure), requiring careful use of advanced tricks available in modern software—no room for trained monkeys here! In the final publication of such results, readers will often be quite unaware that the work has required orders of magnitude more time and effort than other structures churned out in a spare couple of hours while thinking mainly about something else. There is often little correlation between the effort involved and the chemical impact and interest of the result.

Sadly, such effort is sometimes quite unnecessary. On the "garbage in, garbage out" principle, a poor quality crystal is usually going to result, even after much hard work, in a structure of limited precision. Although there are certainly many times when this is the only practical approach, the whole difficulty can often be avoided by obtaining better crystals, either through appropriate crystallization techniques or through the synthesis of a closely related compound. Even a small improvement in crystal quality can make a big difference in the data collection and structure determination. On a number of occasions, I have struggled with a marginal data set, but a subsequent new batch of crystals has led to a simple structure determination from much improved data; a little extra effort and time in sample preparation can save much agony later.

What is meant by "appropriate crystallization techniques"? The usual preparative chemist's approach is with the principal objectives of purity, high yield, and speed, rather than with crystal size and quality, and a microcrystalline powder of mediocre quality is often the result. For single-crystal diffraction, we need only one single crystal of suitable size, so yield is an unimportant issue. A suitable size may be roughly defined as somewhere between a typical grain of salt and a grain of sugar, both familiar household substances, though considerably smaller crystals, down to micron dimensions, can often be tackled with high-intensity x-ray sources, as mentioned earlier. The key to good crystal quality is usually slow crystallization, whether this occurs by solvent evaporation, cooling, or the gradual diffusion of two liquids together to reduce solubility (this may occur at a direct liquid interface, or via the vapor phase). Detailed descriptions of methods and practical tips for their use are available in various references.[2,3,21–24]

In this context it is worth noting that changing the crystallization method or conditions may lead to a different polymorph or solvate (in some instances, better crystals may also be obtained by making a different chemical derivative, such as by changing an unimportant substituent or a counter-ion). Polymorphism, a topic recognized as highly important in the pharmaceutical industry for financial, regulatory, legal, and medical reasons, is an area of research in its own right, and has been a great beneficiary of the huge increase in speed of crystallography. Its existence should not be overlooked, however, by chemists in general, who need to recognize that a crystal structure of a compound they have prepared is not necessarily *the* crystal structure; other forms may be possible, in which there are different conformations, different details of geometry, or even grossly different structural features, such as overall coordination geometry or degree of association of molecular units in monomers, dimers, up to polymers.

Mention of polymorphism and different solid-state structural forms and isomers is a reminder of the possible consequences of unrecognized problems such as disorder, which can produce unfortunate artifacts if it is not correctly treated. At least some of the examples of "bond-stretch isomerism" discussed vigorously a number of years ago were unreal, resulting from substitutional disorder of ligands in metal complexes, especially involving partial replacement of different halogen atoms, which have different covalent radii.[25]

Among the important and valuable tools available to the crystallographer in dealing with difficult structure refinements are flexible and powerful constraints and restraints that can be applied to molecular geometry features (such as target bond lengths and angles, ideal ring shapes, and the imposition of planarity on groups of atoms) or to atomic displacements (approximating isotropic behavior, rigid bonds, and various aspects of similarity for atoms that lie close together in disordered fragments). Like tools in any other trade, they need to be used appropriately, preferably with suitable training and understanding. Inappropriate use can lead to distortions in structural results and, perhaps most dangerously, the imposition of preconceived ideas on the structure. More than once I have seen authors of manuscripts draw conclusions from the observed degree of planarity around an atom (especially involving NH groups in organic molecules or ligands), after they have actually imposed this planarity through constraints in the refinement! This is usually blatant (though it can escape unobservant referees), but other cases may be more subtle, leading to unjustified conclusions.

Even when data are successfully obtained from a good quality crystal and the structure has been correctly solved and appropriately refined, mistakes can still be made (frequently by "trained monkeys," but also sometimes by experts) in interpreting the results. Some of these mistakes are due to a failure to understand the meaning of precision measures. Virtually every numerical result obtained from crystallography comes with a standard uncertainty (also known as estimated standard deviation (esd), but the other term is now preferred); these may be rigorously derived by a proper statistical analysis, or they may be estimates made by computer programs or users. One common belief is that crystallographic standard uncertainties are over-optimistic and should always be inflated (multiplication by 3 is frequently recommended) before any conclusions are drawn about the significance of results, such as differences in bond lengths. While there is a basis of truth in this view (many uncertainties are estimated on the assumption of only random and no systematic errors, though this is less true than it used to be in the statistical treatment of diffraction intensity data), the real reason for scaling up crystallographic uncertainties lies in the fact that they are supposed to be estimates of standard deviations of the results (i.e., estimates of what the standard deviation would be for the complete distribution of results obtained if the experiment were to be repeated many times). In making comparisons of results, such as deciding whether two or more bond lengths are significantly different or not, proper significance tests should be made at appropriate confidence levels, based on assumptions of normal distributions. The standard statistical procedures use Student's *t*-test, as described in all basic statistics and analytical chemistry textbooks. Such tests multiply estimated standard deviations by factors depending on sample sizes and on the desired confidence levels, and lead to confidence intervals (conventionally written with ± symbols, e.g., 1.423 ± 0.009, different from the crystallographic notation with parentheses, e.g., 1.423(3), where these parentheses enclose the standard uncertainty in the last quoted figure, here 0.003). The multiplication of crystallographic uncertainties by 2 or 3 is thus a conversion to a rough confidence interval at the 95% or 99% confidence level, as commonly used for significance testing. In my experience, many chemists do not really understand what the numbers in brackets mean, or have only a vague idea ("It's the error in the bond length"—to which I then ask, "If it's an error, why don't we correct it?"). Consequently, inappropriate statements are made about the supposed significance and reality of small differences in bond lengths and angles, or about the deviation of a group of atoms from a plane, often with complicated arguments to provide reasons for the differences, which actually have no statistical significance at all.

Other mistakes made in interpretation arise from a failure to recognize connections between atoms in the asymmetric unit of a structure (the smallest repeat unit, a fraction of the unit cell, from which the complete crystal structure is generated by symmetry operations of the space group) and those in adjacent symmetry-related units. These connections may be covalent, hydrogen bonding, etc., just the same (and just as important chemically and structurally) as those within the asymmetric unit, the choice of which is quite arbitrary when it is not a distinct single connected molecule. Thus, dimerization or other oligomer or polymer formation by symmetry extension of the asymmetric unit may not be recognized, and may lead to inappropriate use of terms such as *intramolecular* and *intermolecular*. This can have major impact on the assessment of coordination geometry, bridging ligands, polymeric network architecture, hydrogen bonding patterns, and other important structural features. Experienced use of modern computer graphics and other structure analysis programs is vital here.

The final problem resulting from fast modern crystallographic techniques is the fact that many results never get into the public domain. For straightforward structures, formal traditional publication takes far longer than the complete experiment and calculations. Failure to publish results may be caused by the sheer volume of results obtained, from time pressures and the effort involved in publication, from the need (real or imagined) to carry out further related studies, from the fact that the results were not as expected, from patent or other commercial considerations, or from a variety of funding, personal, and other reasons, good or bad. It has been estimated by some prolific research groups that perhaps 80% or more of fully refined crystal structures remain unpublished, and the stockpile is growing fast. Not only is the work done largely wasted (and, many argue, a misuse of research funding), but the community at large is deprived of the benefits (which may be quite unpredictable, through future database analysis looking for structural trends and patterns), those who have carried out the work receive no recognition (particularly unfortunate for those at the beginning of their research career), and there is a danger that someone else will waste yet more time redetermining (and probably also not publishing) the same structure unnecessarily.

CURRENT CHALLENGES AND FUTURE PROSPECTS

Perhaps predicting the future in this subject really is a case of crystal gazing, but we can be reasonably confident about some likely developments. It will probably not be long before we have the next generation of x-ray detectors in routine use. They will be larger than current CCD detectors (for which small size has always been a major disadvantage), probably more sensitive, faster in operation, and (eventually, but not at first) cheaper.[26] X-ray source intensities will be further enhanced, both in the laboratory and at synchrotron facilities, and x-ray free-electron lasers will come along later, though there is still much to be done in this area to make the dream a reality. Computers will continue to increase in power while shrinking in size and cost, of course, but there is unlikely to be anything like the same scale of development in crystallographic software. A major challenge is the attraction of keen and able young scientists into crystallography, and their training and development.

The production of good quality crystals remains a largely unpredictable process, but techniques of multiple screening and automation could be adapted to some extent from macromolecular crystallography. Experiments in robotics for data collection are still rather limited in small-molecule crystallography, and there is scope for much more development here.

Developments in x-ray sources and detectors, speeding up data collection, are beginning to make various kinds of real-time and time-resolved crystallography possible. These include studies of excited states in so-called photocrystallography (where the pulsed nature and time structure of synchrotron radiation can be exploited to make rapid interleaved measurements of diffraction patterns from ground and excited states of molecules in crystalline solids),[27] and real-time monitoring of solid-state reactions.

Low-temperature data collection is now routine down to liquid nitrogen temperatures, and it is becoming easier and more economical to achieve even lower temperatures reliably. Still more adventurous, but undergoing rapid development, is data collection at high pressures, mainly using diamond anvil cells, though gas pressure cells are also being explored.[28]

Area detectors have opened up the feasibility of tackling major structural problems such as twinning, in which two or more diffraction patterns occur simultaneously. Even more challenging are incommensurate structures, for which not only careful experimental measurements but also special refinement techniques are needed.[29]

Particular difficulties in structure determination, requiring expertise and advanced software tools, include major structural disorder and structures with several molecules in the asymmetric unit; these often display tricky features such as pseudosymmetry. In the worst cases of disorder, individual atomic sites can not be modeled, and there is a need for group and whole molecule orientationally averaged scattering factors, which are not widely available in the most commonly used structure refinement programs.

Success with smaller and smaller crystals, down to micron sizes in favorable cases, has blurred the boundary between single-crystal and powder diffraction as techniques. There is a need for methods of structure refinement that can sensibly combine, with appropriate weighting, data from a range of techniques, not only x-ray single-crystal and powder diffraction, but also neutron scattering, some spectroscopic methods, and computer modeling studies. Crystal structure prediction is still successful in only a limited number of cases, and this is tied up with questions of polymorphism.

The sheer volume of crystallographic results is a challenge for the curators of databases, the structure, contents, maintenance, and dissemination of which will need some radical changes if a greater percentage of structures reach the public domain. The publication (in its widest sense) of crystal structures is currently one of the biggest challenges facing us, and various electronic technologies, such as grid developments and local repositories, will have to be developed and embraced (there are pressures from funding agencies for this currently), with proper recognition of the provenance, reliability, and significance of the published results.

Small-molecule crystallography has undergone steady development and various revolutions in its century of history. In the future it is certainly not going to stand still.

REFERENCES

1. Blake, A. J., Clegg, W., Howard, J. A. K., Main, P., Parsons, S., Watkin, D. J., and Wilson, C. 2002. *Z. Kristallogr.*, 217, 427–428.
2. Clegg, W., Blake, A. J., Gould, R., and Main, P. 2001. *Crystal Structure Analysis: Principles and Practice*. Oxford: Oxford University Press.
3. Blake, A. J., Clegg, W., Cole, J. M., Evans, J. S. O., Main, P., Parsons, S., and Watkin, D. J. 2009. *Crystal Structure Analysis: Principles and Practice*. 2nd ed. Oxford: Oxford University Press.
4. Goeta, A. E. and Howard, J. A. K. 2004. *Chem. Soc. Rev.*, 33, 490–500.
5. Clegg, W. 2001. *J. Chem. Soc. Dalton Trans.*, 3323–3332.
6. Cernik, R. J., Clegg, W., Catlow, C. R. A., Bushnell-Wye, G., Flaherty, J. V., Greaves, G. N., Burrows, I., Taylor, D. J., Teat, S. J., and Hamichi, M. 1997. *J. Synchrotron Rad.*, 4, 279–286.
7. http://www.ncl.ac.uk/xraycry
8. Flack, H. D. 1983. *Acta Crystallogr. Sect. A*, 39, 876–881.
9. Flack, H. D. and Bernardinelli, G. 2001. *J. Appl. Crystallogr.*, 33, 1143–1148.
10. Thompson, A. L. and Watkin, D. J. 2009. *Tetrahedron Asymmetry*, 20, 712–717.
11. Brown, I. D. and McMahon, B. 2002. *Acta Crystallogr. Sect. B*, 58, 317–324.
12. Belsky, A., Hellenbrandt, M. Karen, V. L., and Luksch, P. 2002. *Acta Crystallogr. Sect. B*, 58, 364–369.
13. Allen, F. H. 2002. *Acta Crystallogr. Sect. B*, 58, 380–388.
14. Allen, F. H. and Taylor, R. 2004. *Chem. Soc. Rev.*, 33, 463–475.
15. Öztürk, S., Akkurt, M., Demirhan, N., Kahveci, B., Fun, H.-K., and Öcal, N. 2003. *Acta Crystallogr. Sect. E*, 59, i107–i109.
16. Ferguson, G. and Glidewell, C. 2003. *Acta Crystallogr. Sect. E*, 59, m710–m712.
17. von Schnering, H. G. and Vu, D. 1983. *Angew. Chem. Int. Ed. Engl.*, 22, 408.
18. Spek, A. L. 2003. *J. Appl. Crystallogr.*, 36, 713.
19. Barr, D., Clegg, W., Mulvey, R. E., and Snaith, R. 1984. *J. Chem. Soc. Chem. Commun.*, 469–470.
20. Barr, D., Clegg, W., Mulvey, R. E., and Snaith, R. 1984. *J. Chem. Soc. Chem. Commun.*, 700–701.
21. Jones, P. G. 1981. *Chem. Br.*, 222–225.
22. Clegg, W. 2003. Crystal Growth Methods, in *Comprehensive Coordination Chemistry II*, ed. A. B. P. Lever, 579–583. Vol. 1. Amsterdam: Elsevier.
23. http://www.xray.ncsu.edu/GrowXtal.html
24. http://www.cryst.chem.uu.nl/growing.html
25. Parkin, G. 1993. *Chem. Rev.*, 93, 887–911.
26. Kraft, P., Bergamaschi, A., Broennimann, Ch., Dinapoli, R., Eikenberry, E. F., Henrich, B., Johnson, I., Mozzanica, A., Schlepütz, C. M., Willmott, P. R., and Schmitt, B. 2009. *J. Synchrotron Rad.*, 16, 368–375.
27. Cole, J. M. 2004. *Chem. Soc. Rev.*, 33, 501–513.
28. Katrusiak, A. 2008. *Acta Crystallogr. Sect. A*, 64, 135–148.
29. http://www-xray.fzu.cz/jana/jana.html

2 X-Ray Powder Diffraction Characterization of Polymeric Metal Diazolates

Norberto Masciocchi, Simona Galli, and Angelo Sironi

CONTENTS

Abstract: Polymeric metal diazolates typically appear as insoluble and intractable powders, the structure of which can only be retrieved by the extensive use of *ab initio* x-ray powder diffraction (XRPD) methods from *conventional* laboratory data. A number of selected examples from the metal pyrazolate, imidazolate, pyrimidin-2-olate, and pyrimidin-4-olate classes are presented, highlighting the specific crystallochemical properties, material functionality, and methodological aspects of the structure determination process. Linear and helical one-dimensional polymers, layered systems, and three-dimensional networks are described, with particular emphasis on polymorphism and on their thermal, optical, magnetic, and sorption properties. A brief outline of the method, as it has been tailored in our laboratories during the last two decades, is also offered.

Keywords: Powder diffraction, coordination polymers, metal diazolates, polymorphism, crystal structure

INTRODUCTION

Knowledge of the molecular structure, in whatever aggregation state, is at the basis of the comprehension of physicochemical properties, such as stability, inertness, magnetism, and optics, and therefore can be successfully employed to tune the mentioned properties when tailored systems are sought. The amount of structural information gathered in the last two decades by conventional x-ray single-crystal techniques has allowed the estimation of important stereochemical features (geometrical parameters) as well as the interpretation of energetic barriers, the assessment of the correct chiralities, the development of bonding theory, and above all, the discovery of new classes of compounds, which chemical analysis or spectroscopy alone could not have revealed. Thus, single crystals (studied either by x-ray or neutron diffraction) are, and will continue to be, the primary source of this information.

However, many materials only appear as polycrystalline species, and their structural characterization must rely on powder diffraction data, with the obvious difficulties inherent in the partial and complex sampling of the scattered intensity of a 3D reciprocal lattice squeezed onto a 1D space.[1] Only in the last two decades have instrumentation and software for x-ray powder diffraction (XRPD) analysis reached a level of quality permitting the *ab initio* structural analysis of several hitherto unknown structures of inorganic, organic, and organometallic species, all together adding to less than a mere 0.5% of the deposited structures Cambridge Structural Database (CSD) and Inorganic Crystal Structure Database (ICSD). Indeed, while refinement of predetermined models (of a few independent atoms) was already within the structural chemist's hands after the Rietveld and the so-called two-stage methods were proposed for x-rays in the eighties, it was not until recently that more complex structures could be studied by XRPD, opening the way to the crystallographic analysis of several new classes of compounds.

About 15 years ago, our group began to be interested in the chemistry and structural characterization of several metal complexes containing polydentate aromatic nitrogen ligands for their specific reactivity toward simple heterocumulenes and, later, toward organic moieties catalytically transformed by these (typically polynuclear) systems. During these studies, we learned of the existence of yet uncharacterized intractable species, which opened the way to an extensive project in which

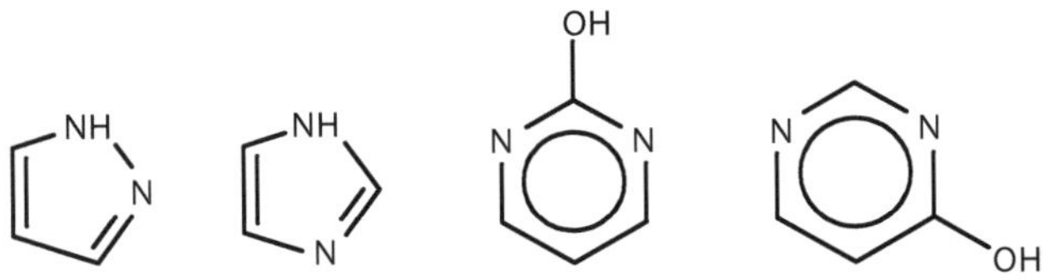

CHART 2.1 Schematic drawing of the heterocyclic ligands, in their neutral forms.

several polymeric compounds, of different nature, topology, and physical properties, were prepared and fully characterized. Their brief structural description, reported in the following paragraphs, will be coupled to the most significant findings, which span from surprisingly high thermal inertness, to polymorphism, intercalation, sorption, and magnetic and nonlinear optical properties.

Worthy of note, all these systems have been characterized by XRPD on conventional laboratory data, showing the viability of the method for the determination of (otherwise inaccessible) structural features and, in several cases, even of the correct stoichiometry. The amount of work reported in the following pages covers the last 15 years and nicely shows that a *complete* class of homologous (insoluble, but polycrystalline) species (in this case, metal diazolates, see Chart 2.1) could be successfully studied by XRPD, opening the way to new and important chemistry, initially unforeseen, until the stereochemical rules and the proper engineering of the materials were devised.

Note: Shortened names will be used throughout this paper, such as pz for pyrazolate, im for imidazolate, and X-pymo for pyrimidin-X-olate (X = 2, 4).

BRIEF DESCRIPTION OF THE *AB INITIO* XRPD TECHNIQUE, AS IMPLEMENTED AND DEVELOPED IN OUR LABORATORIES (1993–2009)

In the last 15 years we have been active in adapting, and extensively using, experimental and computational methods that allowed us, *from conventional powder diffraction data only*, to retrieve that minimal amount of structural information that, coupled with other ancillary data, can satisfactorily answer questions such as: What is the stoichiometry of the sample? What is its crystal and/or molecular structure (in terms of connectivity pattern)? What are the supramolecular features and the structural relations among the different asymmetric units and/or molecular fragments?

Obviously, these (and other) questions can be much more accurately addressed by conventional single-crystal diffraction analysis, which is faster, cheaper, easier to perform, and nowadays included in several practical classes at different university levels and careers. Thus, if a single crystal of the species under study is available, it should be (at least initially) used as such in diffraction experiments. In the unfortunate cases in which monocrystals cannot be grown, then XRPD is the method of choice, provided that the structural complexity (addressed by the number of nonhydrogen atoms in the asymmetric unit, N_{at}) is not too high ($N_{at} < 40$). Consistently, in

May 2009, i.e., about 15 years after the introduction, and diffusion, of the method, the number of crystal structures of molecular compounds (organic and organometallic species) determined by this method barely reaches a few thousand, i.e., a mere 0.4% of the entries of the Cambridge Structural Database. However, the growing interest in this technique is witnessed by the appearance, in the recent literature, of several review papers, focusing on the different methodological aspects of *ab initio* XRPD, as well as a few dedicated monographs or books: For the sake of simplicity, the reader is redirected to the most recent books on the subject[2] and to a collection of minireviews on XRPD studies of molecular functional materials.[3]

Without going into the details of the overall structure analysis process, the following points should complement the information inserted in the original structural papers cited below and trace a line of good practice, which, different from single-crystal analyses, is often abandoned, or twisted, depending on the specific problems encountered in specimen preparation, size, stability, and (para)crystallinity.

- About 30 mg (or more) of a monophasic polycrystalline species should be gently crushed, and eventually sieved, down to the 1–10 μm range; sample holders with minimal background contribution (zero background plates) should be employed and filled, up to the reference rim, by the powder, in either dusting or side-loading mode. The use of inert binders is sometimes helpful.
- Well-aligned conventional diffractometers with monochromatized radiation (typically Cu-Kα) and narrow optics must be used for data collection (typically in step scan mode, $\Delta 2\theta = 0.02°$, $5 < 2\theta < 100°$; $t = 10–30$ s step^{-1}). If the sample is transparent to x-rays, very accurate peak positions for indexing can be obtained by a second tailored data collection on a very thin film.
- Peak search and/or profile fitting should normally afford accurate positions ($\Delta 2\theta < 0.02°$) of at least 20 peaks, later fed to automatic indexing programs. Usually, a combination of different strategies should be used to increase the confidence on the (often multiple and unclear) results. Whole pattern structureless profile matching (Le Bail or Pawley's methods) must be used for cell confirmation, space group determination, and (integrated) intensity extraction.
- Structure solution can be performed by conventional (Patterson or direct) reciprocal space methods, although real space techniques (scavengers, Monte Carlo, simulated annealing, and genetic algorithms) are now prevailing.[4]
- Completion and refinement of the final structural, microstructural, and *instrumental* models are normally performed by whole profile matching (Rietveld method). Inclusion of geometrical restraints (bond distances, angles, torsions, planarity, etc.) can be required in order to stabilize convergence to physically sound values.
- More than ever, careful reconsideration of the results obtained at each step should be done, making the whole process not straightforward, but rather, iterative. Not infrequently, new sample preparations, purifications, depositions, etc., are required in order to "correct" the nonideality of the original sample and to include a minimal set of correction terms.

- The different profile and integrated intensity agreement factors, which, different from conventional single-crystal analyses, are not straightforwardly interpreted, should be estimated. Visual analysis of the matched profile and of the structure derived therefrom, with particular emphasis on *inter*molecular contacts as an estimate of the structure correctness, must be carried out.

Much like a culinary recipe that does not have to be strictly followed in order to prepare excellent gourmet dishes, the art of structural analysis also greatly benefits from a balanced mixture of experience, intuition, curiosity, perseverance (*if not stubbornness*), and diversity of the tools and methods employed. This is particularly true when constantly occurring new instrumental artifacts afford less than optimal data, and preferred orientation, absorption, and optical aberration effects, or sample instability, thermal drifts, beam misalignment, and other somewhat unpredictable, but subtle, source of errors are present.

METAL PYRAZOLATES

The first class of compounds that will be discussed hereafter is based on the pyrazolate anion (pz), the chemistry of which has been thoroughly investigated in the last 30 years, with particular emphasis on its ability to act as a prototypical N,N'-*exo*bidentate ligand, thus favoring the construction of polynuclear metal complexes (with the metals kept at distances typically below 3.5 Å);[5] although more exotic coordination modes (monodentate, η^2, η^5, and even more complex ones) have been recently observed for substituted pz species (particularly when bulky substituents are present in the 3,5-positions[6] or when lanthanides are employed[7]), the following examples, in which the simple, *unsubstituted* pyrazole is used as starting material, all *invariably* show bridging pz anions. These species were among the first we characterized by *ab initio* XRPD methods and, beyond being benchmark cases for the development of this structural technique, allowed the discovery of subtle structural effects such as polymorphism and stability patterns, later confirmed during our studies on the imidazolate and pyrimidinolate polymers discussed below.

Interestingly, among the heterocyclic anions presented here, pz is the only one that can possibly lead to multiple bridges, since virtually no atom is present between the metal ions bridged by a N,N'-*exo*bidentate pz ligand, thus opening the possibility of forming one-dimensional chains, even if metals in +II or +III oxidation states are employed. Worthy of note, a rhodium dimer bridged by *four* pyrazolates is known,[8] but this *unique* occurrence was not found among our samples.

BUCHNER'S "SILBER SALZ"[9]

The Cu(pz) and Ag(pz) species have been long known,[10] but until 1994, their oligomeric or polymeric nature had never been clearly demonstrated. In our seminal paper,[11] we were able to recognize four different phases, namely α- and β-Cu(pz) and α- and β-Ag(pz), the structures of which are drawn in Figures 2.1 and 2.2. Our XRPD analysis clearly showed that α-Cu(pz) and α-Ag(pz) are isomorphous species,

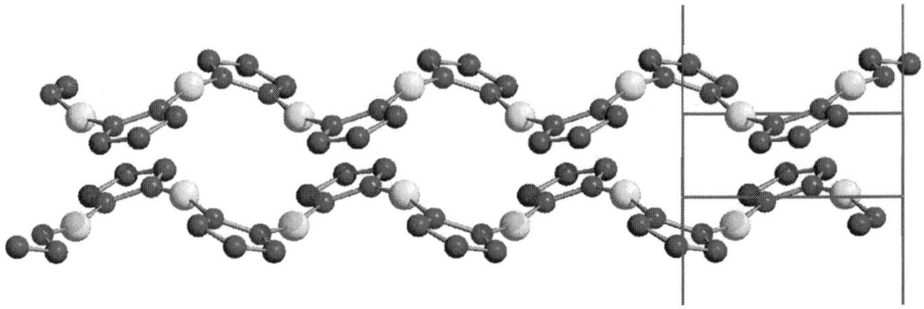

FIGURE 2.1 Schematic drawing of the M(pz) polymeric chains present in α-Cu(pz) and α-Ag(pz). In β-Cu(pz), the chains come slightly closer, in pairs, but possess nearly identical conformation.

containing polymeric 1D chains based upon linearly coordinated M(I) ions and bridging pz ligands, zigzagging through the crystal in an *all-trans* conformation; β-Cu(pz) itself was also found to be based on the same structural motif, although crystallizing in a different packing environment. However, the determination of the molecular structure of β-Ag(pz) (which was later formulated as $[Ag(pz)]_3$) proved to be more difficult since, while we were working on this problem, we believed for a long time that β-Cu(pz) and β-Ag(pz) were also a couple of isomorphous compounds: Indeed, the XRPD patterns of β-Ag(pz) and β-Cu(pz) show some similarities, particularly when a few unindexed peaks in the former are attributed to an impurity. Thus, we repeatedly attempted to refine the structural model obtained for β-Cu(pz) using the spectrum of β-Ag(pz). Being unable to achieve a reasonable agreement and after many failed attempts to obtain a "pure" $[Ag(pz)]_n$ phase, we decided to reconsider the problem by allowing for a larger cell in the indexing procedure. We soon realized

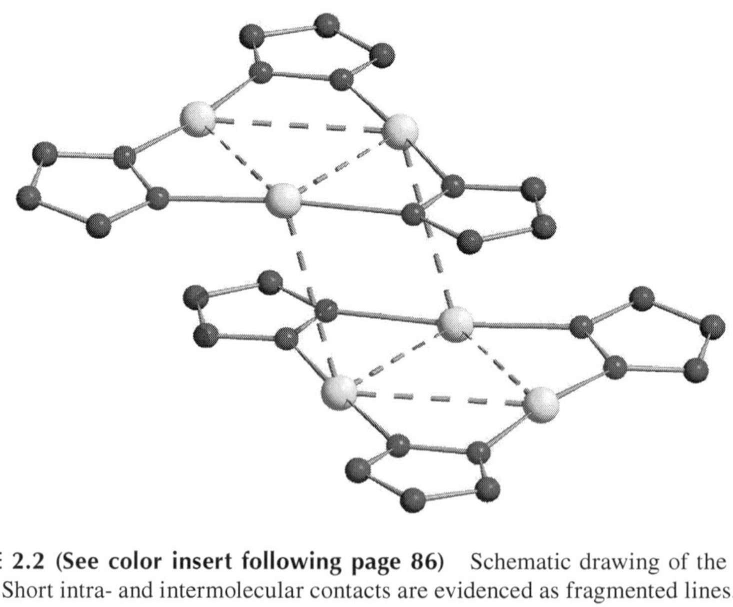

FIGURE 2.2 (See color insert following page 86) Schematic drawing of the $[Ag(pz)]_3$ trimers. Short intra- and intermolecular contacts are evidenced as fragmented lines.

that there was a cell, with a volume ca. 3/2 larger than that of β-Cu(pz), accounting for all the observed peaks of the spectrum[12] (obviously, other diffraction techniques, e.g., selected-area electron diffraction, could help in assigning the correct lattice metrics if a (suitably oriented) microcrystal is found). Direct methods[13] eventually led to the correct formulation of β-Ag(pz) as a cyclic, trimeric compound that was readily modeled in the structure completion and refinement stages by imposing the proper geometrical restraints.

The polymorphic nature of samples α- and β-Cu(pz), as shown in Figure 2.3, is related to the pairing of adjacent copper pyrazolate chains in β-Cu(pz), and to the formation of a pseudohexagonal packing of these "pairs," markedly different from the rectangular motif observed in the α-phase.

The structural analysis briefly presented above already contains a few important methodological aspects, inherent in the XRPD method: (1) the strict[14] requirement on the availability of a crystallochemically pure species (or the belief in the monophasic nature of the sample),[15] (2) the misconceptions raised by wrong perspectives or hypotheses, and finally, (3) the necessity of restraining the bonding parameters of stiff fragments of known geometry to chemically sound values (given the absence of significant accurately measured intensities at medium and high 2θ values). From this example it is clear that XRPD alone first allowed a strong chemical and structural belief (the isomorphism of β-Cu(pz) and β-Ag(pz)) to be discarded, and then led to the proposal of a totally different structural model for $[Ag(pz)]_3$.

Magnetically Active One-Dimensional Polymers

Polymeric materials possessing spin-crossover (SC) or spin-transition (ST) behaviors at, or near, room temperature, which are particularly appealing in the development of nanodevices, sensors, etc., have been recently synthesized and spectroscopically, thermally, and magnetically characterized; within this class of compounds, Fe(II) complexes, bridged by polyazaheterocycles, appeared as versatile and promising species.[16,17] Their interesting functional properties have been attributed to the cooperativity of subtle effects, such as the *collinear* geometry of the active centers and the rigidity of the one-dimensional framework, greatly enhancing (by electron-phonon coupling) their magnetic response, spin-transition features,[18] and even photomechanical effects.[19] However, their chemical and thermal stability is limited by the presence of counter-ions balancing, in the solid state, the net positive charges of the $[(FeL_3)^{2+}]_n$ chains (typically, L = 4-substituted triazoles). Moreover, since these species generally afford materials with low crystallinity[20] and tend to behave very differently as the result of even small variations in their preparation, their structural characterization could not be adequately performed. Thus, only indirect evidences, such as infrared (IR), Raman, Mössbauer, and Extended X-Rays Absorption Fine Structure (EXAFS) spectroscopy, or structural models of the corresponding oligomeric species[21] or of vaguely related polymers,[22] are available.[23]

Aiming to the comprehension of the structural features of these complex polymers, we prepared the $Fe(pz)_3$, $Co(pz)_3$, $Co(pz)_2$, and $Ni(pz)_2$ *model* systems, as well as a novel vapochromic $Cu(pz)_2$ phase.[24] Their XRPD characterization showed that binary metal pyrazolates tend to afford highly crystalline species where *strictly*

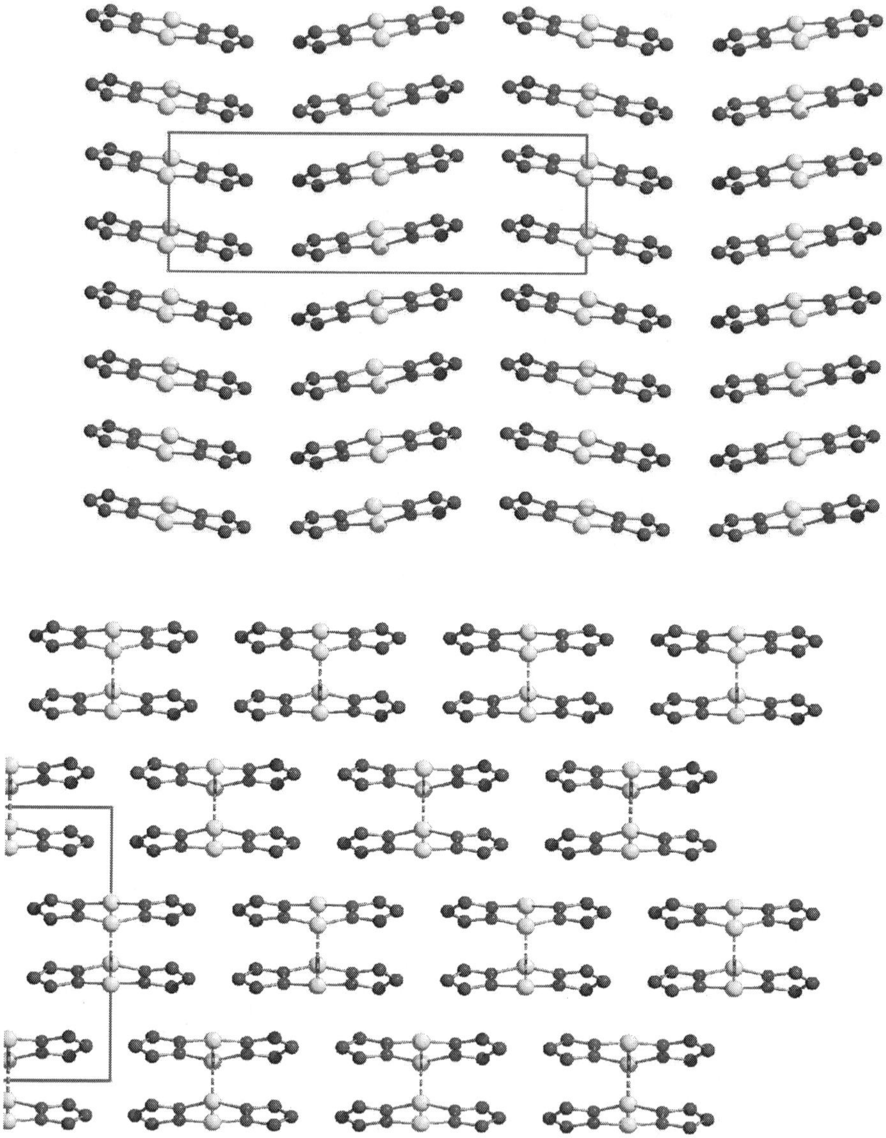

FIGURE 2.3 (See color insert) Schematic drawing of the crystal packing of the Cu(pz) polymeric chains in the α- (top) and β- (bottom) phases. The 1D chains run perpendicularly to the plane of the drawings, thus misleadingly appearing as dimeric entities. Short contacts (vertical fragmented lines) are evidenced in β-Cu(pz).

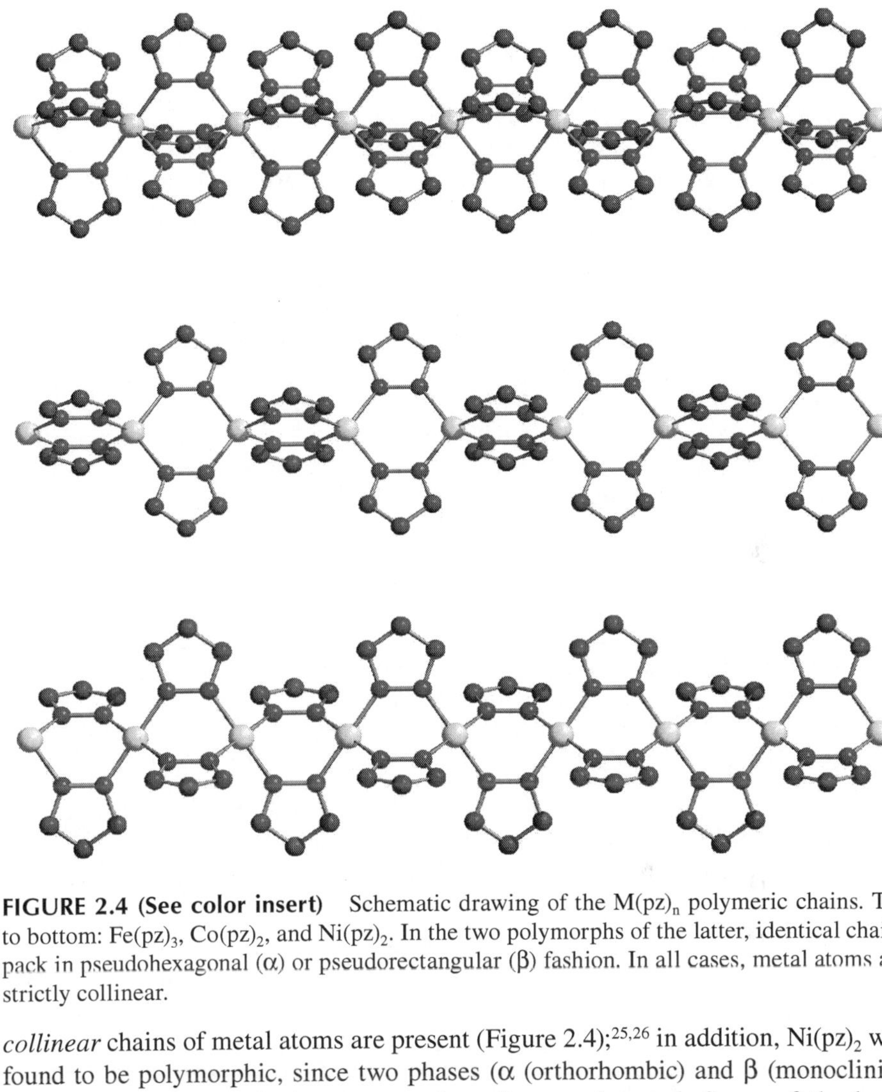

FIGURE 2.4 (See color insert) Schematic drawing of the $M(pz)_n$ polymeric chains. Top to bottom: $Fe(pz)_3$, $Co(pz)_2$, and $Ni(pz)_2$. In the two polymorphs of the latter, identical chains pack in pseudohexagonal (α) or pseudorectangular (β) fashion. In all cases, metal atoms are strictly collinear.

collinear chains of metal atoms are present (Figure 2.4);[25,26] in addition, $Ni(pz)_2$ was found to be polymorphic, since two phases (α (orthorhombic) and β (monoclinic), depending on the synthetic strategy adopted) were detected. Some of the basic topologies of these $M(pz)_n$ chains are not new and can be related to the archetypic β-$RuBr_3$ (or anhydrous scandium acetate) and $BeCl_2$ phases (M = Fe(III)/Co(III) and M = Fe(II)/Co(II)/Cu(II)/Zn(II), respectively); on the contrary, we are not aware of simple binary species adopting, in the solid state, the wavy chain structure of both $Ni(pz)_2$ phases.

The structures of these species, although not unexpected, are the direct proof of the multiple linkages sustained by the polypyrazolate bridges in all analogous, related polymeric chains. Moreover, since all metal atoms lie (in a variety of space groups) on special positions, the final refinements also allowed the assessment of several stereochemical features with high accuracy. The anisotropic disposition of the metals in the crystal lattice suggests that the strongest electronic and, eventually, magnetic couplings in these systems are imputable to the poly-pyrazolate-bridged M···M contacts, and not to interchain interactions, which might be at work only at very low temperatures.[27] Thus, such direct linkages of the metal centers and their collinear geometrical arrangement are likely to cause the large cooperativity of the magnetic interactions, as witnessed by the considerable hystereses found in a number of structurally related, but more complex, polymeric materials.[28,29]

Worthy of note, lipophilic cobalt(II) triazole complexes bearing long alkyl chains in 4-positions, structurally related to the homoleptic $Co(pz)_3$ polymer, have been prepared, and showed interesting gelification and thermochromic structural transitions;[30] since these features have been attributed to the presence of different equilibria, involving oligomeric and polymeric species, XRPD on (suitably simplified) species could provide a structural basis for the observed behavior.

Obviously, the complexity of the materials that can be studied by XRPD methods must not be too high. If the species under investigation does not easily afford crystals for conventional analysis, or monophasic powders of polycrystalline nature, then the preparation of *model systems*, simplified in order to allow a successful structural characterization, is a mandatory step, *provided that the idealization is not too extreme or does not change the sought structural features*.[31] This approach has already been successfully used by us in the effort of understanding the reaction mechanism of CO_2 insertion in a complex organometallic species, substituting one phenyl group by a smaller $-CH_3$ residue, and fully characterizing the reaction intermediate by XRPD.[32]

Polycationic Polymeric Species

Most of the homoleptic metal pyrazolates described above (and in the related literature) have been synthesized from commercially available metal salts (chlorides, sulfates, nitrates, acetates, etc.) and the heterocyclic ligand (in its neutral form) dissolved in water, upon addition of bases (typically ammonia or organic amines); before precipitation of the $M(pz)_n$ systems, i.e., before raising the pH value above 7, metal complexes of uncertain formulation are likely present in solution, and only the removal of the slightly acidic proton of pyrazole allows the quantitative recovery of the cited polymers. Worthy of note, the preparation of group 11 metal pyrazolates follows a different route, since (1) Ag(I) *alone* induces the deprotonation of the coordinated ligand (pyrazole, imidazole, etc.) through the formation of suitable N-Ag bonds, and it does not require the addition of an external base, and (2) Cu(I), which is unstable in aqueous media, is either used as a tetrakis-acetonitrile complex or prepared *in situ* from solid Cu_2O in molten pyrazole.

Within this context, we surprisingly found that the reactions of group 12 metals (in their +II oxidation state), easily leading to the target $M(pz)_2$ polymers for zinc and cadmium,[26b] *quantitatively* form the $Hg(pz)NO_3$ phase if $Hg(NO_3)_2$ is employed. To our surprise, this was the first mixed-anion metal pyrazolate ever prepared. Eventually, in a later study,[33] we demonstrated that imidazole reacts in the very same manner, leading to a species of $Hg(im)NO_3$ formulation. The reason for such behavior is easily explained if the crystal and molecular structures of both phases (Figures 2.5 and 2.6) are described.

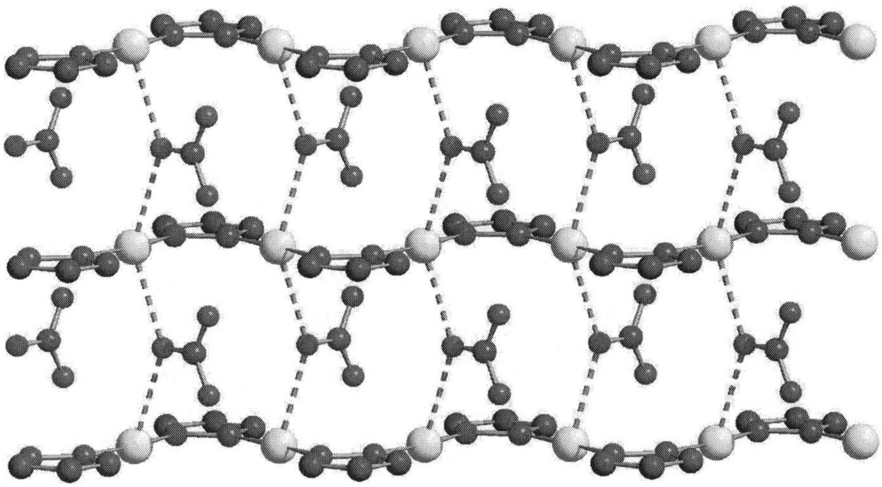

FIGURE 2.5 (See color insert) Schematic drawing of a portion of the $[Hg(pz)]_n^{n+}$ chain, surrounded by loosely interacting (dashed lines) nitrate ions.

Both phases contain linearly coordinated Hg(II) ions, bridged by N,N′-diazolates in their normal exobidentate mode (with Hg-N values slightly above 2.0 Å), thus forming one-dimensional $[Hg(\mu\text{-}pz)]_n^{n+}$ or $[Hg(\mu\text{-}im)]_n^{n+}$ chains running throughout the crystals. The nitrate ions are hosted in the crystal lattice, between these polycationic chains; however, weak Hg···O contacts (above 2.75 Å) are present in the plane normal to the Hg-N vectors, possibly favoring the sudden precipitation of these species during their synthesis, as if they were solids with an extensive three-dimensional nature. Although unexpected, this structural feature is also present in the fully inorganic mercury(II) hydroxo-nitrate, where the NO_3^- anions are completely embedded in a crystalline matrix generated by $[Hg(\mu\text{-}OH)]_n^{n+}$ polycations.[34]

Inter alia, the crystals of the $Hg(im)NO_3$ phase are noncentrosymmetric, P-62c: Metal ions and imidazolates lie in special positions (twofold axes), while two

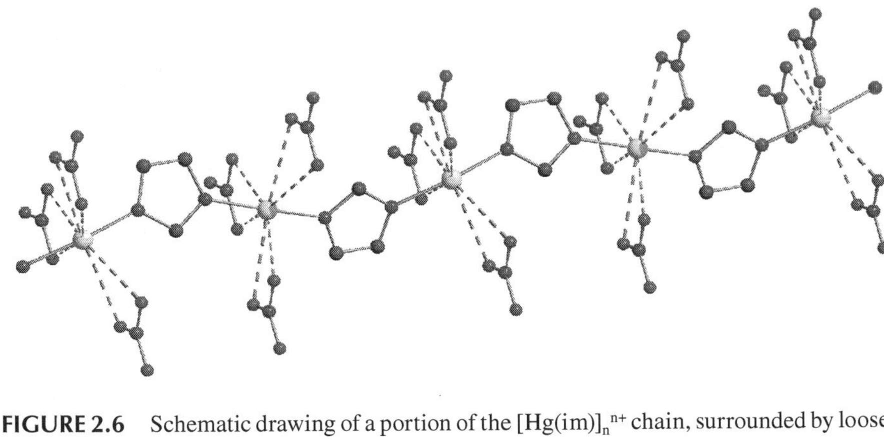

FIGURE 2.6 Schematic drawing of a portion of the $[Hg(im)]_n^{n+}$ chain, surrounded by loosely interacting (dashed lines) nitrate ions.

crystallographically independent NO_3^- anions (of different site symmetries) occupy the voids near the origin and on threefold axes at (2/3, 1/3, z). That this species is acentric was also confirmed by SHG measurements, its activity being (only) 0.1 times that of urea (Kurtz-Perry powder technique).

Methodologically speaking, the structure determination of the $Hg(im)NO_3$ species, crystallizing in the hexagonal system, was not at all straightforward: Indeed, in the high-symmetry crystal systems, not only the Laue class is obscured by the exact coincidence of different classes of reflections, but even the trigonal-hexagonal dichotomy is not geometrically soluble, and thus requires a multitude of parallel structure solution processes, possibly difficult to rank if heavy metals (as in the present case) occupy pseudosymmetrical sites. Thus, more than ever, it is the overall structural consistence, rather than the ranking of the agreement factors, that helps during the whole process of model building. Obviously, at the end of the procedure, R factors reach, for the correct *structural*, *microstructural*, and *instrumental* models, the lowest attainable values.

METAL IMIDAZOLATES

As an obvious extension of our studies on the metal pyrazolato systems, we later turned our attention to the coordination chemistry of the imidazolate anion (im), which has similar bonding ability but a bite angle favoring *exo*bidentate coordination of metals that lie about 5.6–6.0 Å apart. The variety of coordination modes found for imidazolate does not approach the complexity found for the pyrazolate ligand since: (1) the N,N′-*endo*bidentate mode (and its congeners) must be excluded on simple geometrical considerations, and (2) multiple bridges built on μ-im ligands are impossible. However, the limited coordination modes available to the imidazolate "anion" are overwhelmed by the torsional flexibility at the M-N bonds, thus allowing the existence of many different poly- (2 or 3) dimensional frameworks if metal(II) ions are employed. Thus a single coordination (N,N′-*exo*bidentate) mode is likely to afford a wide(r) spectrum of supramolecular arrangements than in the case of pz.

Our studies on the metal pyrazolate and imidazolate polymers are also propedeutic to the comprehension of other five-membered heterocyclic N-ligands: Recently, the coordination chemistry of metal triazolato and tetrazolato species (possessing 1,2- and 1,3-N,N′-connectivity, such as in pz and im, respectively) has been reviewed,[35] and the possibility of isolation of metal pentazolate systems (containing the elusive N_5^- cyclic anion) theoretically predicted.[36]

Moreover, imidazole and its derivatives are widely present in biologically active molecules, such as in the superoxide dismutase enzyme, where im bridges a Cu(II),Zn(II) pair;[37] thus, the understanding of the stereochemical features of this ligand, coupled with its extensive spectroscopic and magnetic characterization, as well as of model systems, allows the study of its biological activity on a molecular basis.

ANTIMICROBIALLY ACTIVE POLYMERS

Before our studies,[38] to our knowledge, the only reports of a species of Ag(im) formulation appeared some decades ago,[39] while Cu(im) was not reported until 2004.[40] Indeed, we attempted the syntheses of both silver and copper analogues,

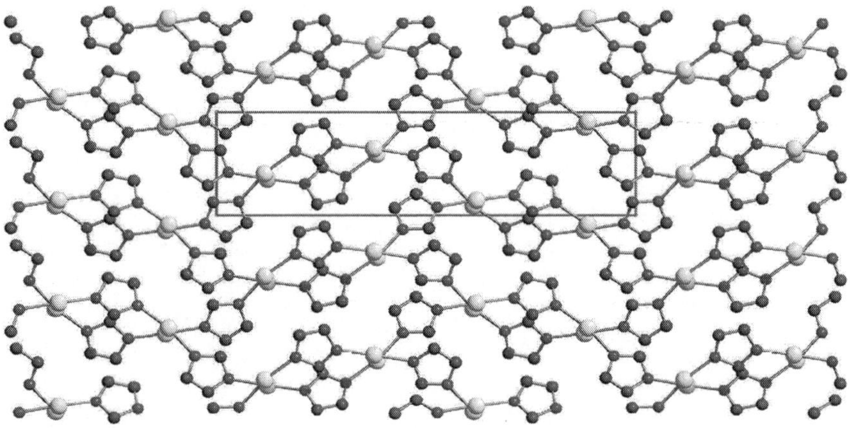

FIGURE 2.7 (See color insert) Schematic drawing of the crystal packing of Ag(im), viewed down [010]. The one-dimensional polymeric chains run nearly parallel to the horizontal cell axis, **c**.

and successfully recovered a monophasic, air-stable Ag(im) species (but not Cu(im), which was a poorly diffracting, easily oxidizable white material). Only at a later stage does the report of novel polymorphic forms, prepared by solvothermal methods, appear, together with the full structural characterization of the Ag(im) and Cu(im) species.[41]

Silver imidazolate was found to crystallize in the acentric $P2_12_12_1$ space group (as also confirmed by powder SHG measurements[42]) and contains wavy one-dimensional chains hinged about short *inter*chain contacts (Ag···Ag ca. 3.16 Å; see Figure 2.7). While in Ag(pz) the conformation of the 1D chain could be described by an *all-trans* pz-Ag-pz sequence, each Ag(im) polymer, still containing linearly coordinated silver ions, shows a (*cis-trans-*) sequence interacting with several adjacent chains, not just two (below and above). Differently from Ag(pz), the crystal packing of Ag(im) can be regarded as a herringbone arrangement of chains, stabilized by a complex network of d^{10}-d^{10} interactions.

After the publication of our paper, we learned of the interesting antimicrobial properties of the Ag(im) compound[43] and of some of its analogues (obtained by substituting the imidazolato group with other heterocyclic anions or introducing ancillary ligands): Indeed, several Ag(I) salts are widely marketed as antifungal drugs, and imidazole-based solutions are often employed in over-the-counter (OTC) pharmaceuticals. The occasional coupling of the two functionalities has therefore prompted for the determination of the antibacterial activity of Ag(im), which, among the many species of the same kind, resulted by far as the most effective. For a detailed analysis and interpretation of these properties, refer to the work of Nomiya et al.[44]

Worthy of note, the poorly soluble Ag(im) species can be dissolved upon reaction with tertiary aromatic phosphines: Subsequent solvent removal or crystallization generates the complex $Ag(im)(PPh_3)$ and $Ag_2(im)_2(PPh_3)_3$ polymers;[45] differently, the Ag(pz) analogue was shown to afford, in the same conditions, dinuclear species only $[Ag_2(pz)_2(PPh_3)_2$ or $Ag_2(pz)_2(PPh_3)_3]$.[46]

Ag(im) crystallizes in the acentric $P2_12_12_1$ space group; thus, only a few peaks should be systematically missing (half of the axial reflections). However, in a rather unpredictable manner, the first clearly evident peak occurs at $2\theta \sim 17°$, while only two small, barely visible reflections occur at lower angles. Without their accurate location, peaks indexing by TREOR[47] failed, the program being unable to address the correct metrics in the absence of suitably high d values. To overcome these difficulties, a variety of strategies can be adopted: (1) the parallel use of different indexing approaches, which, nowadays, is a (much) easier task;[48] (2) the necessity of trusting low-angle small reflections (often attributed to impurities), particularly after reiterated syntheses and measurements; and (3) the accurate detection of low-intensity peaks, favored by highly collimated optics and zero-background sample holders.

Uno, Nessuno, Centomila

During the investigation of the elusive copper(I) imidazolate, we learned of the existence, in the old literature, of a few scattered and incomplete reports on the $Cu(im)_2$ species; this compound immediately appeared to us a very interesting material, from the structural, spectroscopic, magnetic, and bioinorganic points of view. The synthesis itself proved to be a challenging task, since the only structurally characterized species, of blue color (**J**),[49] could not be resynthesized. Instead, a very controlled approach and a fine-tuning of the reaction conditions eventually afforded several other polymorphs, as *monophasic* samples, easily distinguishable by their colors: blue (**B**), green (**G**), olive (**O**), and pink (**P**) (Figure 2.8).[50] An amorphous green species was also prepared. Thus, all together, five polymorphs of $Cu(im)_2$

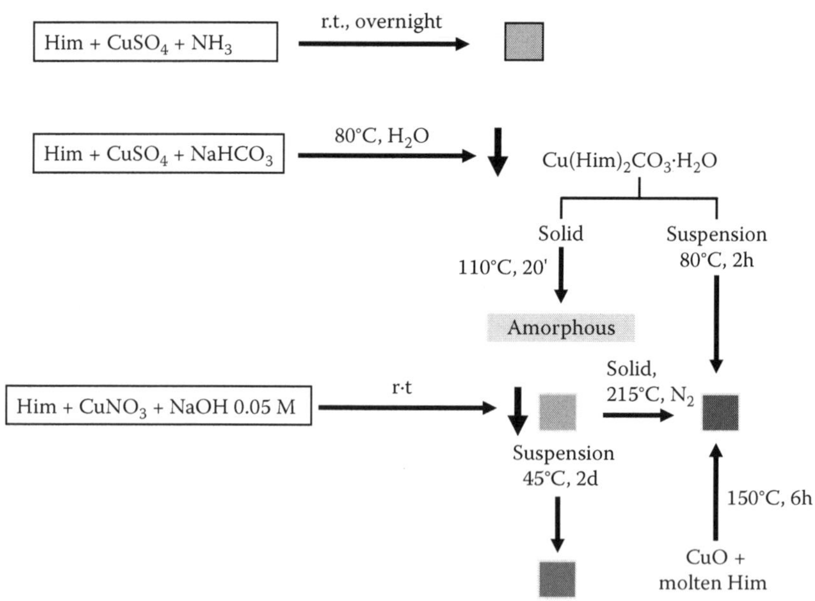

FIGURE 2.8 (See color insert) Reaction conditions employed for the preparation of monophasic samples of the different $Cu(im)_2$ polymorphs, each one indicated by a squared color label.

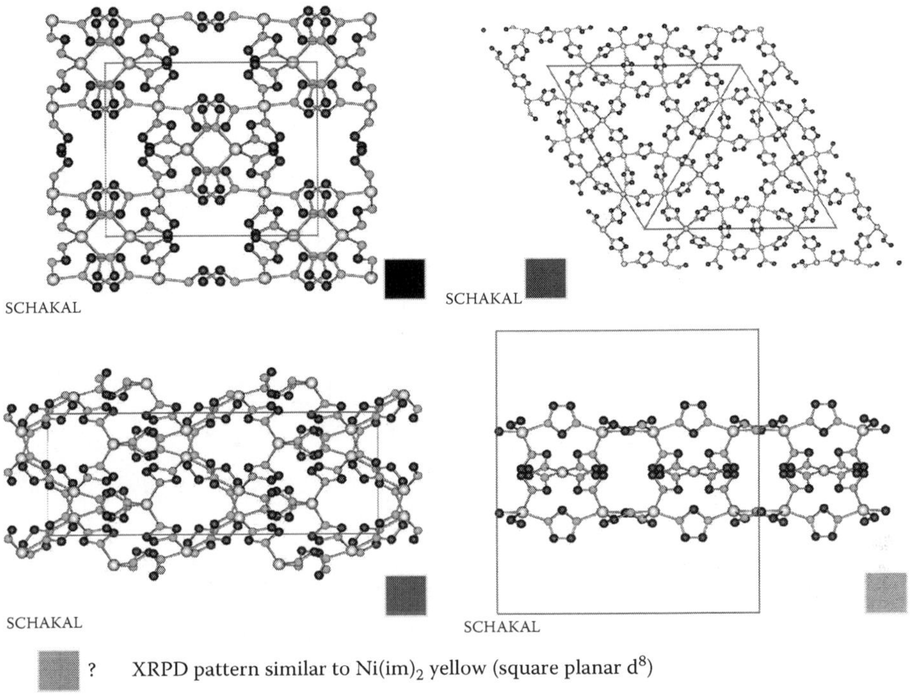

FIGURE 2.9 (See color insert) Schematic drawing of the crystal packing of $Cu(im)_2$: left to right, top to bottom: J, B, O, and G polymorphs. The structure of the pink (P) phase is still unknown.

have been detected, but only four have been fully characterized by diffraction methods (three from XRPD data), the pink phase still awaiting a structural model.[51]

The structures of these polymorphic species are schematically drawn in Figure 2.9, where the presence of cavities, tunnels, or layers can be easily appreciated. For the sake of completeness, a synoptic table showing the many $Cu(im)_n$ phases detected so far (also thanks to the recently published hydrothermal preparations of some Cu(I) containing derivatives) is reported. Table 2.1 nicely illustrates the complex crystal chemistry of the copper-imidazole binary system, which has been recently invoked[52] to form very effective anticorrosion layers protecting the metal from aerial oxidation.

As shown in Figure 2.9, XRPD is the fundamental tool to be used when polymorphic systems are detected (1) for their quick (qualitative) assessment, (2) for their accurate quantitative analysis, and (3) structurally speaking, for retrieving the crystallographic models of samples that, typically, are metastable and, upon manipulation, tend to form (single) crystals of the most stable phase. Consistently, a number of polymorphic systems have been recently studied by us using uniquely laboratory XRPD data: $Pd_2(dmpz)_4(Hdmpz)_4$ (Hdmpz = 3,5-dimethylpyrazole),[53] $Cu(bipy)Cl_2$, (bipy = 4,4′-bipyridine),[54] Alq_3,[55] and several marketed drugs (Acitretin,[56] Linezolid,[57] Sibutramine,[58] Azelastine,[59] and Diflorasone diacetate[60]).

TABLE 2.1
Synoptic Collection of the Main Structural Features of the Various $Cu(im)_n$ Species

Species	$Cu(im)_2$	$Cu(im)_2$	$Cu(im)_2$	$Cu(im)_2$	$Cu(im)_2$	Cu(im)	Cu(im)	Cu(im)	Cu(im)	$Cu_2(im)_3$	$Cu_3(im)_4$
Color	Blue	Blue	Green	Olive	Pink	White	Golden	Orange	Yellow	Green	Mauve
SG	*I2/c*	*R-3*	*Ccca*	*C2/c*	–	–	*C2/c*	*C2/c*	$P2_1/n$	$P2_1/n$	*P-1*
V/Z, Å^3	180	197	187	201	–	–	108	104	102	284	360
ρ, g cm^{-3}	1.824	1.667	1.753	1.632	–	–	2.012	2.069	2.115	1.914	2.118
Topology	3D	3D	2D	3D	–	–	1D	1D	1D	3D	3D
Type	PtS	Sodalite	Slabs	Moganite	–	–	Zig-zag	Zig-zag	Zig-zag	New	NbO
Cavities	NO	Tunnels 7.5 %	No	Closed 11 %	–	–	No	No	No	No	No
Cu site symmetry	−1, 2	2	222, 1	2, 1	–	–	−1	1, 1, 1, 1	1, 1, 1	1, 1, 1, 1	−1, −1, −1
Reference	49	50	50	50	50	[a]	40	41	41	40	98

[a] Powder Diffraction File, ICDD, Swarthmore, PA, PDF No. 52-2401.

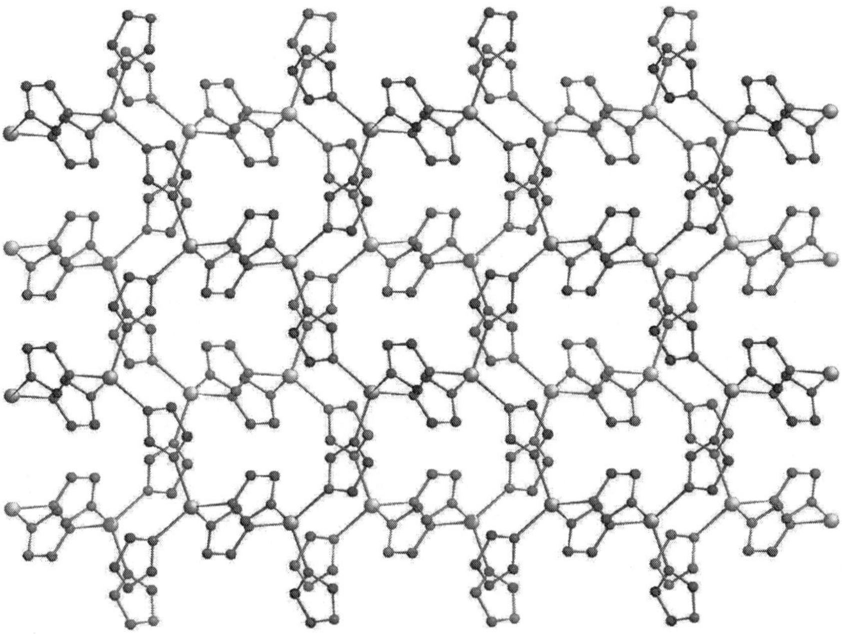

FIGURE 2.10 (See color insert) Schematic drawing of the $M(im)_2$ ($M = Cd, Hg$) species. The two interpenetrating diamondoid frameworks are shown in red and green.

Interpenetrating Diamondoid Frameworks

$Zn(im)_2$ and $Co(im)_2$, two tetragonal but crystallographically distinct phases, are known since long, contain tetrahedrally coordinated M(II) ions and are extremely complex;[61] at variance, the supramolecular arrangement of the two isomorphous orthorhombic Cd and Hg analogues, recently studied by us using XRPD,[33] is surprisingly simple, being based on the 3D diamondoid network, very common for tetrahedral centers and bidentate spacers. However, imidazolate lacks pseudocylindrical symmetry and forces short intermetallic distances: Thus, the observed twofold interpenetration (Figure 2.10) is rather unexpected.[62] Inter alia, $Cd(im)_2$ and $Hg(im)_2$ are the smallest coordination polymers possessing this peculiar structural feature (apart from the inorganic $Zn(CN)_2$ compound). Interestingly, this topological occurrence cannot be achieved if metals of small(er) ionic radii are employed (Zn, Co, and Cu bis-imidazolates).[50,62] Differently, Cd (but not Hg,[26b] due to its peculiar sterochemical requirements) and first-row transition metal (Fe, Co, Cu, and Zn)[25,26] tetrahedral bis-pyrazolates, for which interpenetration is not possible, are found to be isomorphous, although not completely miscible.

The structures of $Cd(im)_2$ and $Hg(im)_2$ were nicely solved by the simulated annealing technique implemented in TOPAS,[63] which, thanks to its powerful graphic routines, allows the real-time estimation of the chemical soundness of the proposed solutions. Worthy of note, the sampling of the whole parameter space is not guaranteed (and never complete) by the random seed generation of the trial structures. Thus, rejection or acceptance of a suitable structural model requires coupling of graphical, numerical, and visual criteria. The structural features of these two samples, together with their easy chemical preparation, make them ideal candidates for classroom demonstrations, where also the indexing, space group determination, and Le Bail intensity extraction steps, as well as Patterson and difference Fourier methods, can be easily tackled.

METAL PYRIMIDINOLATES

Hybrid inorganic-organic microporous materials are of particular interest because their functional properties can be suitably tuned by introducing specific modifications affecting, e.g., polarity, chirality, etc. In this context, we decided to study the class of trifunctional diazaaromatic species, of which the commercially available 2- and 4-hydroxy-pyrimidines (2- and 4-pymoH) are the simplest precursors. These ligands offer a wider set of coordination modes than diazolates, since, in their deprotonated forms, they possess three coordination sites (of nearly equal nucleophilicity, thanks to the semiquinoid nature of one of the mesomeric forms). Moreover, their highly polar nature can be further used to influence the stability and the optical properties (*vide infra*) of the derived frameworks, through interaction with guest molecules (if any).

To our knowledge, the oldest report of a polymeric metal pymo derivative contains the *hydrated* monodimensional polymer Ag(2-pymo)·$2H_2O$, studied by Quirós in 1994.[64] Nowadays, these ligands have been employed in constructing several polynuclear *soluble* systems, capable of solution host-guest chemistry and differential recognition of biologically active species.[65] In the following, only *insoluble*, polymeric compounds will be presented, with particular emphasis on the differences between the species derived from the two structural isomers.

The Lord of the Rings

After having studied the syntheses and crystal structures of several group 11 metal azolates (*vide supra*), we learned[64] of the existence of a hydrated polymer, Ag(2-pymo)·$2H_2O$, whose structure is partially reminiscent of that of the M(pz) (M = Cu, Ag) phases. Accordingly, we decided to attempt the synthesis of the hypothetical Ag(2-pymo) phase either by thermally induced dehydration (monitored by Differential Scanning Calorimetry [DSC], Thermogravimetric Analysis [TGA], IR, and XRPD) or directly from anhydrous components.

Upon heating, Ag(2-pymo)·$2H_2O$ readily loses water in the 80–110°C range (ΔH = 101 kJ mol^{-1}); at about 150°C, a weak exothermic event, not accompanied by weight losses, is observed (ΔH = −4.9 kJ mol^{-1}), well before the complete thermal decomposition (to metallic silver) occurring at about 300°C. XRPD showed that progressive heating generates an amorphous phase that transforms, above 150°C, into a white (poly)crystalline one (Figure 2.11). IR monitoring (nujol mulls) confirmed the loss of water and the formation of a slightly different absorption pattern, which we originally attributed to the anhydrous Ag(2-pymo) polymer. However, rather surprisingly, our XRPD analysis led to the discovery of the novel cyclic, hexameric, chair-like compound $[Ag(2\text{-pymo})]_6$, of crystallographic C_{2h} symmetry.[66]

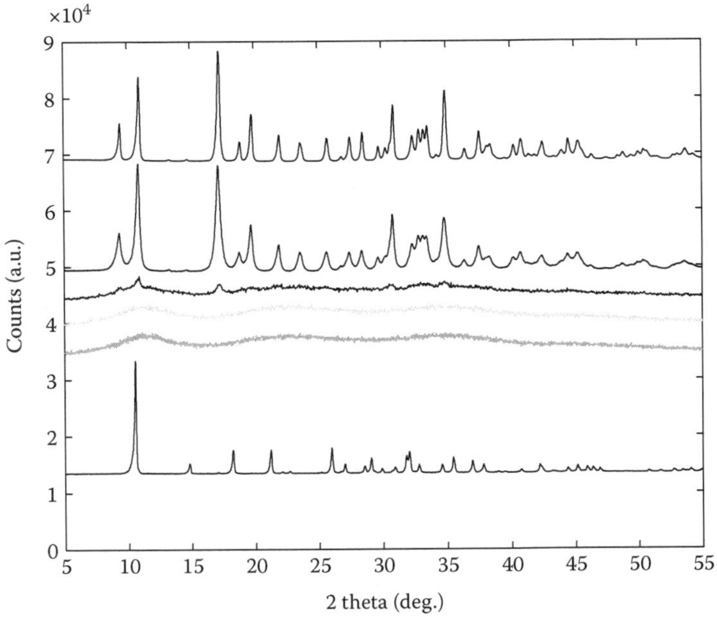

FIGURE 2.11 Raw XRPD data showing the Ag(2-pymo)·$2H_2O$ dehydration (and amorphization) process and subsequent devitrification to crystalline $[Ag(2\text{-pymo})]_6$. Bottom to top: As synthesized; 15', 90°C; 15', 120°C; 15', 140°C; 1 h, 190°C; 2 h, 260°C. Note that the top first and second curves possess full widths at half maximum (FWHMs) of 0.20° and 0.40° (2θ), respectively.

The silver ions, approximately lying at the vertices of a nonbonded hexagon (Ag···Ag 6.117(6) and 5.909(5) Å; Figure 2.12), are linearly coordinated by the nitrogen atoms of the N,N′-bidentate ligands[67] and possess short contacts (of the aurophilic type) with neighboring molecules (Ag···Ag 2.962(2) and 3.061(5) Å).

On attempting the parallel synthesis of the copper analogue, a highly reproducible, but complex XRPD pattern of a yellow polycrystalline species (analyzed as Cu(2-pymo)) was repeatedly obtained. Its (lengthy) crystal structure determination eventually led to a structural model containing separate hexamers (Figure 2.13a) and helical polymers (Figure 2.13b) in a 1:2 ratio, with digonal copper atoms bound to N,N′-bidentate 2-pymo ligands. Therefore, Cu(2-pymo) should be better described as a $1/6[Cu(2\text{-pymo})]_6 \cdot 1/n[Cu_2(2\text{-pymo})_2]_n$ adduct (stoichiometrically consistent with the original Cu(2-pymo) formulation). The helices contain six Cu(2-pymo) monomers per turn (with a pitch of 9.79 Å), i.e., possess a rare (noncrystallographic) 6_1 (or 6_5) character,[68] and are heavily compenetrated, each being trigonally surrounded by three helices of opposite chirality (Figure 2.14). Noteworthy, adjacent helices are displaced by ca. one-third (not half) of their pitch; within the space left by the packing of six adjacent helices, a *closed* cavity is formed (Figure 2.13), which perfectly matches the size and shape of $[Cu(2\text{-pymo})]_6$. Thus, once filled by the hexamers, the packing of helices in Cu(2-pymo) shows no other (solvent-accessible) holes.

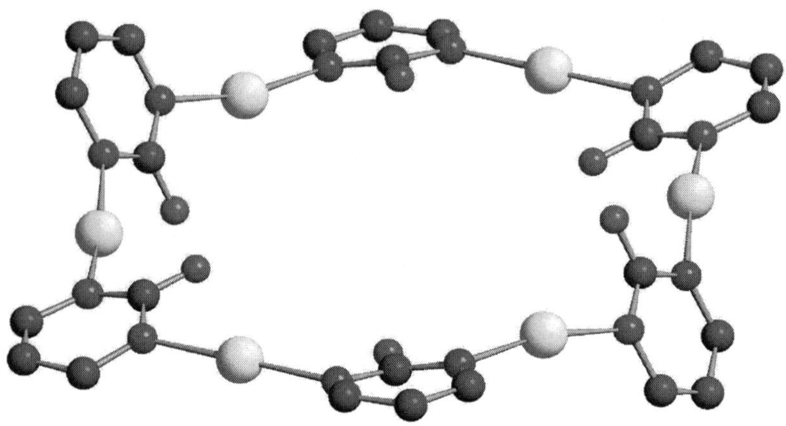

FIGURE 2.12 (See color insert) Schematic drawing of the $[Ag(2\text{-pymo})]_6$ molecule, possessing crystallographic C_{2h} symmetry. Note the folded character of the whole ring.

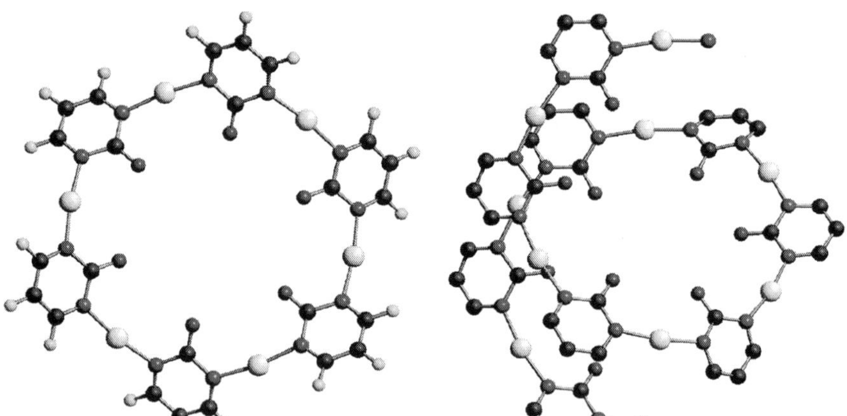

(a) (b)

FIGURE 2.13 (See color insert) Molecular drawing of (a) the $[Cu(2\text{-pymo})]_6$ hexamer (similar to that found in the silver analogue) and (b) a portion of the infinite helix of $[Cu_2(2\text{-pymo})_2]_n$.

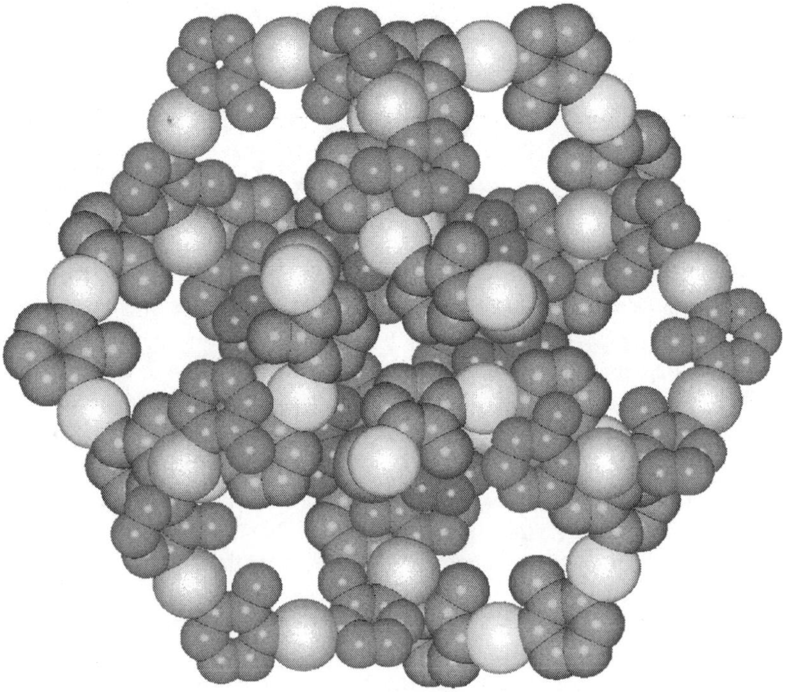

FIGURE 2.14 (See color insert) A packing diagram of the $1/6[Cu(2\text{-pymo})]_6 \cdot 1/n[Cu_2(2\text{-pymo})_2]_n$ species, viewed down [001]; helices of different polarity are depicted by different colors (yellow and green); within this trigonal packing of helices, *closed* cavities about the origin host the $[Cu(2\text{-pymo})]_6$ hexamer (red).

These two structure determinations clearly show that, despite the simple M(2-pymo) formulation, the correct stoichiometry, molecular structure, and overall supramolecular arrangements, all unforeseen beforehand, were satisfactorily determined by XRPD, even in the presence of broad diffraction peaks (widened by the thermal treatment of the hydrated Ag(2-pymo) polymer) or a complex pattern (for Cu(2-pymo), where the Laue class, correct space group, and ligand location were partially obscured by, among other effects, the pseudosymmetric arrangement of the heaviest scatterers).

Some Like It Hot

Thermal stability and chemical inertness are often related, due to the presence of rather strong intra- and intermolecular interactions, to the absence of significantly more stable (oxidation or decomposition) products, and to the lack of accessible reaction paths. In the case of metal diazolates, thermal decomposition generally leads to metal oxides and nitrides (depending on the conditions employed) or, in some cases, to other inorganic species.[69] Often, some amorphous carbonaceous black material accompanies such decomposition, and the formation of gaseous species (water, CO, CO_2, and HCN) is also observed.

Organic molecular species typically melt, or decompose, at temperatures below 300°C, while a few technologically important polymers were found to be stable up to typically 400°C.[70] At first glance, a few metal pyrazolates and imidazolates, particularly those *not* containing metal ions prone to reduction, disproportionation, or even ring metalation, were found to be rather stable, with decomposition temperatures in the 350–450°C range. Accordingly, the very interesting dielectric properties recently discovered for $Zn(im)_2$ were discussed in the frame of potentially useful materials, thanks to the easy method of preparation and its high thermal stability and chemical inertness.[52]

The presence of stiff heteroaromatic rings, with highly delocalized π systems, and of dipolar interactions in the crystal can be invoked to explain the extreme stability of the $M(2\text{-pymo})_2$ species (M = Co, Ni, Zn), which crystallize as diamondoid frameworks and can be heated *without decomposition* (under N_2) up to ca. 550°C.[71]

Compared to the other known metal(II) diazolates, these metal bis-pyrimidin-2-olates show the highest stabilities: Indeed, only $Zn(pz)_2$ was found to sustain such high temperatures. That 2-pymo ligands tend to form rather stable metal adducts was already observed for metal complexes in lower oxidation states, such as Cu(I) and Ag(I): Indeed, the complex Cu(2-pymo) species discussed above decomposes at 70°C higher than α-Cu(pz) (decomposition temperature 270°C[11]), and the *hexameric* $[Ag(2\text{-pymo})]_6$ is stable up to 300°C, about 40°C higher than for the Ag(pz)[11] and Ag(im)[38] polymers.

The acentric nature of the three $M(2\text{-pymo})_2$ species was confirmed by the measurement of SHG activities of a few percent of standard urea powder. Thus, thermal stability, in these species, is nicely accompanied by potentially useful optical properties, opening the way to a class of new polyfunctional materials.

In conventional single-crystal diffractometry, acentric crystals normally require a larger number of independent diffraction data, including Friedel pairs, and the estimate of the correct enantiomorph by anomalous scattering (if applicable); the determination of the correct set of phases in the starting model also requires longer computational times, and fake solutions are slightly more often encountered. Thus, acentricity is normally thought as an annoying, although not removable, effect. While these observations (more or less) apply also to powder diffractometry, there is at least one step, *indexing*, which is (enormously) favored by acentricity: Indeed, for a crystal phase containing N crystallographically independent atoms (say, a full molecule), the cell volume is *smaller* for the acentric vs. the corresponding centric unit cell. Inherently, the indexing procedure is much simpler, and can also benefit from the lower number of systematic absence conditions, normally increasing the low-angle observable lines (and the resulting figures of merit).

Magnetically Active Two-Dimensional and Three-Dimensional Systems

Unsubstituted pyrimidine bridging ligands have been shown to efficiently transmit ferromagnetic interactions in a cobalt(II) layered compound.[72] Thus, in order to prepare magnetically active Co(II) 2D and 3D polymers, we employed the 2-pymo and 4-pymo ligands, which cannot give chain polymers,[25] in the syntheses of extended systems of $Co(2\text{-pymo})_2$[71] and $Co(4\text{-pymo})_2$[73] formulation. The formation of these binary species follows different routes, a stable hydrated intermediate, $Co(4\text{-pymo})_2(H_2O)_4$, being invariably recovered in the latter case. As described above,

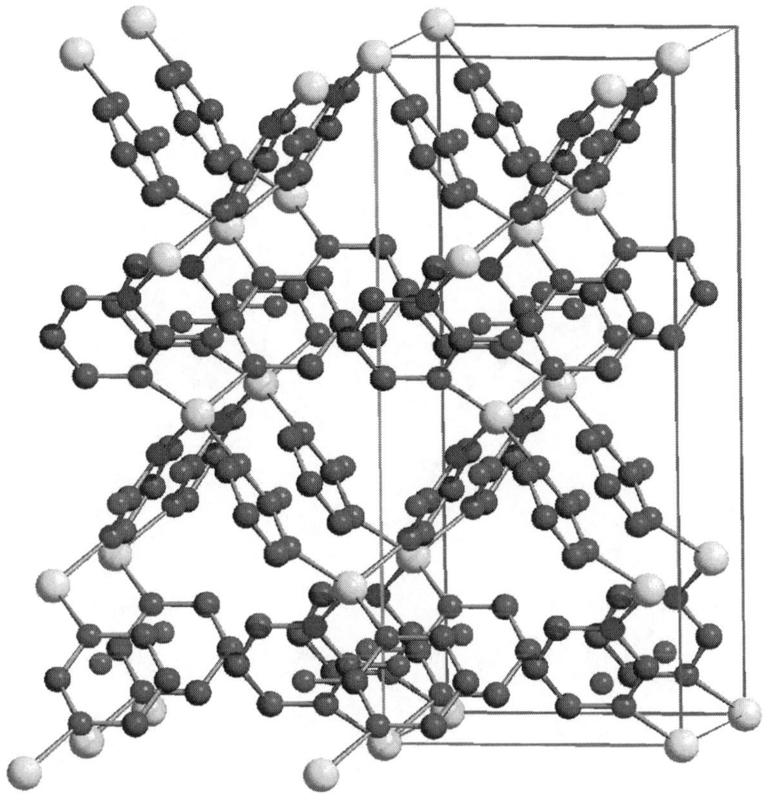

FIGURE 2.15 (See color insert) Partial drawing of the 3D diamondoid network in Co(2-pymo)$_2$.

Co(2-pymo)$_2$ contains N,N′-*exo*bidentate heterocyclic ligands and crystallizes as a 3D diamondoid framework (Figure 2.15).

At variance, Co(4-pymo)$_2$, obtained by heating above 320°C the aqua species (in the 150–320°C temperature range an anhydrous amorphous phase is formed), surprisingly showed a crystal structure based upon stacking of two-dimensional layers of square meshes defined by 4-pymo ligands in their N,O-*exo*bidentate (not chelating) mode (Figure 2.16). Actually, the metal coordination is completed by ancillary, weak(er) Co···N contacts. Incidentally, the chemistry of the nickel analogues is essentially the same, since, under similar conditions, the Ni(2-pymo)$_2$, Ni(4-pymo)$_2$(H$_2$O)$_4$, and Ni(4-pymo)$_2$ compounds,[74] isostructural with the cobalt species, were isolated.[75]

Of these species, the most attractive was found to be Co(2-pymo)$_2$, which can be considered a molecular magnet. Indeed, below a critical temperature (T_c = 23 K), a ferromagnetic ordering takes place (coercive field H_{coer} = 3,900 G, remnant magnetization M_{rem} = 279 cm^3 G mol^{-1}, as determined from the magnetic hysteresis loop, measured at 4.8 K, shown in Figure 2.17).

The remaining cobalt species, including the amorphous material of Co(4-pymo)$_2$ formulation, show only much weaker magnetic interactions; the bridging 4-pymo

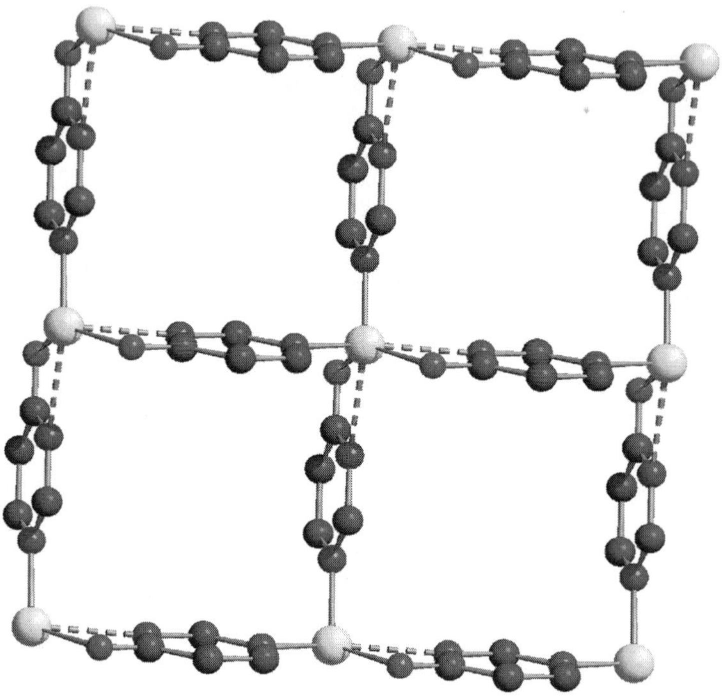

FIGURE 2.16 Partial drawing of the 2D network in Co(4-pymo)$_2$. Hydrogen atoms are omitted for clarity. Fragmented lines address the loose (ca. 2.30 Å) Co···N contacts.

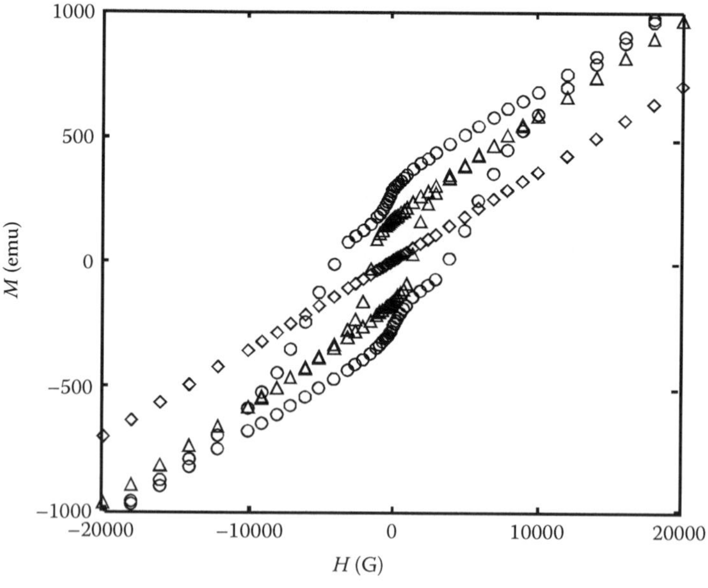

FIGURE 2.17 Magnetic hysteresis loops for Co(2-pymo)$_2$ at 5 K (o), 20 K (Δ), and 30 K (◊).

ligands cause, in Co(4-pymo)$_2$, an antiferromagnetic exchange between the *pseudo*-tetrahedral cobalt(II) ions with |J| = 1.73(1) cm^{-1}, which is increased to about 4.1 cm^{-1} in the nickel analogue; in the aqua-(4-pymo) species, magnetically isolated or weakly interacting M(II) ions (J = −0.313(5) cm^{-1} for Ni(II)) are present.

Mineralomimetic Coordination Frameworks

The last two decades have witnessed the blooming of a profound scientific and economical interest, in academic and industrial fields, for technologically and environmentally important applications such as molecular recognition, gas adsorption, gas storage, and gas or liquid mixtures selective separation. In this respect, zeolites and phyllosilicates have been historically exploited in a wide range of applications, due to their well-defined porous nature, allowing, e.g., selective ion exchange, as well as adsorption and catalytic processes.[76] Recently, to overcome, at least in part, the limits and drawbacks of these and other (crystalline or amorphous) species historically employed for the quoted applications, efforts and resources have been mainly dedicated to find a suitable category of improved functional materials. This quest has identified porous metalorganic frameworks (MOFs) as a promising class.[77] In contrast to inorganic zeolites, phyllosilicates, or activated carbons, MOFs'[78] shape,[79] functionalization, flexibility,[80] and chirality[81] can be finely tuned. However, the rational design and synthesis of MOFs with *zeolitic* networks, which are considered to be the most significant topologies for porous materials, are still a challenge. Metal imidazolates are one of the best known examples of zeolitic MOFs, as demonstrated by us,[50] and others,[82,83] employing different structure-directing agents.

Other angular monovalent anions such as pyrimidinolates[84] can be used as synthons for obtaining zeomimetic materials. Indeed, employing the simple pyrimidin-2- and pyrimidin-4-olate moieties allowed us to isolate and characterize two MOFs of CuL_2 formula (L = 2-pymo[85a,c] or 4-pymo[85b]), in which metallacalix[4]arenes, metallacalix[6]arenes, and hexagonal rings interconnect to give an overall *sodalitic* 3D network (Figure 2.18).

Cu(2-pymo)$_2$ (**CuP**) showed excellent ion pair recognition properties. Indeed, heterogeneous solid-liquid sorption processes are responsible for an unexpected wide variety of guest-induced crystal-to-crystal phase transitions involving the MOF host, and which are highly dependent on liquid phase polarity and the nature of the guests. Thus, the sorption selectivity of **CuP** for ion pairs containing ammonium or alkali cations and cubic anions (ClO_4^-, BF_4^-, PF_6^-) from aqueous solutions[85a] is extended to a wider range of ion pairs containing D_{3h} NO_3^- anions when the solvent polarity is slightly reduced (MeOH/H_2O or EtOH/H_2O mixtures).[85c] The novel sorption processes were found to induce profound structural changes in the original **CuP** MOF (Figure 2.18).[85c]

XRPD showed that, as synthesized, the hydrated form of Cu(2-pymo)$_2$ is *rhombohedral* (**CuPR**). This 3D framework is not rigid but, upon exposition to a solution of MNO_3 (M = NH_4, Li) in aqueous MeOH, a transition to a *cubic* phase, Cu(2-pymo)$_2$·(MNO_3)$_{1/3}$ (**MNO_3@CuPC**), is observed.[86] Single-crystal x-ray studies performed on the **MNO_3@CuPC** systems with M = NH_4 and Li show that they contain a *regular sodalitic* Cu(2-pymo)$_2$ MOF (**CuPC**) with water molecules and $LiNO_3$ or

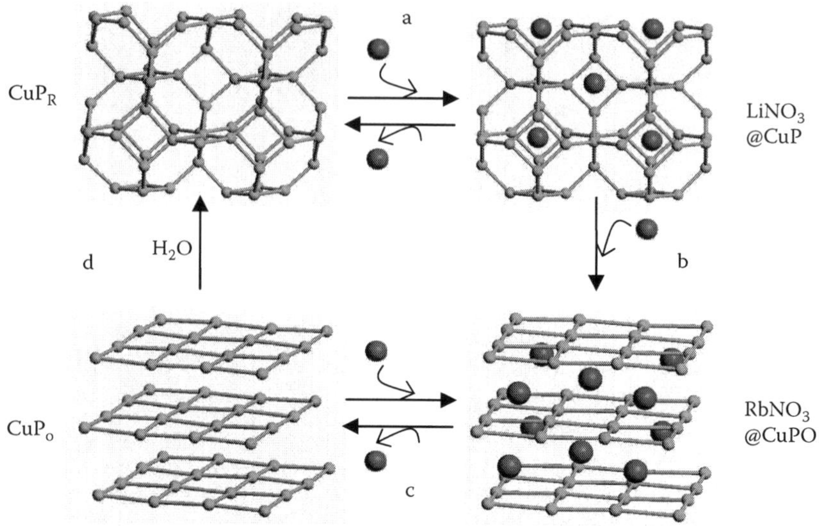

FIGURE 2.18 (See color insert) Guest-induced transformations in the Cu(2-pymo)$_2$ (**CuPR**) framework. (a) Incorporation of n/3 MNO_3. (b) Additional incorporation of 1/6 MNO_3. (c) Removal of 1/2 MNO_3. (d) Water addition. For the M cations, see text. The balls and sticks denote Cu and pyrimidin-2-olate-*N,N′*-bridges, respectively. Coordinates from the crystal structures of **CuPR**, **LiNO$_3$@CuPR**, and **RbNO$_3$@CuPO**.

NH_4NO_3 ionic pairs included in the hexagonal channels (Figure 2.19a). A related structural phase change from the rhombohedral phase to the cubic one also takes place upon dehydration by heating over 60–70°C (Figure 2.20). These processes are fully reversible; i.e., exposition to water of both **CuPC** and **MNO$_3$@CuPC** restores **CuPR**.

Much more pronounced structural changes take place upon exposition of **CuPR** to methanol solutions of nitrate salts of larger cations (Na^+, K^+, Rb^+, Tl^+); the kinetically controlled crystal-to-crystal inclusion process from **CuPR** to **MNO$_3$@CuPC** is, indeed, followed by further incorporation of 1/6 MNO_3, leading to a novel series of isomorphous *orthorhombic layered* materials, Cu(2-pymo)$_2$·(MNO_3)$_{1/2}$ (**MNO$_3$@CuPO**) (Figure 2.18). As determined from single-crystal data, **RbNO$_3$@CuPO** consists of square-grid Cu(2-pymo)$_2$ 2D layers, with the Rb^+ ions coordinated to the pyrimidine exocyclic oxygen atoms of the metallacalix[4]arenes (Figure 2.19b). The MNO_3 guests can be easily removed, giving an *empty layered orthorhombic* Cu(2-pymo)$_2$ species (**CuPO**) that can be further converted into the original **CuPR** phase by exposure to water for a few hours. Thus, the mineralomimetic nature of these systems is related to the topology and functionality of the host MOFs, which, like in sodalite (**CuPR** and **CuPC**) or in talc or muscovite (**CuPO**), can intercalate neutral molecules or metal ions, in the cavities or between the layers,[87] respectively.

We also investigated the adsorption properties of the Cu(2-pymo)$_2$ species, as compared to those of the *cubic sodalitic* Cu(4-pymo)$_2$ counterpart. While their overall structural features are similar and nearly independent of the position of the exocyclic oxygen atom, the shape, size, and hydrophilicity of their cavities are highly

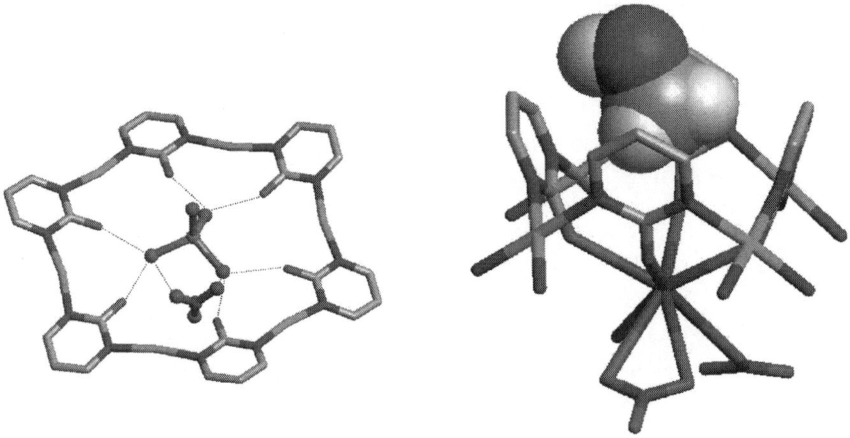

FIGURE 2.19 (left) Li^+ coordination in the hexagonal windows of the sodalite β-cages in **$LiNO_3$@CuPC**. (right) The metallacalix[4]arene motif in **$RbNO_3$@CuPO**: the lower rim recognizes $RbNO_3(H_2O)$, the cone cavity MeOH.

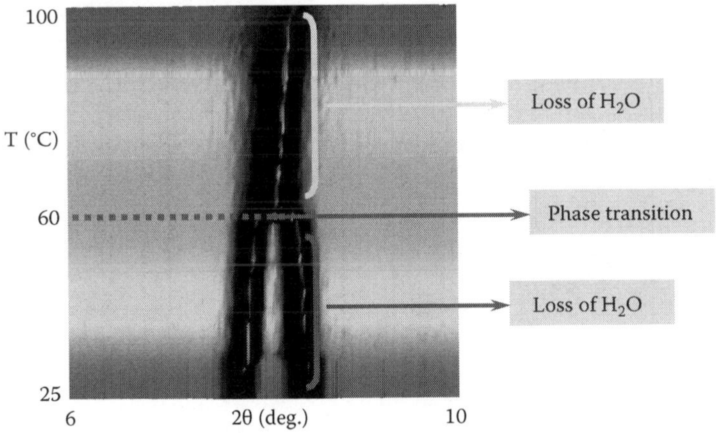

FIGURE 2.20 Variable temperature XRPD traces in the 6–10° 2θ range, showing the progressive merging of the 110 and 012 peaks of **CuPR** (a = 23.040(2), c = 25.140(2) Å) into the 110 reflection of the cubic **CuPC** phase (a_0 = 15.07 Å). Vertical scale, T in the 25–100°C range (bottom to top, 10 scans per 10°C interval). Note that at each T, equilibrium was fully reached. Coalescence occurs at 60°C.

affected. In Cu(2-pymo)$_2$, the internal surface of the cavities is decorated by hydrogen atoms; in Cu(4-pymo)$_2$ one hydrogen out of three is substituted by an oxygen atom. This structural diversity is reflected by the remarkably lower affinity of Cu(4-pymo)$_2$ toward different probe gases, in spite of the comparable empty volumes the two Cu(II) species possess upon activation (Table 2.2).[85b,88] Also, the ion pairs' affinity of Cu(4-pymo)$_2$ is deeply influenced: No stoichiometric species characterized by selective recognition of ion pairs have been isolated.

Breathless

The Pd(2-pymo)$_2$[88] and Pd(F-pymo)$_2$[89] materials (F-pymo = 5-fluoro-pyrimidin-2-olate) were subsequently isolated and characterized, to verify the influence of the metal ion and the ligand substitution on the structural aspects and on the adsorption properties. These derivates possess a *cubic sodalitic* framework, both in the hydrated and in the evacuated form. Dedicated thermodiffractometric studies highlighted that they possess a certain framework rigidity upon dehydration. This behavior is distinct from that of the Cu(II) MOFs (Figure 2.21), and may be explained by invoking, once again, subtle structural differences.

In the hydrated Cu(II) species, interactions exist of the Cu···O_{water} kind. Upon water removal, the decrease in the Cu(II) coordination number brings about a shrinking of the Cu-N bonds, which implies a shortening of the Cu···Cu bridged vectors, with a consequent contraction of the unit cell. This event is not experienced by the Pd(II) materials, due to the limited tendency of the Pd(II) ions to possess higher coordination numbers than 4. Moreover, the Pd(II) species feature longer Pd-N distances, i.e., higher empty volumes and surface areas, concurring to explain their better adsorption performances (Table 2.2).

Blow Up

These outcomes were followed by the isolation and characterization of third-generation,[90] i.e., flexible, porous metal pyrimidinolates, undergoing a reversible framework modification as a response to the size and shape demands of the guest itself.

At variance with the above presented porous MOFs, Cu(F-pymo)$_2$[91] possesses a *gismondinic tetragonal* framework, featuring metallacalix[4]arenes, metallacalix[8]arenes, and helical channels of 2.9 Å diameter running parallel to the crystallographic **c** axis (Figure 2.22).

Apparently striking, if just its kinetic diameter is taken into consideration (3.19 Å), is the remarkable adsorption of CO_2 by Cu(F-pymo)$_2$ (7.6 weight percent at 273 K and 900 torr), following fully reversible type I isotherms, with steep slopes at very low pressures, indicative of pores of small size. Thermodiffractometric experiments under CO_2 showed that, on cooling in the 200–300°C range, the channels expand to accommodate the gas molecules: At 0°C, the thermal energy of the CO_2 molecules and of the framework itself is high enough to widen the channels and permit guests diffusion. Moreover, the Lewis acidic character of CO_2 may allow acid-base interactions with the F-substituents decorating the channels' walls, thus facilitating CO_2 entrance. Rather surprisingly, upon cooling under CO_2, a transient phase is detected

TABLE 2.2
Synoptic Collection of the Main Adsorption Properties of the M(X-pymo)$_2$ Species

Species	Cell V RT, hydrated Å^3	Cell V, % Void Anhydrous Å^3	SA $m^2\ g^{-1}$	Adsorbed N_2 77 K, 76 torr $cm^3\ g^{-1}$ STP	Adsorbed CO_2 273 K, 900 torr $cm^3\ g^{-1}$ STP	Adsorbed H_2 77 K, 900 torr Weight %	Adsorbed H_2 77 K, 900 torr kg H_2 L^{-1}
Cu(2-pymo)$_2$	3,852	3,422, 27.8	350[c]	150	n.a.	0.86	0.013
Cu(4-pymo)$_2$	3,920	3,512, 24.4	65[c]	30	n.a.	0.03	<0.001
Pd(2-pymo)$_2$	4,315	1,801, 41.3	600[c]	150	69	13	0.018
Pd(F-pymo)$_2$	4,340	1,670, 38.8	600[c]	149	84	1.15	0.018
Cu(F-pymo)$_2$	4,337	n.a., 13.0[a]	n.a.	None	39	0.56[f]	0.010
Co(F-pymo)$_2$	5,758	4,985, 13.0	300[d]	None	160[e]	None	
Zn(F-pymo)$_2$	5,906	1,518, negligible[b]	n.a.	None	180[e]	None	

[a] This value should be considered an approximation derived from the room temperature crystal structure after removing the water molecules.
[b] The symmetry change, from trigonal to triclinic, imposed a 3× reduction of the cell volume.
[c] Brunauer, Emmett, Teller (BET) method applied on the N_2 isotherm at 77 K.
[d] Dubinin-Radushkevich method applied on the CO_2 isotherm at 273 K.
[e] At 273 K and ca. 2,800 kPa.
[f] At 90 K.

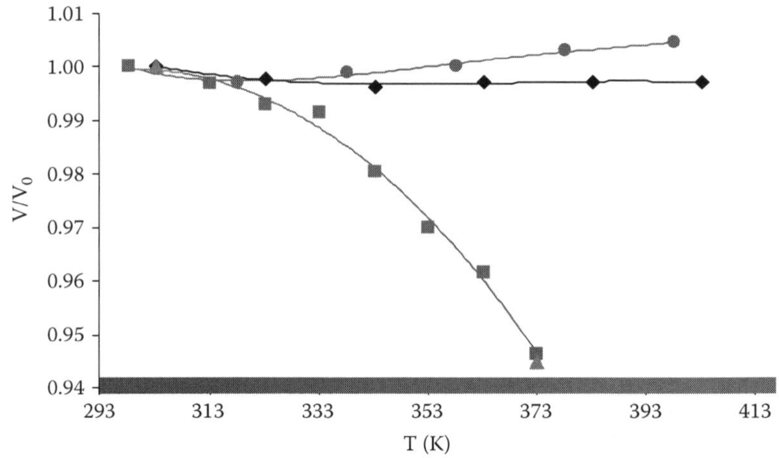

FIGURE 2.21 Thermal behavior of the unit cell volume V, normalized to the 30°C value (V_0), for Cu(2-pymo)$_2$ (green squares), Cu(4-pymo)$_2$ (orange triangles), Pd(2-pymo)$_2$ (red circles), and Pd(F-pymo)$_2$ (brown rhombi).

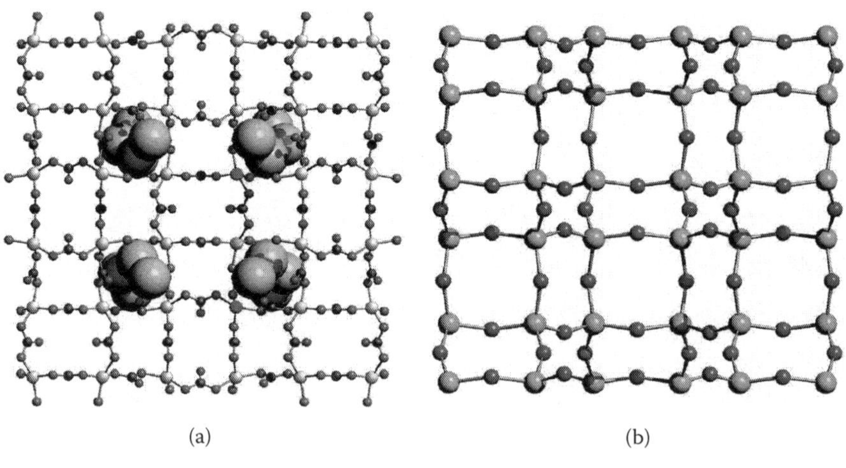

FIGURE 2.22 (See color insert) View, down [001], of (a) the crystal structure of Cu(F-pymo)$_2$, as compared to (b) the inorganic gismondine. Highlighted with PCK style, the four helical channels comprised within one unit cell.

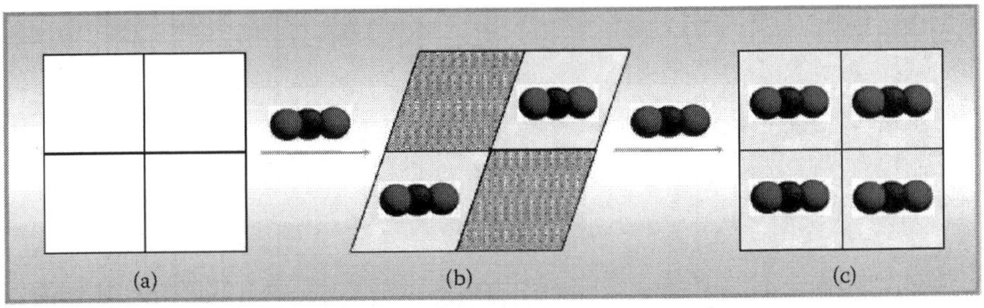

FIGURE 2.23 Structural modifications experienced by the Cu(F-pymo)$_2$ MOF upon CO_2 inclusion: (a) activated, void framework; (b) start of the CO_2 settling within two opposite helical channels out of the four of the unit cell, with concomitant rhombic distortion; (c) settling of the CO_2 molecules in the remaining two helical channels, with restoring of the pristine tetragonal symmetry. The orientation of the CO_2 molecules within the framework has been arbitrarily assigned.

at about 90°C, possessing a lower (*rhombic*) symmetry than the parent phase: At a certain stage of CO_2 adsorption, settling of the guest molecules cooperatively occurs in two opposite channels out of the four of the unit cell (Figure 2.23), promoting a rhombic distortion of the pristine tetragonal lattice. When settling involves the other two channels, then the tetragonal symmetry is restored.

With the M(F-pymo)$_2$ materials (M = Co, Zn)[92] we turn back again to *trigonal sodalitic* MOF. Heated up to 90°C, the M(F-pymo)$_2$·2.5H_2O as synthesized compounds (**α-M**) reversibly dehydrate to *sodalitic* M(F-pymo)$_2$ species (**β-M**) of significantly different lattice metrics. Further heating induces irreversible polymorphic transformations into *layered orthorhombic* M(F-pymo)$_2$ phases (**γ-M**), in which the original MN_4 coordination sphere changes into the MN_3O one (Figure 2.24). The **α-Co** → **β-Co** transformation leads to a *trigonal sodalitic* **β-Co** phase, leaving the main topological features unaltered, and with a concomitant cell volume decrease of 13.4%. The **α-Zn** → **β-Zn** phase change, occurring with a higher cell volume decrease (24.5%), leads to a symmetry loss down to the triclinic system, the overall network still possessing *sodalitic* topology.

At 77 K, in the lower pressures range, the activated **β-M** materials do not adsorb either N_2 or H_2. This behavior may be traced back to the significant shrinkage experienced by the unit cell volumes upon dehydration, worsened by the probable framework rigidity at low temperatures. CH_4 molecules are not adsorbed even in the higher pressures range (up to 2,600 kPa), possibly due to size exclusion reasons and to the apolar nature of the gas molecules.[93] By contrast, at 273 K, CO_2 enters the porous networks of **β-M**. For both materials, above a certain gate-opening pressure (ca. 600 kPa and 900 kPa for **β-Co** and **β-Zn**, respectively), a sudden rise in the amount of adsorbed CO_2 is observed (Table 2.2). Possibly, at 273 K, the higher kinetic energy of the gas molecules and the higher thermal energy of the coordination networks may allow CO_2 to open the gate,[94] i.e., to force the transformation of the **β-M** phases back to the **α-M** open frameworks.

These latter studies have been performed using a cheap temperature-controlled sample holder (Figure 2.25),[95] which, thanks to the Peltier battery technology, can cool (down

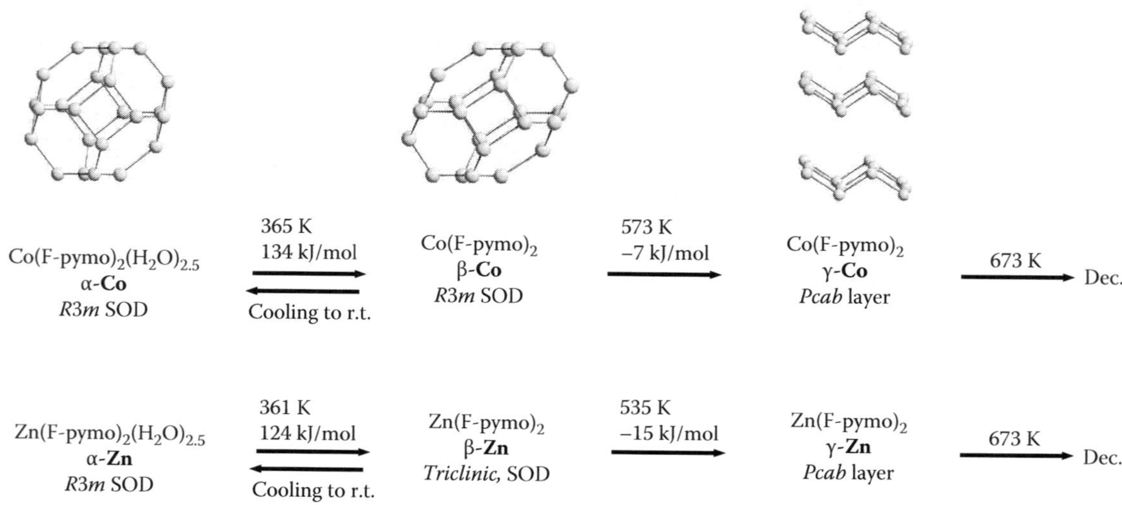

FIGURE 2.24 (See color insert) Schematic representation of the thermal behavior of the M(F-pymo)$_2$(H$_2$O)$_{2.5}$ species (**β-M**, M = Co, Zn), as retrieved from the concomitant usage thermal analysis and thermodiffractometry.

FIGURE 2.25 (See color insert) Custom-made temperature-controlled sample holder mounted on our x-ray powder diffractometer.

to –20°C) or heat (up to 160°C) the deposited powders with programmable linear gradients and ±0.1°C stability, even under a gas flux, when the proper sample chamber is mounted. The detection and characterization of sample dehydration processes (which, inter alia, is an extremely important feature during drug development and formulation in the pharmaceutical industry) is thus straightforward.

CONCLUSIONS AND OUTLOOK

Polymeric metal diazolates (pyrazolates, imidazolates, pyrimidin-X-olates, X = 2, 4) typically appear as insoluble and intractable powders. However, we have shown in this review that the complete retrieval of their structures, using *ab initio* x-ray powder diffraction methods and well-tuned laboratory diffractometers, is now possible. This methodology has been successfully applied to more than 50 species of this class, and not only benchmark case structures. Accordingly, the results schematically presented above on a large class of coordination polymers, structurally characterized by *ab initio* XRPD, clearly demonstrate that powder diffraction is indeed a very efficient and powerful tool in the structural chemist's hand (once properly employed with tailored software resources) for the complete understanding of the origin of a number of functional properties, such as thermal, optical, magnetic, and sorption behavior.

However, it must be continuously stressed that the structural "details" that can be obtained by powder diffraction are rather blurred; nevertheless, XRPD still affords plenty of useful, otherwise inaccessible, information, such as molecular shape, heavy atoms stereochemistry, rough bonding parameters, crystal packing, and nature of paracrystallinity effects. Even if these were the only attainable results,

XRPD would be considered an unavoidable source of structural information and a fruitful complement to other structural techniques. Of even higher importance, as recently reviewed,[96] the availability of ancillary methods (of the experimental or even computational type: selected-area electron or neutron diffraction, NMR, steric energy estimates, etc.) can further improve the "defocused" structural image retrieved by the *ab initio* XRPD process, and therefore should be employed whenever a higher-resolution picture of the structure, which XRPD alone cannot afford, is sought.

According to the results reported above, and to the slow, but steady, increase of crystal structures solved and refined, uniquely, from powder diffraction data,[97] more than ever, we feel that the *ab initio* XRPD method will soon be incorporated in the research activity of a number of laboratories (dealing with, e.g., polymeric functional materials, molecular magnets, catalysts, pharmacologically active species, etc.), as well as in the classrooms of chemistry, physics, and material science students.

ACKNOWLEDGMENTS

We thank all coworkers who have participated in this project for their continuous support and helpful discussions. The Italian MIUR (Prin 2006: Polimeri di Coordinazione con Proprietà Funzionali per l'Ottica Non-Lineare, il Magnetismo, la Catalisi Eterogenea ed il Riconoscimento Molecolare), CNR, University of Insubria (Progetto di Eccellenza Sistemi Poliazotati), Fondazioni Provinciale Comasca, and CARIPLO (Project 2007-5117) are acknowledged for funding.

REFERENCES

1. Generally, 2θ, sinθ, t, or 1/d, depending on the nature of the experiment (angle vs. energy-dispersive diffraction, time of flight, etc.).
2. Pecharsky, V. K., and Zavalij, P. Y. 2003. *Fundamentals of powder diffraction and structural characterization of materials.* Boston: Kluwer Academic Publishers; Clearfield, A., ed. 2002. *Crystal structure determination from powder diffraction.* American Crystallographic Association; Dinnebier, R., and Billinge, S., eds. 2008. *Powder diffraction: theory and practice.* Cambridge: Royal Society of Chemistry.
3. Masciocchi, N., ed. 2004. *Powder diffraction of molecular functional materials.* CPD Newsletter 31. IUCr, Chester, UK.
4. With the notable exception of the so-called charge flipping methods, recently proposed in *Oszlányi, G., and* Süto, A. 2004. *Acta Crystallogr.*, A60, 134.
5. La Monica, G., and Ardizzoia, G. A. 1997. *Prog. Inorg. Chem.,* 46, 151, and references therein.
6. (a) Perera, J. R., Neeg, M. J., Schlegel, H. B., and Winter, C. H. 1999. *J. Am. Chem. Soc.,* 121, 4536; (b) Gust, K. R., Knox, J. E., Heeg, M. J., Schlegel, H. B., and Winter, C. H. 2002. *Eur. J. Inorg. Chem.*, 2327.
7. See, for example, Deacon, G. B., Gitlits, A., Roesky, P. W., Bürgstein, M. R., Lim, K. C., Skelton, B. W., and White, A. H. 2001. *Chem. Eur. J.*, 7, 127, and references therein.
8. Barron, A. R., Wilkinson, G., Motevalli, M., and Hursthouse, M. B. 1985. *Polyhedron,* 4, 1131.
9. Büchner, E. 1889. *Ber. Dtsch. Chem. Ges.,* 22, 842.
10. Trofimenko, S. 1972. *Chem. Rev.*, 72, 497, and references therein.

11. Masciocchi, N., Moret, M., Cairati, P., Sironi, A., Ardizzoia, G. A., and La Monica, G. 1994. *J. Am. Chem. Soc.*, 116, 7768.
12. The increase of the tolerated cell volume is reflected by the simultaneous appearance of extra peaks in calculated positions, not matched by observable diffracted intensity. However, if the space group symmetry imposes several systematic absence conditions, the required set of observable peaks (at low 2θ) is not significally augmented. This was indeed the case for the *Pbcn* space group of the silver trimer.
13. Cascarano, G., Favia, L., and Giacovazzo, C. 1992. *J. Appl. Crystallogr.*, 25, 310.
14. Dinnebier, R. E., Albrich, F., VanSmaalen, S., and Stephens, P. W. 1997. *Acta Crystallogr.*, B53, 153.
15. Unless the few, possibly weak, "unindexable" peaks are attributed to a known contaminant phase.
16. (a) Kahn, O., Kröber, J., and Jay, C. 1992. *Adv. Mater.*, 4, 718; (b) Garcia, Y., van Koningsbruggen, P. J., Codjovi, E., Lapouyade, R., Kahn, O., and Rabardel, L. 1997. *J. Mater. Chem.*, 7, 857.
17. Gütlich, P., Garcia, Y., and Goodwin, H. A. 2000. *Chem. Soc. Rev.*, 29, 419.
18. van Koningsbruggen, P. J., Garcia, Y., Codjovi, E., Lapouyade, R., Kahn, O., Fournès, L., and Rabardel, L. 1997. *J. Mater. Chem.*, 7, 2069.
19. (a) Jung, O.-S., and Pierpont, C. G. 1994. *J. Am. Chem. Soc.,* 116, 2229; (b) Jung, O.-S., Lee, Y.-A., and Pierpont, C. G. 1995. *Synth. Met.,* 71, 2019.
20. Smit, E., Manoun, B., Verryn, S. M. C., and de Waal, D. 2001. *Powder Diffr.*, 16, 37.
21. (a) Vos, J. G., and Groeneveld, W. L. 1997. *Inorg. Chim. Acta*, 23, 123; (b) Vos, G., le Fêbre, R. A., de Graaff, R. A. G., Haasnot, J. G., and Reedijk, J. 1983. *J. Am. Chem. Soc.*, 105, 1682; (c) Garcia, Y., van Koningsbruggen, P. J., Bracic, G., Guionneau, P., Chasseau, D., Cascarano, G. L., Moscovici, J., Lambert, K., Michalowicz, A., and Kahn, O. 1997. *Inorg. Chem.*, 136, 6357; (d) Michalowicz, A., Moscovici, J., Charton, J., Sandid, F., Benamrane, F., and Garcia, Y. 2001. *J. Synch. Rad.*, 8, 701.
22. van Koningsbruggen, P. J., Garcia, Y., Kahn, O., Fournès, L., Koijman, H., Spek, A., Haasnoot, J. G., Moscovici, J., Provost, K., Michalowicz, A., Renz, F., and Gütlich, P. 2000. *Inorg. Chem.*, 39, 1891.
23. With the notable exceptions of the bispyrazolate-iron(II) (Patrick, B. O., Reiff, W. M., Sanchez, V., Storr, A., and Thompson, R. C. 2001. *Polyhedron*, 20, 1577) and bis(1-methyl-2-thioimidazolate)iron(II) (Rettig, S. J., Sánchez, V., Storr, A., Thompson, R. C., and Trotter, J. 1999. *Inorg. Chem.*, 38, 5920) polymers, containing tetrahedral d^6 ions, the structures of which have been recently correlated to their magnetic properties.
24. Cingolani, A., Galli, S., Masciocchi, N., Pandolfo, L., Pettinari, C., and Sironi, A. 2005. *J. Am. Chem. Soc.*, 127, 6144.
25. Masciocchi, N., Ardizzoia, G. A., Brenna, S., LaMonica, G., Maspero, A., Galli, S., and Sironi, A. 2002. *Inorg. Chem.,* 41, 6080, and references therein.
26. See also (a) α-$Cu(pz)_2$: Ehlert, M. K., Rettig, S. J., Storr, A., Thompson, R. C., and Trotter, J., 1989. *Can. J. Chem.,* 67, 1970; (b) $Zn(pz)_2$ and $Cd(pz)_2$: Masciocchi, N., Ardizzoia, G. A., Maspero, A., LaMonica, G., and Sironi, A. 1999. *Inorg. Chem.,* 38, 3657.
27. This is particularly true when substituted polyazoles and noncoordinating anions are used instead of the simple pyrazolate, the packing density of the polymers being lowered by side chains, anions, and in some cases, solvent molecules.
28. Kröber, J., Codjovi, E., Kahn, O., Grolière, F., and Jay, C. 1993. *J. Am. Chem. Soc.*, 115, 9810.
29. van Koningsbruggen, P. J., Garcia, Y., Codjovi, E., Lapouyade, R., Kahn, O., Fournès, L., and Rabardel, L. 1997. *J. Mater. Chem.*, 7, 2069.

30. Kuroiwa, K., Shibata, T., Takada, A., Nemoto, N., and Kimizuka, N. 2004. *J. Am. Chem. Soc.*, 126, 2016.
31. Similarly, in computational chemistry, large residues are often substituted by simpler isolobal fragments, aiming for a significant reduction of computing times.
32. Masciocchi, N., Ragaini, F., Cenini, S., and Sironi, A. 1998. *Organometallics*, 17, 1052.
33. Masciocchi, N., Ardizzoia, G. A., Brenna, S., Castelli, F., Galli, S., Maspero, A., and Sironi, A. 2003. *Chem. Commun.* 2018.
34. Ribar, B., Matkovic, M., Sljukic, F., Gabela, F. 1971. *Z. Kristallogr.*, 134, 311.
35. (a) Yélamos, C., Gust, K. R., Baboul, A. G., Heeg, M. J., Schlegel, H. B., and Winter, C. H. 2001. *Inorg. Chem.,* 40, 6451; (b) Mösch-Zanetti, N. C., Ferbinteanu, M., and Magull, J. 2002. *Eur. J. Inorg. Chem.* 950.
36. Zheng, W., Heeg, M. J., and Winter, C. H. 2003. *Angew. Chem. Int. Ed.* 42, 2761, and references therein.
37. Ohtsu, H., Shinobu, I., Nagatomo, S., Kitagawa, T., Ogo, S., Watanabe, Y., and Fukuzumi, S. 2000. *Chem. Commun.*, 1051, and references therein.
38. Masciocchi, N., Moret, M., Cairati, P., Sironi, A., Ardizzoia, G. A., and La Monica, G. 1995. *J. Chem. Soc. Dalton Trans.* 1671.
39. (a) Bauman, J. E., Jr., and Wang, J. C. 1964. *Inorg. Chem.*, 3, 368; (b) Sigwart, C., Kroneck, P., and Hemmerich, P. 1970. *Helv. Chim. Acta*, 53, 177.
40. Tian, Y.-Q., Xu, H.-J., Weng, L.-H., Chen, Z.-X., Zhao, D.-Y., and You, X.-Z. 2004. *Eur. J. Inorg. Chem.*, 1813.
41. Huang, X.-C., Zhang, J.-P., and Chen, X.-M. 2006. *Cryst. Growth Des.,* 6, 1194, and references therein.
42. Buck, M., private communication.
43. http://www.portfolio.mvm.ed.ac.uk/studentwebs/session2/group29/infdru.htm
44. See, for example, Nomiya, K., Tsuda, K., Sudoh, T., and Oda, M. 1997. *J. Inorg. Biochem.*, 68, 39.
45. Ardizzoia, G. A., Brenna, S., Castelli, F., Galli, S., LaMonica, G., Masciocchi, N., and Maspero, A. 2004. *Polyhedron*, 23, 3063.
46. Ardizzoia, G. A., La Monica, G., Maspero, A., Moret, M., and Masciocchi, N. 1997. *Inorg. Chem.*, 36, 2321.
47. Werner, P. E., Eriksson, L., and Westdahl, M. 1985. *J. Appl. Crystallogr.,* 18, 367.
48. Shirley, R. 2002. *The Crysfire 2002 System for automatic powder indexing: User's manual.* Guildford, Surrey, England: The Lattice Press.
49. Jarvis, J. A. J., and Wells, A. F. 1960. *Acta Crystallogr.*, 13, 1027.
50. Bruni, S., Cariati, E., Cariati, F., Galli, S., and Sironi, A. 2001. *Inorg. Chem.*, 40, 5897.
51. The similarity of its XRPD trace with that of a still uncharacterized yellow $Ni(im)_2$ phase (Masciocchi, N., Castelli, F., Forster, P. M., Tafoya, M. M., and Cheetham, A. K. 2003. *Inorg.Chem.*, 42, 6147) witnesses the isomorphous character and the square planar disposition of the MN_4 chromophores.
52. Gasparač, R., Stupnišek-Lisac, E., and Martin, C. R. 2008. Imidazole and its derivatives as inhibitors for prevention of corrosion of copper. In *Electrochemical approach to selected corrosion and corrosion control studies*, ed. P. L. Bonora and F. Deflorian, 20–36. EFC Book Series 28, Leeds, UK: Maney.
53. Masciocchi, N., Ardizzoia, G. A., La Monica, G., Moret, M., and Sironi, A. 1997. *Inorg. Chem.*, 36, 449.
54. Masciocchi, N., Cairati, P., Carlucci, L., Mezza, G., Ciani, G., and Sironi, A. 1996. *J. Chem. Soc. Dalton Trans.,* 2739.
55. Brinkmann, M., Gadret, G., Muccini, M., Taliani, C., Masciocchi, N., Sironi, A. 2000. *J. Am. Chem. Soc.*, 122, 5147; Muccini, M., Loi, M. A., Keveney, K., Zamboni, R., Masciocchi, N., and Sironi, A. 2004. *Adv. Mater.*, 16, 861.

56. Malpezzi, L., Magnone, G. A., Masciocchi, N., and Sironi, A. 2004. *J. Pharm. Sci.,* 94, 1067.
57. Maccaroni, E., Alberti, E., Malpezzi, L., Masciocchi, N., and Vladiskovic, C. 2007. *Int. J. Pharm.,* 251, 144.
58. Maccaroni, E., Alberti, E., Malpezzi, L., Masciocchi, N., and Pellegatta, C. 2008. *J. Pharm. Sci.,* 97, 5229.
59. Maccaroni, E., Alberti, E., Malpezzi, L., Razzetti, G., Vladiskovic, C., and Masciocchi, N. 2009. *Cryst. Growth Des.*, 9, 517.
60. Maccaroni, E., Giovenzana, G. B., Palmisano, G., Botta, D., Volante, P., and Masciocchi, N. 2009. *Steroids,* 74, 102.
61. (a) Sturm, M., Brandel, F., Engel, D., and Hoppe, W. 1975. *Acta Crystallogr.*, B31, 2369; (b) Lehnert, R., and Seel, F. 1980. *Z. Anorg. Allg. Chem.*, 464, 187.
62. Very recently, $Cd(im)_2$ was also prepared by solvothermal methods, and found to be strongly photoluminescent: Tian, Y. Q., Xu, L., Cai, C. X., Wei, J. C., Li, Y. Z., and You, X. Z. 2004. *Eur. J. Inorg. Chem.*, 1039.
63. *TOPAS-R*, version 3.1. 2003. Karlsruhe, Germany: Bruker AXS.
64. Quirós, M. 1994. *Acta Crystallogr.*, C50, 1236.
65. Navarro, J. A. R., Fresinger, E., and Lippert, B. 2000. *Inorg. Chem.*, 39, 2301; Barea, E., Navarro, J. A. R., Salas, J. M., Quirós, M., Willermann, M., and Lippert, B. 2003. *Chem. Eur. J.*, 9, 4414.
66. Masciocchi, N., Corradi, E., Moret, M., Ardizzoia, G. A., Maspero, A., La Monica, G., and Sironi, A. 1997. *Inorg. Chem.*, 36, 5648.
67. The possibility of a N,O-coordination, similar to that found in the $[Cu(mpyo)]_4$ (Hmpyo = 2-hydroxy-6-methylpyridine) tetramer (Berry, M., Clegg, W., Garner, C. D., and Hillier, I. H. 1982. *Inorg. Chem.*, 21, 1342), was also originally considered, but IR evidences, hard/soft acid/base considerations, and successful refinement led to a more plausible N,N'-link.
68. See, for example, Batten, S. R., Hoskins, B. F., and Robson, R. 1997. *Angew. Chem. Int. Ed. Engl.*, 36, 636; Withersby, M. A., Blake, A. J., Champness, N. R., Hubberstey, P., Li, W. S., and Schroeder, M. 1997. *Angew. Chem. Int. Ed. Engl.*, 36, 2327.
69. For instance, $Zn(CN)_2$ is quantitatively recovered in pure crystalline form from the thermal decomposition of zinc bispyrazolate, with concomitant acetonitrile extrusion.
70. See, for example, the poly(*p*-hydroxybenzoic) acid (PHBA): Yoon, D. Y., Masciocchi, N., Depero, L. E., Viney, C., and Parrish, W. 1990. *Macromolecules*, 23, 1793.
71. Masciocchi, N., Ardizzoia, G. A., La Monica, G., Maspero, A., and Sironi, A. 2000. *Eur. J. Inorg. Chem.*, 2507.
72. Lloret, F., De Munno, G., Julve, M., Cano, J., Ruiz, R., and Caneschi, A. 1998. *Angew. Chem. Int. Ed. Engl.*, 37, 2781.
73. Masciocchi, N., Galli, S., Sironi, A., Barea, E., Navarro, J. A. R., Salas, J. M., and Tabares, L. C. 2003. *Chem. Mater.,* 15, 2153.
74. Actually, another hydrated phase, $Ni(2\text{-pymo})_2(H_2O)_{2.5}$, absent in the cobalt(II) system, was found.
75. Barea, E., Navarro, J. A. R., Salas, J. M., Masciocchi, N., Galli, S., and Sironi, A. 2004. *Inorg.Chem.,* 43, 473.
76. Corma, A. 1997. *Chem. Rev.,* 97, 2373.
77. See, e.g., (a) Yaghi, O. M., O'Keeffe, M., Ockwig, N. W., Chae, H. K., Eddaoudi, M., and Kim, J. 2003. *Nature*, 423, 705; (b) Janiak, C. 2003. *J. Chem. Soc. Dalton Trans.*, 2781.
78. Eddaoudi, M., Kim, J., Rosi, N., Vodak, D., Wachter, J., O'Keeffe, M., and Yaghi, O. M. 2002. *Science*, 295, 469.
79. Keller, S. W., and Lopez, S. 1999. *J. Am. Chem. Soc.*, 121, 6306.
80. Kitaura, R., Seki, K., Akiyama, G., and Kitagawa, S. 2003. *Angew. Chem. Int. Ed.*, 42, 428.
81. (a) Seo, J. S., Whang, D., Lee, H., Jun, S. I., Oh, J., Jeon, Y. J., and Kim, K. 2000. *Nature*, 404, 982; (b) Lee, S. J., and Lin, W. 2002. *J. Am. Chem. Soc.*, 124, 4554.

82. (a) Tian, Y.-Q., Zhao, Y.-M., Chen, Z.-X., Shang, G.-N., Weng, L.-H., and Zhao, D.-Y. 2007. *Chem. Eur. J.*, 13, 4146; (b) Huang, X.-C., Lin, Y.-Y., Zhang, J.-P., and Chen, X.-M. 2006. *Angew. Chem. Int. Ed.*, 45, 1557.
83. (a) Park, K. S., Ni, Z., Côté, A. P., Choi, J. Y., Huang, R., Uribe-Romo, F. J., Chae, H. K., O'Keeffe, M., and Yaghi, O. M. 2006. *Proc. Natl. Acad. Sci. USA*, 103, 10186; (b) Hayashi, H., Côté, A. P., Furukawa, H., O'Keeffe, M., and Yaghi, O. M. 2007. *Nature Mater.*, 6, 501.
84. Navarro, J. A. R., Barea, E., Galindo, M. A., Salas, J. M., Romero, M. A., Quirós, M., Masciocchi, N., Galli, S., Sironi, A., and Lippert, B. 2005. *J. Solid State Chem.*, 178, 2436.
85. (a) Tabares, L. C., Navarro, J. A. R., and Salas, J. M. 2001. *J. Am. Chem. Soc.,* 123, 383; (b) Barea, E., Navarro, J. A. R., Salas, J. M., Masciocchi, N., Galli, S., and Sironi, A. 2003. *Polyhedron*, 22, 3051; (c) Barea, E., Navarro, J. A. R., Salas, J. M., Masciocchi, N., Galli, S., and Sironi, A. 2004. *J. Am. Chem. Soc.,* 126, 3014.
86. Apart from the evacuated **CuPC** and **CuPO** species, all species host, in their MOF channels, a number of water molecules.
87. Putnis, A. 1992. *Introduction to Mineral Sciences*. Cambridge: Cambridge University Press.
88. Navarro, J. A. R., Barea, E., Salas, J. M., Masciocchi, N., Galli, S., Sironi, A., Ania, C. O., and Parra, J. B. 2006. *Inorg. Chem.*, 45, 2397.
89. Navarro, J. A. R., Barea, E., Salas, J. M., Masciocchi, N., Galli, S., Sironi, A., Ania, C. O., and Parra, J. B. 2007. *J. Mater. Chem.*, 17, 1939.
90. Uemura, K., Matsuda, R., and Kitagawa, S. 2005. *J. Solid State Chem.*, 178, 2420.
91. Navarro, J. A. R., Barea, E., Rodríguez-Diéguez, A., Salas, J. M., Ania, C. O., Parra, J. B., Masciocchi, N., Galli, S., and Sironi, A. 2008. *J. Am. Chem. Soc.*, 130, 3978.
92. Galli, S., Masciocchi, N., Tagliabue, G., Sironi, A., Navarro, J. A. R., Salas, J. M., Mendez-Liñan, L., Domingo, M., Perez-Mendoza, M., and Barea, E. 2008. *Chem. Eur. J.*, 14, 9890.
93. Llewellyn, P. L., Bourrelly, S., Serre, C., Filinchuk, Y., and Férey, G. 2006. *Angew. Chem. Int. Ed.*, 45, 7751.
94. Tanaka, D., and Kitagawa, S. 2008. *Chem. Mater.*, 20, 922.
95. Equipped with zero background plates and supplied by Officina Elettrotecnica di Tenno, Italy.
96. Masciocchi, N., and Sironi, A. 2005. *Comptes Rendus Chimie*, 8, 1617.
97. See, for example, Armel Le Bail's Web site: http://www.cristal.org/iniref.html
98. Powder Diffraction File, ICDD, Swarthmore, PA, PDF 52-2401.

3 Single Crystal Neutron Diffraction for the Inorganic Chemist— A Practical Guide

Paula M. B. Piccoli, Thomas F. Koetzle, and Arthur J. Schultz

CONTENTS

Abstract: Advances and upgrades in neutron sources and instrumentation are poised to make neutron diffraction more accessible to inorganic chemists than ever before. These improvements will pave the way for single crystal investigations that currently may be difficult, for example, due to small crystal size or large unit cell volume. This article aims to highlight what can presently be achieved in neutron diffraction and looks forward toward future applications of neutron scattering in inorganic chemistry.

INTRODUCTION

The impact of x-ray diffraction on the development of inorganic chemistry is undeniable. From the organometallic literature we cite Kealy and Pauson's initial report[1] on ferrocene and the resulting structure from x-ray diffraction,[2,3] along with the discovery of agostic bonds in Trofimenko's scorpionate complexes,[4–7] as seminal examples that revolutionized the way in which inorganic chemists think about structure and bonding. The automated diffractometer enabled routine structural characterization of inorganic and organometallic compounds and arguably has contributed more to the advancement of the field than any other structure characterization tool.

However, a major limitation of x-ray diffraction is its insensitivity to hydrogen atoms. This will hold especially when the proton is located close to a metal atom, which will dominate the scattering in the x-ray experiment. Whereas ^{1}H NMR can provide valuable information about the chemical environment of hydrogen atoms in a compound, the particular advantage of diffraction techniques resides in their ability to determine structure at the atomic level. Neutron diffraction, with its ability to accurately locate and characterize hydrogen atoms, is ideal for providing information that cannot be gained from x-ray diffraction alone. Properties of the neutron, particularly including the fact that it scatters from the nucleus of the atom and possesses spin and a magnetic moment, make it a powerful probe for chemical structure determination and a complement to x-ray diffraction.

Fortunately, analysis of single crystal neutron diffraction data is very analogous to that of x-ray data. Widely available software packages, including SHELX[8] and GSAS,[9] can be used to refine neutron structures, thereby allowing x-ray crystallographers to conveniently perform their own analyses. As facilities at existing sources such as the ILL (Grenoble, France), SINQ (Switzerland), ISIS (RAL, UK), Lujan Center (Los Alamos, New Mexico), and IPNS (Argonne, Illinois, decommissioned, 2008) are joined over the next several years by those at more advanced neutron sources, including the Spallation Neutron Source (SNS) in Oak Ridge, Tennessee, the KEK-JAEA Japan Spallation Neutron Source (JSNS) of J-PARC Center, the new reactor OPAL at ANSTO in Australia, and the second target station at ISIS, opportunities for chemists to engage in single crystal neutron diffraction will increase dramatically, as will our ability to handle smaller crystal sizes and larger unit cell volumes, two factors that currently limit the problems we can explore. These new facilities, with higher beam intensities and new instrumentation, will expand on the excellent science already being done at established facilities around the globe. These developments are poised to bring single crystal neutron diffraction to the doorstep of the majority of inorganic and organometallic chemists and promise greater availability than ever before.

In recent years several comprehensive reviews on the topic of single crystal neutron diffraction in relation to chemical crystallography have been published.[10–12] It is the goal of this chapter to inform the greater community of inorganic chemists how neutron scattering can be utilized, now and in the future, to answer the research questions not currently addressable by conventional x-ray crystallographic techniques. We will highlight some important results from neutron diffraction of small molecules, focusing particularly on results of the last several years. Hopefully we can address here many of the questions that new users will have when contemplating a single crystal neutron experiment. Neutron scattering can be useful to the chemist for a number of applications other than single crystal diffraction; for an overview we refer the reader to Roger Pynn's "Neutron Scattering: A Primer," which provides an excellent general introduction.[13]

PRACTICAL MATTERS: THE EXPERIMENT

Properties of the Neutron

Neutrons are neutral, subatomic particles possessing a magnetic moment that interact with matter in a different manner than do x-rays. Neutrons scatter from the atomic nuclei, whereas x-rays or electrons scatter primarily from the electrons surrounding the nuclei. As we see in Figure 3.1, although there is a general tendency for neutron scattering lengths to increase with atomic number, there is a quite random variation of scattering lengths between elements and, for that matter, among isotopes of the same element.

The contrast between hydrogen and deuterium, with their respective negative and positive scattering lengths, allows for good discrimination between the two isotopes. The contrast can be a very useful feature, for example, when determining the percentage of deuterium in a partially deuterated complex where the fractional occupancies of H and D on the same site can be refined. This was demonstrated in the investigation of an equilibrium isotope effect in the metal hydride $H_2Os_3(CO)_{10}CH_2$.[14] Neutron diffraction data showed a preferential distribution of deuterium in the methylene ligand and hydrogen in the hydride sites, as expected on the basis of zero-point energy considerations, findings also supported by NMR. This contrasting of H and D in neutron diffraction provides especially valuable information when investigating biological structures. One ubiquitous manifestation of the negative scattering length

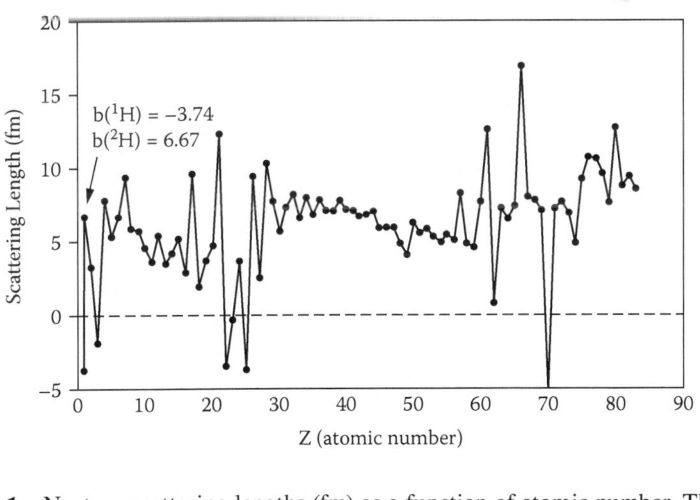

FIGURE 3.1 Neutron scattering lengths (fm) as a function of atomic number. The neutron scattering length varies quite randomly across the periodic series and, if scattering lengths are sufficiently different, neutron scattering can distinguish between neighboring elements and among isotopes of the same element.

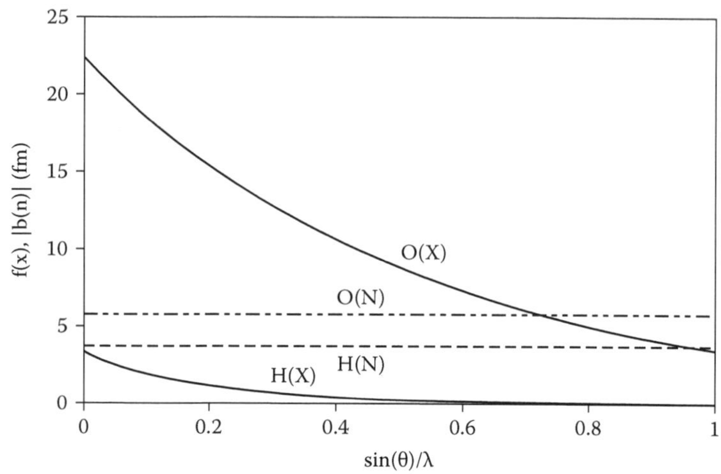

FIGURE 3.2 Scattering factors (fm) as a function of sin(θ)/λ for x-rays (f(x), solid lines) compared to the corresponding neutron scattering lengths (|b(n)|, dashed lines). Destructive interference due to the scattering of x-rays from the diffuse electron cloud surrounding the atomic nucleus causes a dramatic fall-off in intensity as data are collected at higher angles. In contrast, the nucleus acts as a point scatterer, and neutron scattering lengths accordingly do not vary with scattering angle.

of hydrogen is that it appears as a hole in neutron Fourier maps, making it extremely easy to distinguish hydrogen from other types of atoms in the structure.

Because x-rays scatter from the electron density surrounding the nucleus, heavy atoms tend to dominate the total scattering. Furthermore, the diffuse nature of the electron distribution causes destructive interference among rays scattered from a given atom, resulting in a decrease of the x-ray scattering factor as a function of scattering angle (this function is commonly known as the x-ray form factor). In contrast, the nucleus behaves as a point scatterer, and neutron scattering lengths accordingly do not vary with scattering angle (Figure 3.2).

The neutrality and highly penetrating nature of neutron beams, along with the almost complete absence of radiation damage to the sample, combine to make neutrons an ideal probe for determining chemical structure. The properties enumerated above allow neutrons to:

- Find light atoms in the presence of heavy atoms
- Distinguish among isotopes of the same element, and atoms of similar atomic number
- Determine magnetic structure
- Provide accurate nuclear positions and mean square atomic displacement parameters (ADPs)

In the following sections we will examine these applications in detail along with specific examples, but first we will explore some practical considerations of the single crystal neutron experiment.

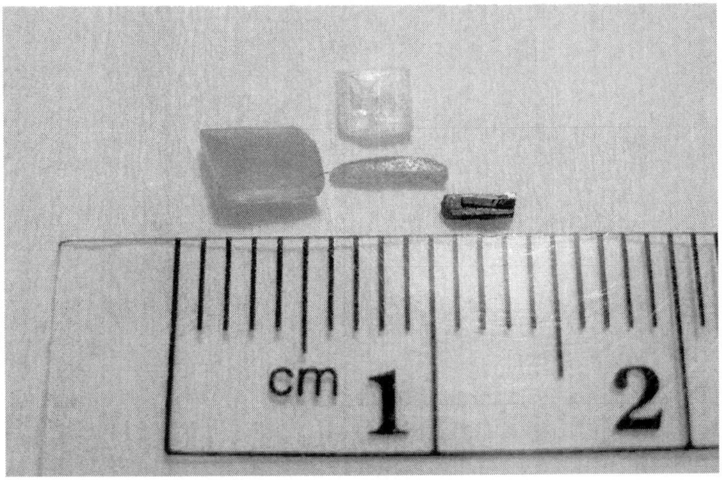

FIGURE 3.3 (See color insert following page 86) Some representative samples from the IPNS. Although a minimum sample size of approximately 1 mm^3 in volume is required for the typical single crystal experiment, in practice larger samples are often desirable.

Sample Size

Traditionally, the limiting factor in single crystal neutron diffraction has been sample size. Typically a crystal of not less than 1 mm^3 in volume has been essential to obtain a suitable signal for proper structure refinement, and the process of growing a crystal at least this size can be challenging, to say the least. In practice, larger crystals than the recommended minimum are often desirable to obtain an adequate signal-to-noise ratio. (See Figure 3.3, which depicts some representative samples used in single crystal experiments at the IPNS.) The reason for the need for large crystals is the relatively low flux available from neutron sources. Also, a factor is the resolution of the instrument, which has often limited the structures that can be investigated to those having unit cell axes of approximately 25 Å in length or less. Low flux has also resulted in fairly long experiment times, although this situation has improved in recent years with the availability of higher-intensity sources and the application of area detectors. The low flux can result in a low data-to-parameter ratio, particularly for larger structures, which in turn can make a full refinement difficult. One solution to this difficulty may be to collect x-ray diffraction data at the same temperature as the neutron data, and to refine both data sets jointly. In this case the heavy atoms are determined and refined primarily with the x-ray data, and hydrogen atoms are refined using the neutron data.

The advent of more intense neutron sources and new instrumentation promises to decrease data collection times as well as the minimum sample size required for the experiment. Better resolution will allow samples with larger unit cell volumes to be tackled routinely; as problems in inorganic chemistry are becoming more complex, this will be absolutely vital to addressing current research issues. Two instruments that have made major progress on this front in recent years are the LADI and VIVALDI diffractometers at the ILL in Grenoble, France. The success of LADI and VIVALDI in measuring smaller crystals and larger unit cells has led to a backlog of experiment proposals for these instruments, and clearly the demand for neutron crystallography is on the rise. To illustrate the pace of progress that has been achieved, consider that a crystal of $[N(CH_3)_4]_3[H_2Rh_{13}(CO)_{24}]$ (1.8 mm^3, a = 16.239(6) Å, b = 17.887(7) Å, c = 20.080(8) Å, β = 94.62(3)°, V = 5,814 Å^3)[15] required eleven weeks of data collection time at the conventional four-circle diffractometer at the Brookhaven High Flux Beam Reactor (HFBR; since decommissioned) in 1997. By contrast, a 0.4 × 0.4 × 0.6 mm^3 (0.096 mm^3) crystal of hydrogen-loaded $Zn_4O(BDC)_3$ (*vide infra*; cubic, a = 25.88 Å, V = 17,334 Å^3) required less than one day of data collection per full data set on the VIVALDI instrument in 2006.[16]

Neutron Absorption

Absorption of radiation by the sample is generally less of a problem for neutron diffraction than for x-ray diffraction. Absorption of x-rays increases with increasing atomic number, which can create a real challenge in materials containing heavy atoms. With neutrons, just as each isotope has a particular cross section for scattering, it also has a unique cross section for absorption of the neutron. A good example of the difference in absorption of x-rays versus neutrons can be seen in the polyoxometalate $K_7Na_9[Pt(O)(H_2O)(PW_9O_{34})_2]\cdot 21.5H_2O$, which has many heavy atoms in the structure. With linear absorption coefficient values of μ(X, Mo*K*α) = 257 cm^{-1} and μ(N, λ = 0.7107 Å) = 0.69 cm^{-1}, transmissions through a 1 mm crystal are essentially 0 and 0.93, respectively. Even for a small crystal of approximately 0.3 mm, transmission is about 0.05 for x-rays. From these numbers we see that absorption is usually far less significant for neutrons than for x-rays.[17]

A small number of isotopes do possess a high cross section for absorption of neutrons, and their presence can pose potential problems with data collection. For example, boron-containing composites and materials are used in neutron shielding because of boron's high absorption cross section (767 barns for natural abundance boron; 1 barn = 10^{-24} cm^2). This precludes mounting a crystal for neutron scattering in a tube containing Pyrex glass, as the neutrons will be absorbed by the boron atoms in the glass, and few neutrons will make it to the sample or the detector. Crystals that must be mounted in a capillary for the neutron experiment typically are mounted in tubes made of quartz, lead glass, or soda glass, and stabilized with plugs of quartz wool on either side of the crystal (Figure 3.4). Samples that are only moderately air or moisture sensitive, and accordingly can withstand a short exposure time to air, are often simply coated in fluorocarbon grease prior to mounting on the diffractometer.

How does having an elemental composition containing highly absorbing components affect the neutron structure refinement? For a material containing only a few atom percent of boron in the overall structure, absorption is not an enormous problem; there are many examples of neutron structures of borohydrides or other boron-containing complexes in the literature. For compounds containing a high percentage of boron, for example, decaborane, it may be necessary to substitute ^{11}B for natural abundance boron.[18] (An inspection of the neutron scattering tables[19] reveals that ^{11}B has a minimal cross section for absorption, 0.055 barns.) Some members of the lanthanide series, for example, gadolinium and samarium, have even higher cross sections for absorption than boron. Without resorting to isotopic substitution, neutron

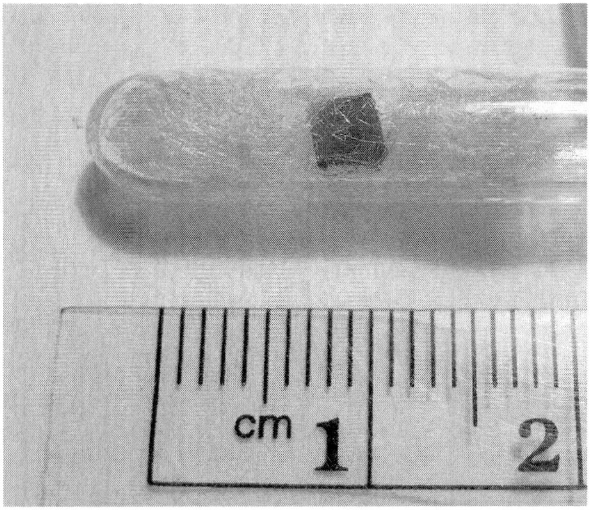

FIGURE 3.4 Air-sensitive single crystal sample sealed in a quartz tube. Plugs of quartz wool on each side of the sample help prevent it from moving in the tube during data collection.

diffraction may not be feasible for problems featuring these elements. Of course, in the future, with smaller sample sizes absorption may not pose as large a problem.

Sample Environment

The highly penetrating nature of the neutron makes it well suited to experiments under non-ambient conditions. Neutron data for small molecules are routinely collected at very low temperatures using closed-cycle helium refrigerators. Experiments utilizing high-temperature furnaces, diamond anvil and gas pressure cells, and applied magnetic and electric fields are all highly feasible for neutron diffraction. Additionally, polarizers can be placed in the beam path to provide a polarized beam for analysis of magnetic structures (*vide infra*).

Neutron Sources

The two common categories of neutron sources are steady-state reactor sources and spallation (pulsed) sources. With a reactor source, such as the High Flux Isotope Reactor (HFIR) at Oak Ridge National Laboratory, or the High Flux Reactor (HFR) at the ILL, the beam is generally monochromated (either by a single crystal or by choppers) to give a small range of wavelengths of radiation, similar to the normal procedure for x-ray scattering. Traditionally, a conventional four-circle diffractometer was often used in which each reflection is recorded individually using a single point neutron detector. Currently, all instruments at reactor and pulsed spallation sources use some type of position-sensitive area detectors, such as multiwire ^3He gas-filled detectors or neutron sensitive ^6Li-containing scintillation detectors. These detectors provide large solid angle coverage and therefore shorter data collection times. Also, in recent years a modified Laue technique using image plate detectors and a band of wavelengths (adjustable to minimize peak overlap) has been developed at reactor sources and is utilized at the ILL with the LADI and VIVALDI diffractometers that were mentioned above. These instruments have been shown to be capable of handling smaller crystals and larger unit cells of small molecules, as well as some protein structures. The novel detector of cylindrical design allows for coverage of a large solid angle of reciprocal space and greatly reduces the time necessary for data collection. A new instrument, the KOALA diffractometer, has been commissioned at the ANSTO OPAL reactor and is designed after VIVALDI.

The first spallation, or pulsed neutron source, the IPNS at Argonne, was commissioned in 1981. The LANSCE facility at Los Alamos National Laboratory and ISIS at the Rutherford Appleton Laboratory in the UK followed closely behind IPNS, and most recently the SNS at Oak Ridge National Laboratory and the JSNS in Japan have begun production of neutrons. In addition, several spallation sources are currently under construction or in the planning stages, including the ISIS second target station and the proposed European Spallation Source (ESS). Whereas reactor sources have reached a plateau with respect to practical intensity output, the spallation source design can achieve considerably higher flux (Figure 3.5). In the spallation process a pulsed proton beam that has been accelerated to high energies strikes a heavy-element target. Neutrons are expelled from the target with each pulse of the proton beam, creating approximately 15–30 neutrons per proton. The data are analyzed using the time-of-flight (TOF) technique, in which neutrons are sorted by velocity v, which is related to wavelength λ by the de Broglie equation $\lambda = h/mv = (h/m)(t/L)$, where h is Planck's constant, m is the neutron mass, and t is the TOF for path length L. There is no need to monochromate the neutron beam, and so a broad spectrum

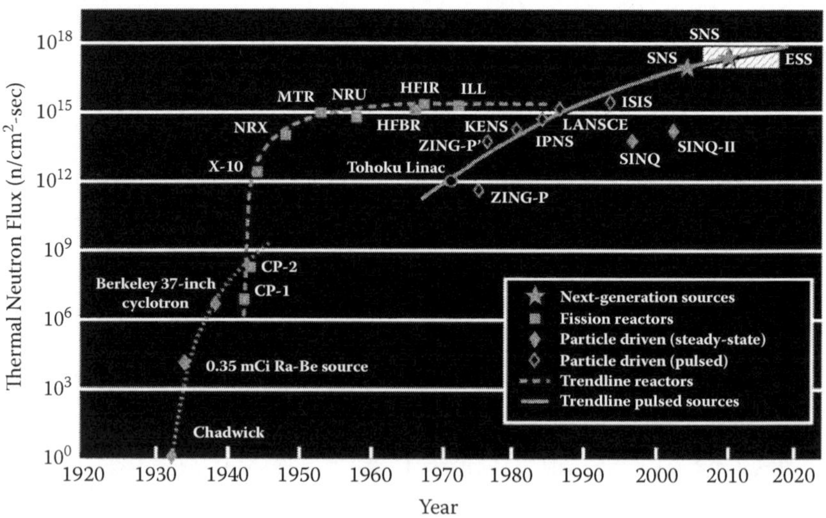

FIGURE 3.5 Neutron flux at various facilities versus year of operation. (Based on Figure 1 in: J. M. Carpenter and W. Yelon (1986). *Neutron Sources. Neutron Scattering*, Part A. K. Sköld and D. L. Price. Orlando, Florida, Academic Press. 23:339–367.)

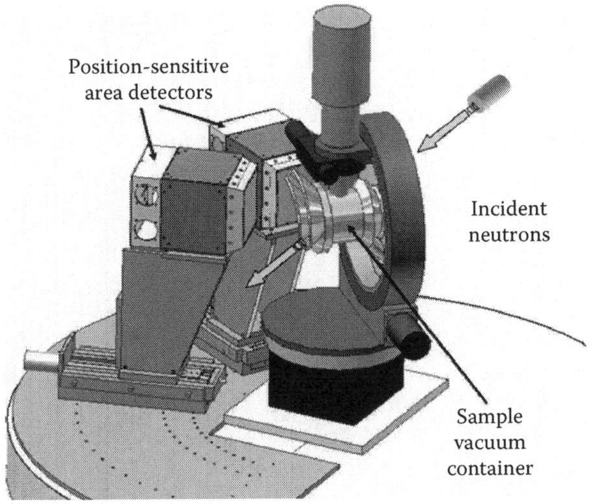

FIGURE 3.6 (See color insert) Diagram of the IPNS SCD instrument now at Los Alamos. The sample is mounted in the center of the vacuum chamber and can be rotated 90° about the χ circle and 360° about φ. The closed-cycle refrigerator is mounted vertically on the φ axis. Two position-sensitive area detectors are centered at 75° and 120° scattering angles from the sample, and can cover a large volume of reciprocal space. The crystal is stationary during the collection of each data frame; approximately 22 settings of the diffractometer are required to cover one hemisphere of reciprocal space.

of thermal neutrons can be used in data collection. With two-dimensional position-sensitive detectors and a stationary crystal, this technique allows for large volumes of reciprocal space to be covered in a single sample orientation. Sampling this large volume is especially useful for the investigation of superlattice peaks and diffuse scattering. Figure 3.6 shows a schematic of the IPNS SCD instrument, which is currently installed at the Lujan Neutron Center in Los Alamos. The SXD instrument at ISIS works on the same principle as the SCD, but with a larger detector array that correspondingly reduces data collection time.

Data collected at any neutron source will be corrected for absorption from the sample, the incident spectrum, and detector efficiency prior to structure refinement. Extinction corrections are made during the refinement of the structure by refining an extinction parameter in a program such as GSAS or SHELX. Each facility typically has its own in-house set of programs that collect, integrate, and reduce the data to structure factor amplitudes (or their squares).

NEUTRONS FIND LIGHT ATOMS IN THE PRESENCE OF HEAVY ATOMS

Hydride and Hydrogen Complexes

As we have seen before, the neutron is diffracted from the nucleus of the atom, and neutron scattering lengths vary quite randomly with atomic number; neutron diffraction determines nuclear positions directly and is not influenced by the electronic density, except in the special case of magnetic materials. These properties make neutron diffraction ideal for locating light atoms in the presence of heavy atoms. The classical application of neutron diffraction in inorganic and organometallic chemistry has concerned the location of hydrogen atoms in metal hydrides or complexes containing an agostic bond. Complexes with these features can be difficult to characterize by NMR owing to their sometimes highly fluxional nature. Numerous hydride complexes have been characterized by single crystal neutron diffraction over the years; of the 460 organometallic neutron structure entries listed in the Cambridge Structural Database (CSD; 2009),[20] 146 of these are hydride or borohydride complexes of transition metals.

Hydride complexes of the transition metals are quite diverse, with coordination numbers of the hydride ligand in molecular complexes ranging from the 1-coordinate terminal hydrides to the 6-coordinate interstitial hydrides at the centers of octahedral clusters. The precise location of hydride ligands bound to metal centers is not reliable with conventional x-ray diffraction methods. Bau and coworkers, leaders in the structural characterization of hydride complexes, have compiled neutron structures of hydrides through 1996 in two excellent reviews.[21,22]

The first 4-coordinate interstitial hydride was characterized only recently by the Bau group with data collected at ILL.[23] The Y_4H_8 cluster of $(Cp'')_4Y_4H_8(THF)$ ($Cp'' = C_5Me_4(SiMe_3)$) is noteworthy not only in that it contains the first example of an unusual and elusive bonding mode for a hydride ligand, but, as illustrated in Figure 3.7, because the cluster contains six edge-bridging 2-coordinate and one face-sharing 3-coordinate hydride ligands as well. It is thought that the presence of these additional bridging hydrides enhances the stability of the cluster with the interstitial hydride H1 occupying the tetrahedral cavity.

The aforementioned cluster $[N(CH_3)_4]_3[H_2Rh_{13}(CO)_{24}]$ was found to possess two hydride ligands of 5-coordinate geometry (Figure 3.8).[15] Each of these hydrides is sited virtually coplanar with the base of a Rh_5 square pyramid, pulled slightly out of the basal plane toward the center of the square pyramidal cavity formed by the five rhodium atoms. The hydride ligands are clearly localized on two sites and are not distributed over the six available equivalent square faces of the Rh_{13} polyhedron. Rather than having hydrides located in interstitial cavities, as is found in the preceding example, the structure of the $[H_2Rh_{13}(CO)_{24}]^{3-}$ cluster anion is more suggestive of hydrogen atoms chemisorbed on a Rh(100) surface.

A recent unusual example of 6-coordinate hydrogen not involving transition metals is the interstitial hydride found at the center of the $[(t\text{-}Bu_2AlMe_2)_2Li]^-[\{Ph(2\text{-}C_5H_4N)N\}_6HLi_8]^+$ cluster, a main group complex with potential importance concerning fuel cell technology (Figure 3.9).[24] While a peak at 583 cm^{-1} in the infrared data indicated the possible presence of the hydride, ^1H NMR gave no such confirmation. Single crystal neutron diffraction, in this case, was the only technique able to identify and precisely characterize the nature of the interstitial hydride.

Neutron diffraction has been crucial in making the distinction between classical dihydrides and nonclassical dihydrogen complexes. Characterization of dihydrogen complexes is especially important as the dihydrogen ligand represents a potential transition state in the activation of H_2 by transition metals. A recent publication from

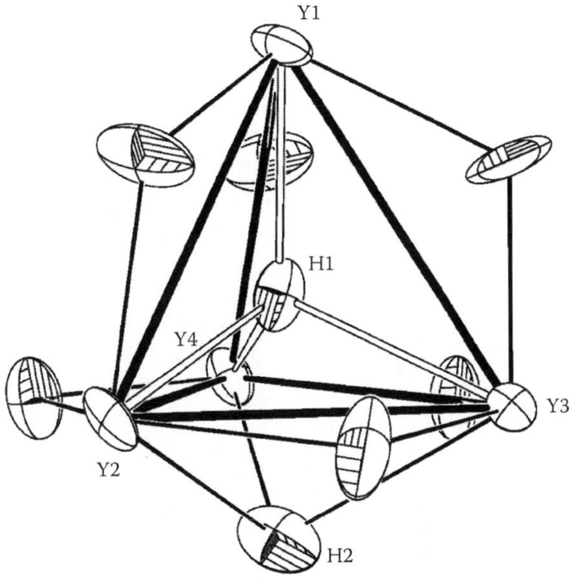

FIGURE 3.7 ORTEP plot of the Y_4H_8 core of the [Cp''YH_2]$_4$(THF) molecule. H1 is the first example of a 4-coordinate interstitial hydride. Also worthy of note is the presence of six 2-coordinate edge-bridging hydrides and the face-sharing, 3-coordinate hydride ligand H2.

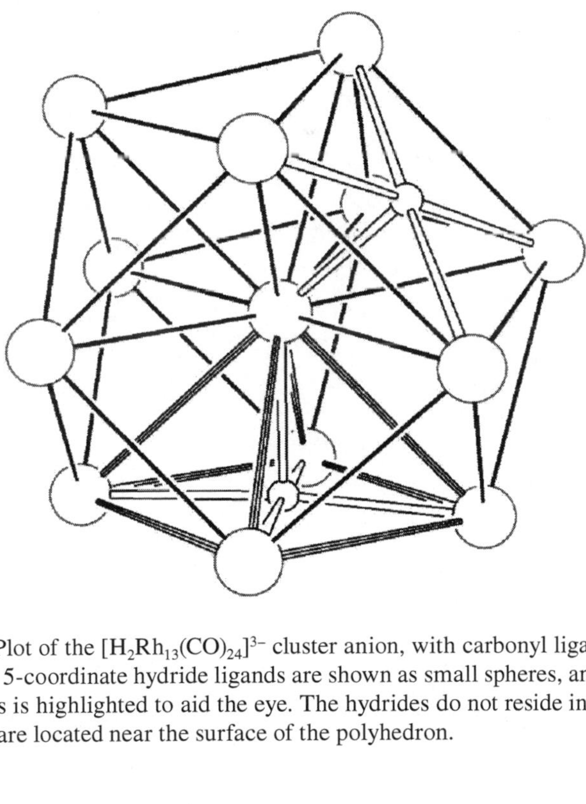

FIGURE 3.8 Plot of the $[H_2Rh_{13}(CO)_{24}]^{3-}$ cluster anion, with carbonyl ligands removed for clarity. The two 5-coordinate hydride ligands are shown as small spheres, and one of the Rh_5 square pyramids is highlighted to aid the eye. The hydrides do not reside in interstitial cavities but instead are located near the surface of the polyhedron.

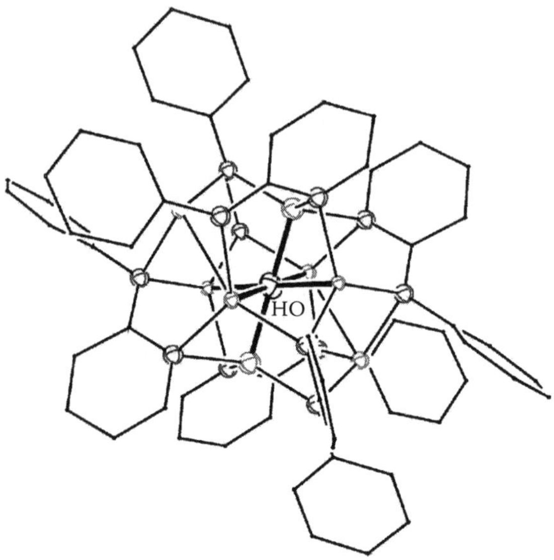

FIGURE 3.9 $[\{Ph(2\text{-}C_5H_4N)N\}_6HLi_8]^+$ cation, showing the interstitial 6-coordinate hydride ligand H0 at the center of the $(Li^+)_8$ cluster.

Webster and coworkers on a series of osmium hydrides elegantly illustrates the effect of varying the sterics and electronics of the ancillary ligands.[25] As illustrated in Figure 3.10, changing the ligand group from PPh_3 to $AsPh_3$ results in a lengthening of the H-H bond of the dihydrogen ligand by 0.07 Å. The dihydrogen ligand is oriented parallel to the plane formed by the center of the Cp* ligand (Ct), Os, and L. Substituting PCy_3 for L (Cy = cyclohexyl) dramatically increases the H-H distance to 1.31 Å and reorients the dihydrogen ligand to lie perpendicular to the aforementioned Ct-Os-L plane. This lengthening of the H-H bond serves as a snapshot of H-H bond activation in progress, inasmuch as H-H distances of longer than 1.50 Å are typical for classical dihydride complexes.[26]

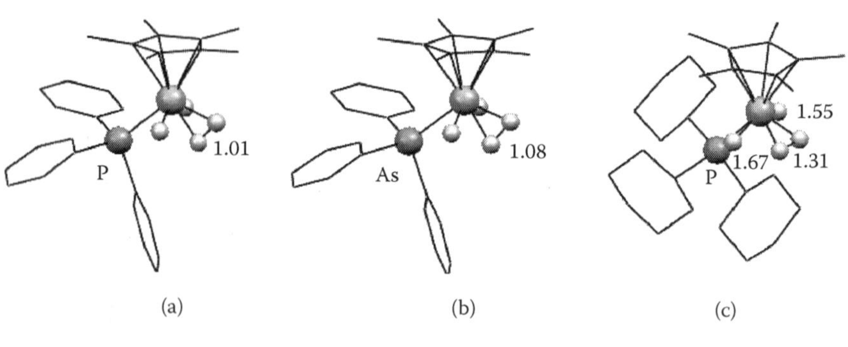

FIGURE 3.10 (See color insert) Neutron structures of $[Cp^*Os(H)_2(\mu\text{-}H_2)L]^+$ complexes (L = PPh_3, $AsPh_3$, PCy_3), showing variation in the H-H bond as a function of sterics and electronics. Bond distances in Å.

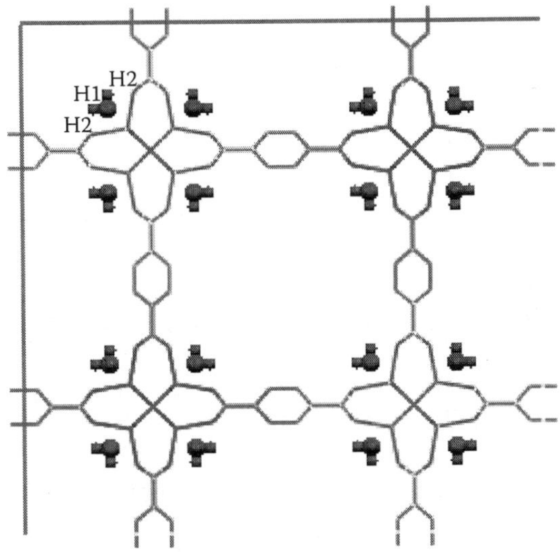

FIGURE 3.11 Plot of the unit cell for $Zn_4O(BDC)_3{\cdot}4H_2$ at 30 K. Atom H2 is disordered over three sites as atom H1 lies on a crystallographic threefold axis. This H_2 position lies close to a framework node and is 100% occupied at 30 and 50 K. A second site, populated by H_2 at 5K, is not shown. Hydrogen atoms on the carbon ligands have been removed for clarity.

Another important application of neutron diffraction is the location of absorption sites of hydrogen in porous, crystalline storage materials, as has been reported for Yaghi's metal-organic framework (MOF) system Zn_4O(1,4-benzenedicarboxylate)$_3$.[16,27] Inelastic neutron scattering experiments[28] have suggested that the hydrogen binding sites for a series of MOF compounds vary among compounds even though chemical makeup is quite similar, and that the nature of the organic linker in these compounds plays a significant role in how and where hydrogen is stored in these materials. In $Zn_4O(BDC)_3$ (BDC = 1,4-benzenedicarboxylate), the H_2 site is clearly localized at a single framework node at temperatures between 30 and 120 K (see Figure 3.11). At 5 K, a second framework node is also populated at 98%. As was mentioned earlier, data collection was possible at VIVALDI, despite the very small crystal size (0.4 × 0.4 × 0.6 mm^3), which represents a significant point in the future of single crystal neutron diffraction. As many MOF compounds form small crystals and have moderate to large unit cells, clearly this is an area for exciting growth in the very near future.

Complexes Containing Agostic Bonds

The agostic interaction,[29] where a pendant X-H bond comes into close contact with a metal center, has been a very important concept for chemists over the last twenty years. Molecules with this type of bond, also called transition metal σ-complexes,[30] appear to be intermediates in the oxidative addition of X-H to the metal. Oxidative addition is an important step in such catalytic processes as the hydrogenation of olefins and other unsaturated hydrocarbons, hydroformylation, and hydrosilation.[31]

X can be one of a number of main-group elements including hydrogen (technically, in this formulation, dihydrogen species are a special class of agostic complexes). The agostic bond is typically characterized by an elongation of the X-H distance, which is easily recognized by neutron diffraction. Agostic C-H-M interactions may be seen in ^{1}H NMR, as well as, the signal of the proton shifts to higher field ($\delta = -5$ to -15 ppm). Reduced coupling constants (^1J (C,H) = 75–100 Hz) can also be an indicator of an agostic interaction, although the potential fluxional nature of the interaction can obscure these signs. Vibrational frequencies at low wave numbers (υ_{CH} = 2,700–2,300 cm^{-1}) may also suggest the presence of an agostic bond. The most compelling data that we have available for such interactions are from crystal structures, and especially from neutron diffraction.[32]

The agostic interaction was first explored in the early 1970s by Trofimenko and his collaborators. NMR spectra of purported 16-electron scorpionate complexes, $[H_2B(pz)_2]Mo(\eta^3\text{-}C_3H_5)(CO)_2$ and $[Et_2B(pz)_2]Mo(\eta^3\text{-}C_3H_4Ph)(CO)_2$,[4,5] indicated a shift of the methylene proton of $[Et_2B(pz)_2]Mo(\eta^3\text{-}C_3H_4Ph)(CO)_2$ into the hydridic region. The x-ray structures[7,33] revealed a close approach of the B-H or ethyl groups to the Mo center, stabilizing them as 6-coordinate, 18-electron complexes where the agostic interaction is counted as a 3-center, 2-electron bond. Recent neutron diffraction studies[34] of isostructural scorpionates $[Bp^x][Mo(CO)_2(\eta^3\text{-}C_3H_4Me)]$ (Bp = pyrazol-1-yl borate; x = 3,5-Ph$_2$, 3,5-Me$_2$,4-Br, 3,4,5-Me$_3$, 3,4,5-H$_3$) show, as expected, that a B-H bond is elongated by 0.05–0.08 Å, compared to an unactivated distance of 1.2 Å, when it is complexed to the metal center. A general trend is found where the strength of the agostic bond, as reflected in the Mo-H distance, is correlated with the electron withdrawing strength of the scorpionate ligand substituents. This finding is also reflected in the *trans* influence of the opposing carbonyl ligand (longer Mo-CO distances correspond to shorter Mo-H distances; see Figure 3.12).

The neutron structure of $TiCl_3(dmpe)Me$ (dmpe = $Me_2PCH_2CH_2PMe_2$),[35] one of the first early transition metal complexes for which an agostic M-C-H interaction was suggested,[36] shows that the internal geometry of the methyl group bound to Ti is not distorted, but that the methyl group is canted toward the metal center with a Ti-C-H angle of 93.5(2)° (Figure 3.13a). However, no elongation of the C-H bond was found. This result is consistent with the ^{1}H NMR spectrum, which finds no significant variation in the chemical shift that would indicate an agostic interaction.

An elongated C-H bond of 1.153(6) Å (versus a nonactivated C-H bond distance of approximately 1.09 Å) is found for the nitrosyl complex $Cp^*W(NO)(CH_2CMe_3)_2$ (Figure 3.13b).[37] As with the previous example, a reduced W-C-H bond angle of 80.6(3)° is once again found. Interestingly, though, in this case ^{1}H NMR spectroscopy does indicate the presence of an agostic interaction with the methylene proton signal at –1.43 ppm.[38] The complex has been described as doubly agostic, where one methylene proton on each alkyl ligand has an interaction with the metal center, although one interaction is stronger than the other. On the NMR timescale, both of these agostic methylene protons are equivalent; by contrast, they are rendered inequivalent in the solid state.

A more recent example of a doubly agostic complex is found in $RuCl_2[PPh_2(2,6\text{-}Me_2C_6H_3)]_2$, where the *ortho* methyl groups on the phenyl ligands have a close approach to the metal center. As with the aforementioned tungsten example, C-H bonds are found to be somewhat elongated to 1.119(11) and 1.111(14) Å, respectively.

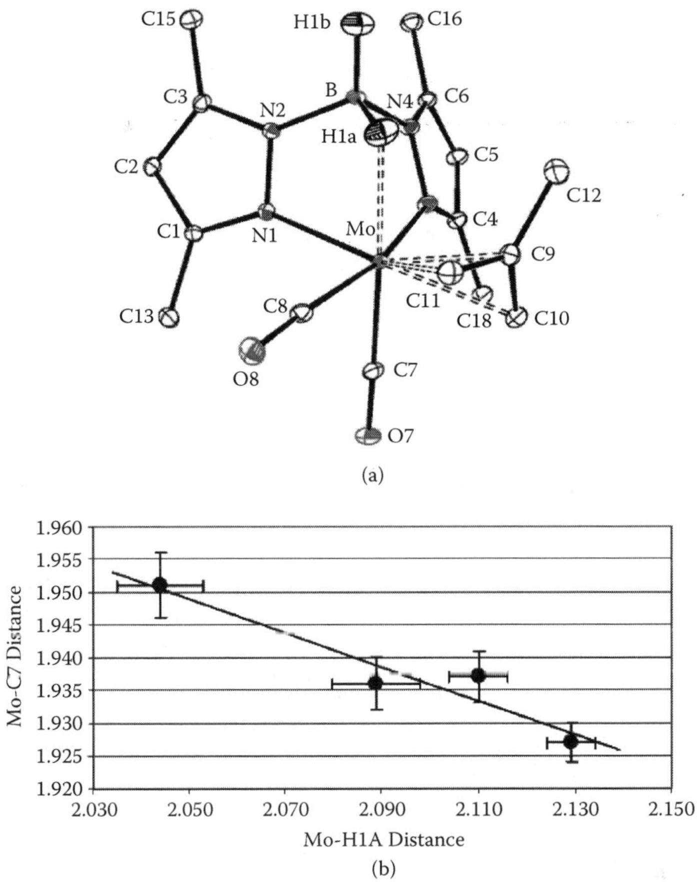

FIGURE 3.12 (See color insert) (a) The neutron structure of Bp*[Mo(CO)$_2$(η^3-C_3H_4Me)]. (b) Plot of Mo-C7 distance versus Mo-H1A distance for a series of substituted scorpionates. As the substituents on the pyrazolylborate ligands become more electron withdrawing, the agostic Mo-H interaction becomes stronger (shorter Mo-H distance). This is reflected in the *trans* influence, where Mo-CO distances increase with decreasing Mo-H distance.

While the ^{1}H NMR spectrum does not show the hydridic character of the agostic protons, presumably due to free rotation of the methyl groups, the ^{13}C NMR exhibits both chemical shifts and $^1J_{CH}$ coupling constants that are consistent with an agostic bond.[39] The agostic interactions stabilize what would otherwise be an unsaturated, 14-electron complex.

Just as the search for stable, multiple-bonded, transition metal-carbon complexes (alkylidene complexes) was of great interest 20 to 30 years ago, Mork et al. have recently been investigating complexes with multiple bonding between silicon and transition metals.[40] One reaction route that they have developed to prepare silylene complexes includes the activation of a Si-H bond in a $L_nM(SiHR_2)$ intermediate complex to yield a silylene hydride complex of the type $L_n(H)M{=}SiR_2$. One of these intermediate complexes, (η^5-C_5Me_5)Mo($SiHEt_2$)($Me_2PCH_2CH_2PMe_2$), was isolated,

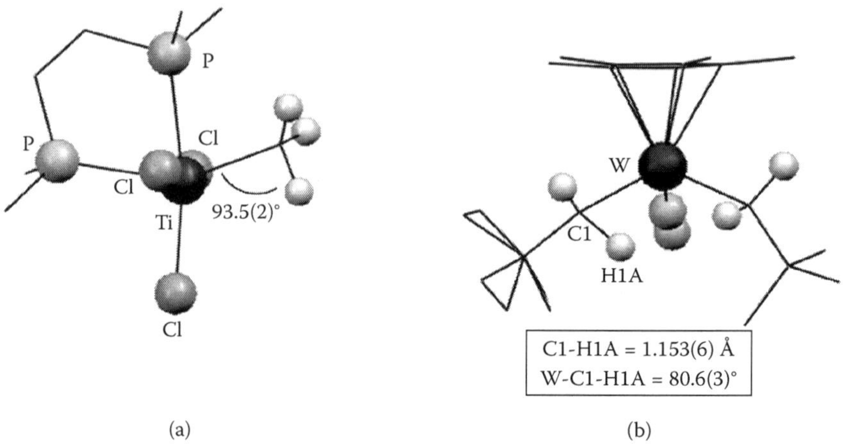

FIGURE 3.13 Neutron diffraction structures of agostic complexes: (a) $TiCl_3$(dmpe)Me and (b) Cp*W(NO)(CH_2CMe_3)$_2$. In (a) a reduced Ti-C-H angle of 93.5(2)° is the only indication of a possible agostic interaction, as the internal geometry of the methyl group and C-H bond lengths are in the normal range. Analysis of complex (b) clearly shows an elongation of the agostic C-H bond from a typical length of 1.09 to 1.153(6)Å. The reduced W-C1-H1A angle seen in complex (a) is also seen here. Also of note for complex (b) is that the two methylene groups are inequivalent in the solid state, but in solution they both appear to have equivalent agostic interactions with the metal center. Hydrogen atoms not of interest have been omitted for clarity.

and its neutron structure provides a snapshot of the intermediate in the oxidative addition of a Si-H bond to a metal. The structure, shown in Figure 3.14, is that of a three-legged piano stool with one leg consisting of a Si-H σ-bond complex with the molybdenum. The molecule formally contains a 16-electron system in the absence of the Si-H σ-bond interaction with the metal center. As observed previously, formation of a 3-center, 2-electron, σ-complex bond achieves a stable 18-electron configuration. The Mo-Si bond length of 2.34(1) Å is longer than the values of 2.219(2) and 2.288(2) Å in complexes with multiple Mo-Si bonds.[41] The Si-H bond length, 1.68(1) Å, is indicative of a significant degree of activation in comparison to a normal Si-H bond length of 1.48 Å in tetrahedral silanes. The third leg of the piano stool points in between the Si and H(1) atoms, as exhibited by the Si-Mo-P(1) and Si-Mo-P(2) angles of 87.1(3)° and 101.9(4)°. The plane of the Si atom and the two α-C atoms of the ethyl groups is nearly coplanar with the Mo atom and nearly perpendicular to the plane of the Cp* ligand. This orientation may be indicative of π-bonding with the metal.

While only a few agostic systems have been studied by neutron diffraction, it is clear that spectroscopic techniques are not always adequate to address agostic bonding issues. Neutron diffraction remains the most powerful method to locate an agostic hydrogen atom.[42]

Absence of a Hydrogen Atom

Neutron diffraction is useful in situations where it may be important to confirm the absence of a proton. This type of problem arose recently in the case of late transition

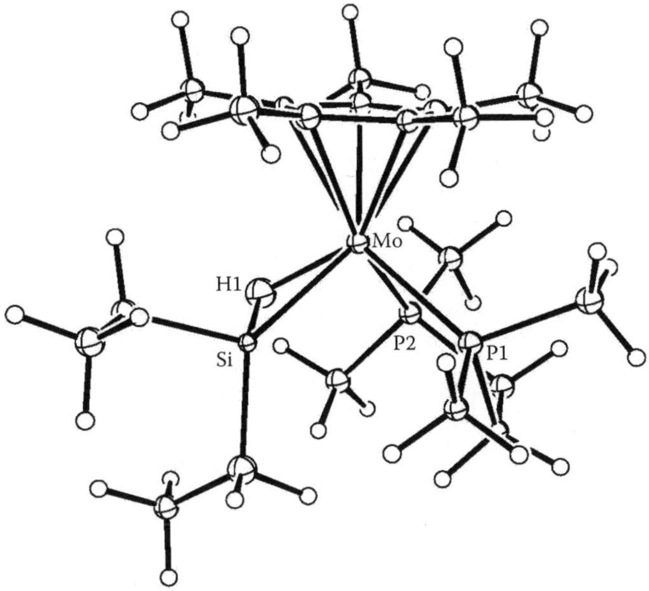

FIGURE 3.14 Neutron structure of (η^5-C_5Me_5)Mo($SiHEt_2$)($Me_2PCH_2CH_2PMe_2$) illustrating the Si-H-Mo agostic-type interaction. Except for H1, the hydrogen atoms have been displayed as small spheres for clarity, although they were refined with isotropic displacement parameters.

metal oxo (LTMO) complexes. LTMO species, the existence of which has long been doubted, have recently been synthesized by the Hill group.[17] In conjunction with other characterization techniques, neutron diffraction was an essential tool in helping to preclude terminal hydroxyl complexes of Pt and Au, which have better precedents (see Figure 3.15). Lattice water molecules in these LTMO polyoxometalates (POMs) were clearly visible in the neutron difference Fourier maps, and the refinement of some of these water molecules was stable; this showed that hydrogen atoms located from the neutron diffraction experiment were both real and refineable. No features in the vicinity of the terminal oxo ligand were consistent with hydrogen bound to the oxygen atom, and so the results are consistent with a terminal oxo ligand. Molecular orbital calculations on model LTMO complexes indicate that the POM ligands act as an electron sink, reducing destabilizing electron-pair repulsions.

Hydrogen Bonding

Hydrogen bonding has long been a subject explored in great detail by neutron scattering in work dating back, for example, to the pioneering studies of Peterson and Levy on the structure of ice.[43] Hydrogen bonding in small molecule organic systems[44] and in macromolecules[45] has attracted much attention. The paramount importance of the hydrogen bond in biological systems has recently motivated the development of new and improved instrumentation for neutron protein crystallography.[46–48] For the inorganic chemist, particularly with compounds synthesized hydrothermally or

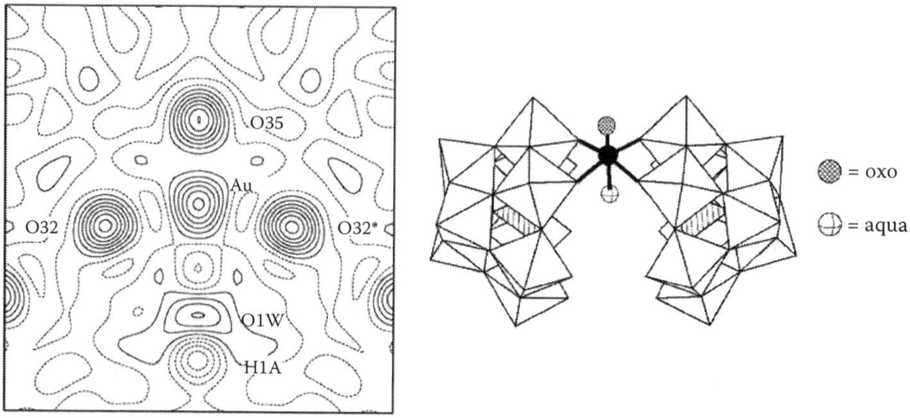

FIGURE 3.15 Fourier synthesis map derived from the observed neutron structure factor amplitudes (left) of the Au-oxo-POM-aqua plane in $K_{15}H_2[Au(O)(OH_2)P_2W_{18}O_{68}]\cdot 25H_2O$ (anion shown at right). Solid contours indicate positive neutron scattering density and broken contours represent negative scattering density (indicative of hydrogen atoms). Negative scattering density that models as hydrogen atoms is seen close to the disordered aqua ligand OW1, but no such density is seen in the vicinity of the O35 oxo ligand. This map is typical for both the Pt and Au terminal oxo complexes. (Figure at right reprinted from Anderson, T. M., et al. 2004. *Science* 306:2074–2077. With permission from AAAS.)

solvothermally, or supramolecular structures utilizing hydrogen bonding as their method of self-assembly and organization, neutron diffraction would prove invaluable for complete characterization of the system. Since such compounds usually possess a large unit cell volume or grow as small crystals, much science in this rapidly growing field has yet to benefit from single crystal neutron diffraction.

Characterization of the hydronium ion $(H_5O_2)^+$ and aquo complexes of metals is readily achievable with neutron diffraction, which can provide structural information not accessible with x-ray methods alone. Taking into consideration the location of hydrogen atoms, the true point group symmetry for an aquo metal complex can be resolved only by locating the water hydrogens. This will also indicate whether the water ligands are pyramidal or planar with respect to the M-OH_2 bond. In the case of $[V(H_2O)_6][H_5O_2](CF_3SO_3)$, for example, the VO_6 framework has O_h symmetry, which is lowered to D_{3d} when considering the geometry of the essentially planar metal-bound water molecules.[49] The neutron data thus support the lowering of the degenerate ground state of O_h or T_h symmetry for the d^2 metal center as required by the Jahn-Teller theorem. The hydronium ion in $[V(H_2O)_6][H_5O_2](CF_3SO_3)$ is characterized by pyramidal waters with a strong, centered hydrogen bond where the proton lies on a crystallographic center of inversion.

Neutron diffraction has been used to characterize the role of the extended hydrogen bonding network in the cooperative Jahn-Teller switch in ammonium copper sulfate Tutton salts. In 1984, it was reported from neutron powder diffraction data that the Jahn-Teller elongation of two *trans* Cu–O bonds in the perdeuterated salt, $(ND_4)_2[Cu(D_2O)_6](SO_4)_2$, is orthogonal to the elongated bonds in the fully hydrogenated salt.[50] A subsequent single crystal neutron investigation of

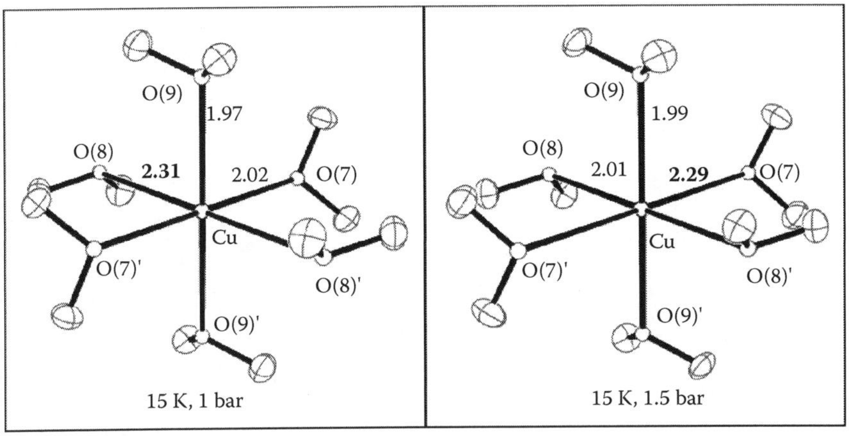

FIGURE 3.16 Results from a single crystal neutron diffraction study of $(ND_4)_2[Cu(D_2O)_6]$-$(SO_4)_2$ at ambient pressure (left) and under an applied pressure of 1.5 kbar. Note the changes in the Cu-O bond distances.

$(ND_4)_2[Cu(D_2O)_6]$-$(SO_4)_2$ under applied pressure[51] demonstrated that the switch is reversible (Figure 3.16) and led to several powder diffraction studies exploring the pressure-temperature phase diagram.[52] Because the scattering lengths for hydrogen (–3.74 fm) and deuterium (6.67 fm) differ by sign and magnitude, it is possible to obtain the H/D ratio and distribution by refinement of the scattering from each hydrogen site. Thus, from single crystal data, it was determined that a 42% deuterated crystal exhibited no indication of a phase transition, even though the structure had been shown to switch at 50% deuteration from electron paramagnetic resonance (EPR) measurements.[53] All of these studies highlight the ability of neutron diffraction to precisely locate and refine hydrogen atoms with anisotropic thermal parameters, to determine H/D ratios, and to easily utilize pressure cells due to the high penetrating ability of neutrons.

The hydrogen bond, or hydrogen bridge,[54] is an immensely useful noncovalent interaction in crystal engineering and supramolecular chemistry,[55] and neutron diffraction is uniquely suited to probe how these relatively weak interactions influence the overall structure of molecules in the solid state. D-H···A systems can involve anything from the classic example where the acceptor A is oxygen or nitrogen, to systems where A is an aromatic π-system or a halogen.[56] For example, the close intramolecular C-H···X interactions of the terminal halide ligands in linear hydride complexes $[(dippm)_2Ni_2X_2](\mu\text{-H})$ (dippm = 1,2-bis(diisopropylphosphino)-methane; X = Br, Cl)[57,58] grip the halide into a locked position, suggesting that these kinds of interactions may directly influence the geometry of the bridging hydride. The M-H-M angles of 177.9(10)° and 176.7(6)° are far greater than those for all other M-H-M bridging systems having a *bent* geometry (angles typically less than 160°).[22] Related complexes $[(dcpm)_2Ni_2X_2](\mu\text{-H})$ (dcpm = 1,2-bis(dicyclohexylphosphino) methane; X = Br, Cl)[58,59] do not possess the same close approaches of C-H to X due to the steric bulk of the ligand, and the geometry of the bridging hydride appears to be bent. While the neutron structures of these bent $[(dcpm)_2Ni_2X_2](\mu\text{-H})$ complexes

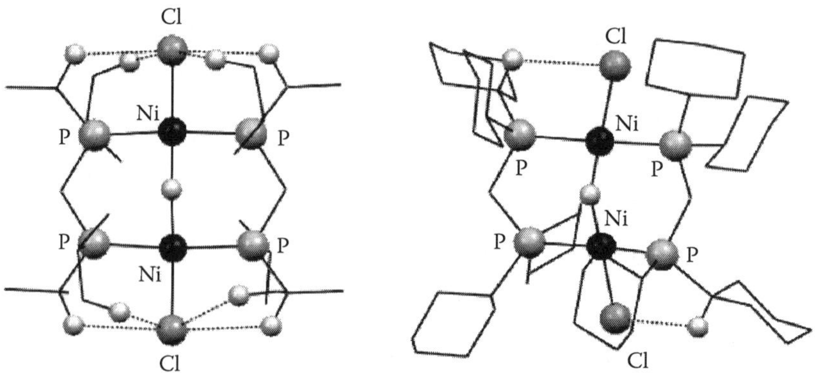

FIGURE 3.17 (See color insert) Comparison of the unsupported bridging hydride complexes [(dippm)$_2$Ni$_2$Cl$_2$](μ-H) (neutron structure, left) and [(dcpm)$_2$Ni$_2$Cl$_2$](μ-H) (x-ray structure, right). C-H···Cl contacts shorter than the sum of the van der Waals radii (2.95 Å) are shown as dashed lines. The sterics of the phosphine ligand are suggested to be the influencing factor on the geometry of the bridging hydride. In the linear example, the close approach of the isopropyl groups "locks" the chlorides into place, resulting in a linear hydride complex. Hydrogen atoms not involved in the hydrogen bridges have been omitted for clarity.

have yet to be determined, the published result is a fine example of how neutron crystallography can probe the effect of steric and electronic factors for a series of related compounds (Figure 3.17).

NEUTRONS CAN DISTINGUISH AMONG ATOMS OF SIMILAR ATOMIC NUMBER

While location of atoms heavier than hydrogen is generally not a problem for the x-ray diffraction experiment, it is extremely difficult for x-rays to distinguish between two atoms of similar atomic number that reside in the same structure or occupy the same crystallographic site. The difference between x-ray scattering factors for adjacent atoms in the periodic table is often not sufficient to make this distinction. The largely random variation in neutron scattering lengths over the periodic series or between isotopes of the same element (Figure 3.1) makes neutron diffraction well suited to answer questions regarding chemical identity or fractional occupancy of similar atoms, provided that the difference in scattering length between the atoms or isotopes in question is sufficiently large.

The quaternary aluminum silicide $Pr_8Ru_{12}Al_{49}Si_9(Al_xSi_{12-x})$, grown from an aluminum melt, was studied by neutron diffraction to determine not only which sites were occupied by Al and Si, but also the degree of fractional occupancy over one of the sites. Neutron results determine the value of x to be approximately 4, in agreement with the energy-dispersive spectroscopy (EDS) measurements.[60] The Kanatzidis group conducted a similar study on a crystal of $Tb_4FeGa_{12-x}Ge_x$ in which it was found that Ga partially occupies the Ge *12e* site in the structure.[61] The neutron diffraction result in this study was also consistent with EDS analysis. Neutron diffraction on the solid solution clathrate $Ba_8Al_{14}Si_{31}$ revealed that Al is substituted over

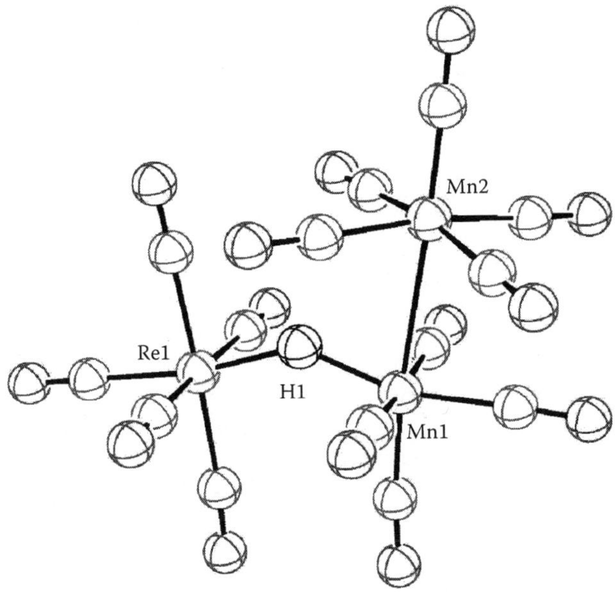

FIGURE 3.18 (See color insert) Neutron diffraction structure of $(CO)_5Re(\mu\text{-}H)Mn(CO)_4Mn(CO)_5$. The Mn2 site was found to be 9.2% occupied by Re due to a co-crystallization of the isomorphous $(CO)_5Re(\mu\text{-}H)Mn(CO)_4Re(CO)_5$.

all of the framework sites in the crystal, a finding also supported by ^{27}Al magic angle spinning (MAS) NMR.[62]

During the analysis of neutron diffraction data from a single crystal of $(CO)_5Re(\mu\text{-}H)Mn(CO)_4Mn(CO)_5$ (Figure 3.18), the isotropic displacement parameters of the Re and Mn(2) atoms refined to unusual values.[63] Subsequent refinement of the scattering lengths of these atoms revealed that the terminal Mn(2) site was partially occupied by approximately 9.2% Re due to a co-crystallization of the isomorphous $(CO)_5Re(\mu\text{-}H)Mn(CO)_4Re(CO)_5$. The scattering length of the Re site refined to more than that of its reported value, but was consistent with the scattering length obtained from the refinement of $[ReH_7\{P(p\text{-tolyl})_3\}_2]$.[64] Once the occupancy of the Mn(2) site was adjusted to reflect the site substitution of Mn by Re, and the scattering length for Re was refined and corrected, anisotropic displacement parameters had reasonable values for all metal atoms. The findings from the neutron diffraction study prompted a reevaluation of the NMR data and also helped to postulate a mechanism for metal-metal exchange to form the co-crystal. While this is not an example of diffraction concerning near neighbors, in this case the contrast between the negative scattering length of Mn (–3.73 fm) and the highly positive scattering length of Re (9.2 fm) proved to be crucial to recognizing the site substitution, which had been missed in the x-ray structure.[65]

NEUTRONS DETERMINE MAGNETIC STRUCTURE

One of the interesting features of the neutron is that it has a magnetic moment, meaning that it behaves as a very small bar magnet. When the neutron encounters an atom with unpaired spin in an ordered magnetic material, such as we find in ferromagnets, antiferromagnets, and in paramagnetic materials when placed in a strong orienting field, the magnetic interaction gives rise to magnetic scattering, which may or may not coincide with the ordinary nuclear Bragg scattering. An important difference between magnetic and nuclear scattering is that the magnetic component is proportional to the sine of the angle between the diffraction vector and the spin. For magnetic materials, the neutron scattering is thus dependent on the direction and spatial distribution of magnetization.[13]

With an unpolarized beam the neutrons arrive at the sample with their spins in random orientations. This type of beam can still be very useful in characterizing ferro- and antiferromagnets. The use of a polarizer enables the separation of neutrons that are "spin-up" and "spin-down" for use in more complex magnetic scattering experiments. A magnetic guide field between the polarizer and sample ensures that neutrons of one spin type arrive at the sample without reverting to the random orientation of an unpolarized beam. For paramagnetic materials the spin density can be determined by aligning the spin of the sample in a magnetic field, usually at low temperature, and measuring the resulting magnetic Bragg scattering. The magnetic structure factors are Fourier components of the spin density, and so by combining measurements with neutrons that are in both the spin-up and spin-down orientations, the spin density can be reconstructed, for example, by means of a multipolar refinement of the spin density around the nuclear positions. This method was used successfully in 1994 by Zheludev et al. in the investigation of $[Bu_4]^+[TCNE]^{\bullet-}$ (TCNE = tetracyanoethylene) at 1.8 K in an applied field of 4.65 T.[66] After determination of the nuclear structure with both x-rays and neutrons, 211 independent Bragg reflections were collected with neutrons in both orientations. The crystal was mounted in two separate orientations to maximize the number of reflections scattered from the plane of the molecule; free rotation of the sample for full coverage of reciprocal space is not yet achievable for this type of experiment. The results after multipolar refinement of the spin density on the radical anion are shown in Figure 3.19, in which the electronic spin density is localized on the sp^2-carbon atoms and also on the nitrogen atoms.

The rare earth complex $Y(HBPz_3)_2(DTBSQ)$ ($HBPz_3$ = hydrotrispyrazolylborate and DTBSQ = di-*tert*-butylsemiquinonate) was investigated using polarized neutrons to determine the spin density of the anion in this paramagnetic complex.[67] The S = 1/2 ground state is attributed to antiferromagnetic coupling between the Y^{3+} ion and the DTBSQ radical anion. The polarized neutron data were taken at 1.9 K under an applied field of 9.5 T at the ILL. Figure 3.20 shows the spin density as reconstructed by multipolar refinement based on the polarized neutron diffraction data. The spin density from the anion is partially delocalized over the Y^{3+} site, which is primarily 5*s* in character and has virtually no 4*d* character (within error). This delocalization from the radical anion onto Y is most likely due to significant σ-character on the oxygen atoms of the radical anion. This is consistent with results determined from the experimental charge density obtained from the x-ray experiment, which revealed highly polarized oxygen atom lone pairs that carry a large negative charge on the semiquinone radical. Spin density on the carbon atoms of the ring is π-type in nature, which overlaps with the empty valence orbitals on Y^{3+},

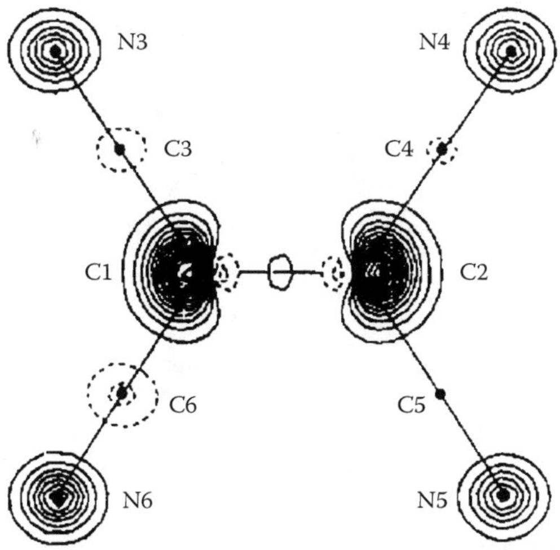

FIGURE 3.19 Magnetic spin density reconstructed by multipole model refinement, projected onto the [TCNE]$^{\bullet-}$ molecular plane. (Reprinted from Zheludev et al., *J. Am. Chem. Soc.*, 116, 7243–7249, 1994. Copyright 1994, American Chemical Society. With permission.)

supporting the conclusion that the magnetic interaction between the rare earth ion and radical anion results from overlap of the magnetic orbitals of these two components of the complex.

Tutton salts of the first-row transition metals, with formulas $A_2[M(H_2O)_6]X_2$ (A = alkali metal or ammonium, M = transition metal, X = sulfate or selenate), have been extensively studied by single crystal and powder diffraction, and by polarized neutron diffraction (PND). Chandler and coworkers have investigated the spin density delocalization in the ammonium sulfate series of salts with transition metals V, Cr, Mn, Fe, and Ni using the ratio of the Bragg intensities from spin-up and spin-down polarized neutron diffraction. It is possible using this method to extract the nuclear and the magnetic contribution to each peak, and to then use the magnetic structure factors to analyze the spin density in real space. From these studies, the occupancies of the metal orbitals can be derived and spin density out to the hydrogen atoms on the water ligands can be observed.[68]

For crystals in noncentrosymmetric space groups, reconstruction of the spin density by the multipole refinement method is not possible due to the fact that the imaginary portion of the magnetic structure factor is nonzero, and therefore the Fourier inversion needed to extract the spin density information is also nonlinear. In this case, a model-free maximum entropy reconstruction based on spin density maps is possible; however, the technique is qualitative at best.[73]

Molecular magnets like the TCNE radical anion mentioned above are a popular research topic as of late. Polarized neutron diffraction will continue to be a very powerful tool in the complete characterization of these materials. During 2005, approximately 20% of all neutron scattering experiments at the ILL concerned magnetic materials[69]; over the 1994–2005 period a majority (55%) of single crystal scattering experiments concerned magnetism. This area of single crystal neutron

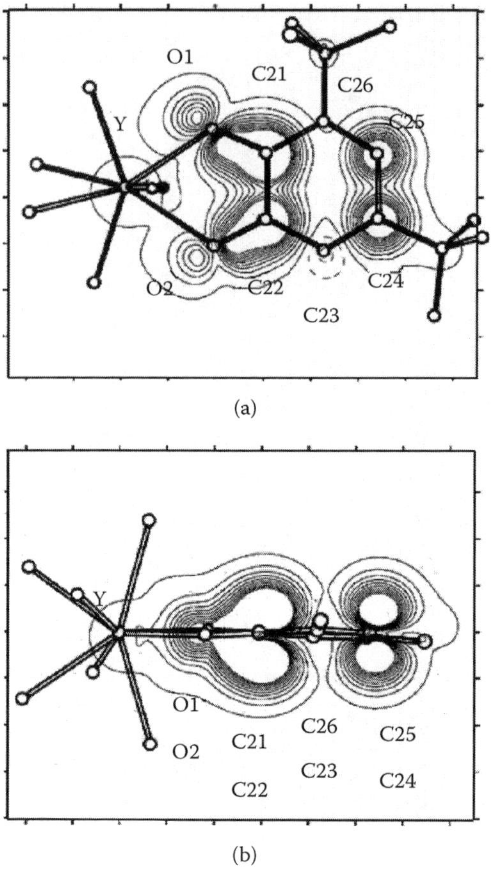

FIGURE 3.20 Projection of the induced spin density at 1.9 K under 9.5 T in Y(HBPz$_3$)$_2$(DTBSQ) obtained by multipole model A reconstruction: (a) along the perpendicular to the mean plane of the semiquinonate ring, and (b) along the C21–C22 direction. (Reprinted from Claiser et al., *J. Phys. Chem. B*, 109, 2723–2732, 2005. Copyright 2005, American Chemical Society. With permission.)

scattering is poised to see a dramatic increase in interest as more advanced technologies become available.

NEUTRONS PRODUCE DATA FREE OF THE INFLUENCE OF ELECTRONIC EFFECTS

Because neutrons are diffracted from nuclei and generally do not interact with electrons, except in the case of magnetic materials, single crystal neutron diffraction data become valuable in electron charge density studies. The neutron data can accurately determine the atomic position and anisotropic displacement parameters of hydrogen atoms, information that is not available even from high-resolution x-ray data. When x-ray and neutron diffraction data are collected at the same temperature, the hydrogen positions and anisotropic displacement parameters can be properly scaled and subsequently fixed during the multipole refinement of the charge density from the x-ray data. The application of this procedure can be seen, for example, in the experimental electron density study of the peroxo complex $MoO(O_2)$(HMPA)-(dipic).[70] With neutron data determining the nuclear positions of atoms, bonding electron density can be accurately described using the x-ray data. The use of neutron parameters for hydrogen atoms in multipolar refinement of the experimental electron density has seen far more utilization in the study of small organic compounds, but with the availability of low-temperature data from synchrotron sources and anticipated smaller crystal sizes for neutron diffraction, hopefully we will see an increase in the number of inorganic and organometallic complexes being studied.

A NOTE ON POWDER DIFFRACTION

Neutron powder diffraction has been used extensively over many years by materials scientists, physicists, and chemists alike. Every major neutron scattering center has one or more beamlines dedicated to powder diffraction. Sample volumes tend to be even larger than for single crystal diffraction; powder studies often require gram quantities of sample. Powder diffraction is thus problematic for materials containing highly absorbing elements; the large volume of sample will be absorbing and very little scattering will be observed. In this case, substitution to remove the highly absorbing isotope is the only way to get around the problem. Furthermore, powder diffraction is limited by unit cell size and by hydrogen-containing materials (see below).

Powder studies are sensitive to impurities in the sample; therefore, sample purity and size can be a challenging part of the experiment. Neutron powder diffraction has been invaluable in the investigation of oxide materials, where characterization with x-rays can be difficult due to the relatively low scattering power of oxygen compared to that of transition metals or heavier main group elements. The sensitivity to impurities can also be useful in those experiments following a chemical reaction over time, where products from complex reactions may be identified. Questions concerning site occupancy have also been a staple in powder diffraction experiments with neutrons.

Samples for powder neutron diffraction containing hydrogen are typically deuterated due to the high incoherent scattering of the proton. In single crystal diffraction, deuteration is generally not required because the entire sample contributes both to coherent (Bragg scattering) and incoherent scattering. In the powder experiment, the random orientation of many small crystallites creates the problem in which the entire sample in the neutron beam is contributing to incoherent scattering, but only a small portion of the sample is contributing to the coherent scattering. This results in a very high background for hydrogenated materials. By substituting deuterium for hydrogen, the scattering power of the sample increases (the scattering length of deuterium is approximately twice that for hydrogen, in absolute value) and the incoherent scattering is greatly reduced. For organometallic chemists, unfortunately, the cost of perdeuterating a sample often will outweigh the benefit of pursuing neutron powder diffraction, in which case the only reasonable option is to attempt to grow a large single crystal.

THE FUTURE

What, then, does the future of single crystal neutron diffraction hold for us? Many of the challenges that we face in neutron crystallography of inorganic and organometallic compounds are currently being addressed. With the design of more intense neutron sources and new instrumentation optimized to handle the larger unit cells and complex structures that are becoming far more common in small molecule characterization, many of the barriers that in the past have limited the application of neutron diffraction are being removed.

A general-purpose single-crystal diffractometer, TOPAZ (Figure 3.21), is under construction at the SNS at the time of this writing and is scheduled for completion by mid-2010, with user operations to commence later in 2010. The instrument design is optimized to handle sample sizes approaching that typically found for x-ray diffraction, 0.1 mm^3 in volume or 0.5 mm on an edge. Unit cell lengths of 50 Å should be routine, thus expanding greatly on what we can currently address at existing sources. Plans for the instrument include a large number of detectors that, coupled with the increased flux on the sample, have the potential to decrease data collection times from several days or weeks to one day, or perhaps even a few hours, depending on the sample.

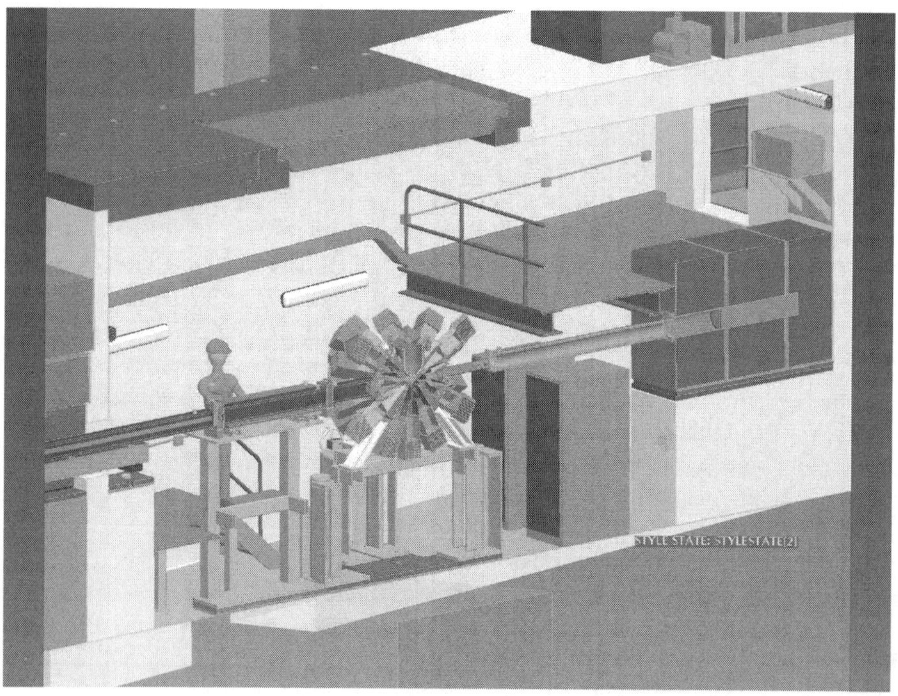

FIGURE 3.21 (See color insert) A cutaway diagram of the TOPAZ single-crystal diffractometer currently under construction at the SNS. This instrument will be ideal for smaller sample sizes and larger unit cells, both of which limit the number of currently feasible problems in single-crystal neutron diffraction.

A real strength of the TOPAZ instrument in small molecule crystallography, aside from the ability to collect data on smaller crystals and larger unit cell volumes, will likely be in parametric studies. Rapid data collection times will enable studies at variable temperature or pressure, or systematic studies of a series of related compounds, to compare features in the manner that we have outlined above, while discussing the structures from the Girolami group (Figure 3.10). Problems involving hydrogen or other gas storage issues should also be relatively straightforward. More complex materials utilizing hydrogen bonding as a means of self-assembly, which often crystallize with large unit cell parameters, may be better understood with complete characterization by neutrons. Given the importance of hydrogen bonding in biological systems, bioinorganic chemistry problems are particularly well suited for study with an instrument such as TOPAZ due to the small size of crystals that typically can be grown.

TOPAZ may enable researchers to use the same small crystal for both x-ray and neutron studies, which would eliminate systematic sources of error in charge density studies. Plans for a flexible sample environment that include a magnet and polarizer for polarized neutron studies should enable users to engage in magnetic characterization of materials, if not as a "day-one" capability on TOPAZ, then sometime in the near future. Tests of a compact ^{3}He polarizer and spin flipper at the IPNS over the last five years are important steps in achieving this capability.[71,72]

CONCLUSIONS

In this chapter we outlined some of the major cases for utilizing single crystal neutron diffraction in research concerning inorganic and organometallic chemistry. In upcoming years, with advances being made in neutron source brightness and new instrumentation, access to single crystal neutron diffraction should become easier. The capability to handle smaller sample sizes will dramatically increase the number of problems we are able to examine. All of this is good news, especially for those researchers unable to grow crystals of suitable size for neutron diffraction in the past. Faster data collection times will enable researchers to perform parametric studies similar to those being done with x-ray diffraction, perhaps over a wider temperature range than can be performed at a laboratory x-ray source, as well as systematic studies of a series of compounds exhibiting subtle changes due to sterics or electronics.

Getting started in neutron diffraction is as simple as contacting your friendly neighborhood instrument scientist and writing a proposal for beam time. Facility Web sites are the first place to look for contact information, details about instrument and sample environments, and proposal submission guidelines. Questions concerning the feasibility of an experiment can be addressed directly to the instrument scientist, and as a group, we are always happy to answer questions!

ACKNOWLEDGMENT

Work at Argonne National Laboratory was supported by the U.S. Department of Energy, Office of Science, Office of Basic Energy Sciences, under contract DE-AC02-06CH11357.

REFERENCES

1. Kealy, T. J., and Pauson, P. L. 1951. *Nature* 168:1039–1040.
2. Wilkinson, G., Rosenblum, M., Whiting, M., and Woodward, R. B. 1952. *J. Am. Chem. Soc.* 74:2125–2126.
3. Fischer, E. O., and Pfab, W. 1952. *Zeitschrift Fur Naturforschung B* 7:377–379.
4. Trofimenko, S. 1968. *J. Am. Chem. Soc.* 90:4754.
5. Trofimenko, S. 1970. *Inorg. Chem.* 9:2493.
6. Kosky, C. A., Ganis, P., and Avitabile, G. 1971. *Acta Cryst. B* 27:1859.
7. Cotton, F. A., LaCour, T., and Stanislowski, A. G. 1974. *J. Am. Chem. Soc.* 96:754.
8. Sheldrick, G. M. 2008. *Acta Cryst.* A64:112–122.
9. Larson, A. C., and Von Dreele, R. B. 2000. *General Structure Analysis System—GSAS.* Los Alamos, NM: Los Alamos National Laboratory.
10. Wilson, C. C. 2000. *Single Crystal Neutron Diffraction From Molecular Materials.* Singapore: World Scientific Publishing Co. Pte. Ltd.
11. Wilson, C. C. 2005. *Z. Kristallogr.* 220:385–398.
12. Koetzle, T. F., and Schultz, A. J. 2005. *Topics in Catalysis* 32:251–255.
13. Pynn, R. 1990. *Los Alamos Science* 19:1–31. Online at http://neutrons.ornl.gov/research/ns_primer.pdf
14. Calvert, R. B., Shapley, J. R., Schultz, A. J., Williams, J. M., Suib, S. L., and Stucky, G. D. 1978. *J. Am. Chem. Soc.* 100:6240.
15. Bau, R., Drabnis, M. H., Garlaschelli, L., Klooster, W. T., Xie, Z., Koetzle, T. F., and Martinengo, S. 1997. *Science* 275:1099.
16. Spencer, E. C., Howard, J. A. K., McIntyre, G. J., Rowsell, J. L. C., and Yaghi, O. M. 2006. *Chem. Commun.* 278–280.
17. Anderson, T. M., Neiwert, W. A., Kirk, M. L., Piccoli, P. M. B., Schultz, A. J., Koetzle, T. F., Musaev, D. G., Morokuma, K., Cao, R., and Hill, C. L. 2004. *Science* 306:2074–2077.
18. Tippe, A., and Hamilton, W. C. 1969. *Inorg. Chem.* 8:464–470.
19. http://www.ncnr.nist.gov/resources/n-lengths/. 1992. *Neutron News* 3:29–37.
20. Allen, F. H. 2002. *Acta. Cryst. B.* 58:380–388.
21. Teller, R. G., and Bau, R. 1981. *Struct. Bonding* 44:1.
22. Bau, R., and Drabnis, M. H. 1997. *Inorganica Chimica Acta* 259:27–50.
23. Yousufuddin, M., Baldamus, J., Tardif, O., Hou, Z., Mason, S. A., McIntyre, G. J., and Bau, R. 2006. *Physica B* 385–386:231–233.
24. Boss, S. R., Cole, J. M., Haigh, R., Snaith, R., Wheatley, A. E. H., McIntyre, G. J., and Raithby, P. R. 2004. *Organometallics* 23:4527.
25. Webster, C. E., Gross, C. L., Young, D. M., Girolami, G. S., Schultz, A. J., Hall, M. B., and Eckert, J. 2005. *J. Am. Chem. Soc.* 127:15091–15101.
26. Kubas, G. 2001. *Metal Dihydrogen and Sigma-Bond Complexes: Structure, Theory, and Reactivity.* New York: Kluwer Academic/Plenum Publishers.
27. Li, H., Eddaoudi, M., O'Keeffe, M., and Yaghi, O. M. 1999. *Nature* 402:276.
28. Rowsell, J. L. C., Eckert, J., and Yaghi, O. M. 2005. *J. Am. Chem. Soc.* 127:14904–14910.
29. Brookhart, M., and Green, M. L. H. 1983. *J. Organomet. Chem.* 250:395.
30. Crabtree, R. H. 1993. *Angew. Chem. Int. Ed. Engl.* 32:789.
31. Collman, J. P., Hegedus, L. S., Norton, J. R., and Finke, R. G. 1987. *Principles and Applications of Organotransition Metal Chemistry.* Mill Valley, CA: University Science Books.
32. Elschenbroich, C., and Salzer, A. 1992. *Organometallics: A Concise Introduction.* New York: VCH Publishers.
33. Cosky, C. A., Ganis, P., and Avitabile, G. 1971. *Acta Cryst.* B27:1859.

34. Piccoli, P. M. B., Cowan, J. A., Schultz, A. J., Koetzle, T. F., and Trofimenko, S. 2008. *J. Mol. Struct.* 890:63.
35. Dawoodi, Z., Green, M. L. H., Mtetwa, V. S. B., Prout, K., Schultz, A. J., Williams, J. M., and Koetzle, T. F. 1986. *J. Chem. Soc. Dalton Trans.* 1629.
36. Dawoodi, Z., Green, M. L. H., Mtetwa, V. S. B., and Prout, K. 1982. *J. Chem. Soc. Chem. Commun.* 1410.
37. Bau, R., Mason, S. A., Patrick, B. O., Adams, C. S., Sharp, W. B., and Legzdins, P. 2001. *Organometallics* 20:4492–4501.
38. Debad, J. D., Legzdins, P., Rettig, S. J., and Veltheer, J. E. 1993. *Organometallics* 12:2714.
39. Baratta, W., Mealli, C., Herdtweck, E., Ienco, A., Mason, S. A., and Rigo, P. 2004. *J. Am. Chem. Soc.* 126:5549–5562.
40. Mork, B. V., Tilley, T. D., Schultz, A. J., and Cowan, J. A. 2004. *J. Am. Chem. Soc.* 126:10428–10440.
41. Mork, B. V., and Tilley, T. D. 2003. *Angew. Chem. Int. Ed. Engl.* 42:357–360.
42. Scherer, W., and McGrady, G. S. 2004. *Angew. Chem. Int. Ed.* 43:1782–1806.
43. Peterson, S. W., and Levy, H. A. 1952. *J. Chem. Phys.* 20:704–707.
44. Steiner, T. 1998. *J. Phys. Chem. A* 102:7041–7052.
45. Schoenborn, B. P., and Knott, R. B. 1997. *Neutrons in Biology*. New York: Plenum Publishing Corporation.
46. Langan, P., Greene, G., and Schoenborn, B. P. 2004. *J. Appl. Cryst.* 37:24–31.
47. Schultz, A. J., Thiyagarajan, P., Hodges, J. P., Rehm, C., Myles, D. A. A., Langan, P., and Mesecar, A. D. 2005. *J. Appl. Cryst.* 38:964–974.
48. Brammer, L., Helliwell, J. R., Wilson, C. C., Keen, D. A., and Radaelli, P. G. 2005. LMX—A diffractometer for large molecule crystallography. http://www.isis.rl.ac.uk/targetstation2/instruments/phase2/lmx2005Web.pdf
49. Cotton, F. A., Fair, C. K., Lewis, G. E., Mott, G. N., Ross, F. K., Schultz, A. J., and Williams, J. M. 1984. *J. Am. Chem. Soc.* 106:5319–5323.
50. Hathaway, B. J., and Hewat, A. W. 1984. *J. Solid State Chem.* 51:364–375.
51. Simmons, C. J., Hitchman, M. A., Stratemeier, H., and Schultz, A. J. 1993. *J. Am. Chem. Soc.* 115:11304–11311.
52. Schultz, A., Henning, R., Hitchman, M. A., and Stratemeier, H. 2003. *Crystal Growth Design* 3:403–407.
53. Henning, R., Schultz, A. J., Hitchman, M. A., Kelly, G., and Astley, T. 2000. *Inorg. Chem.* 39:765–769.
54. Desiraju, G. R. 2002. *Acc. Chem. Res.* 35:565–573.
55. Braga, D., and Grepioni, F. 2000. *Acc. Chem. Res.* 33:601–608.
56. Brammer, L., Bruton, E. A., and Sherwood, P. 2001. *Crystal Growth Design* 1:227–290.
57. Vicic, D. A., Anderson, T. J., Cowan, J. A., and Schultz, A. J. 2004. *J. Am. Chem. Soc.* 126:8132–8133.
58. Tyree, W. S., Vicic, D. A., Piccoli, P. M. B., and Schultz, A. J. 2006. *Inorg. Chem.* 45:8853–8855.
59. Kriley, C. E., Woolley, C. J., Krepps, M. K., Popa, E. M., Fanwick, P. E., and Rothwell, I. P. 2000. *Inorg. Chim. Acta* 300–302:200–205.
60. Sieve, B., Chen, X. Z., Henning, R., Brazis, P., Kannewurf, C. R., Cowen, J. A., Schultz, A. J., and Kanatzidis, M. G. 2001. *J. Am. Chem. Soc.* 123:7040–7047.
61. Zhuravleva, M. A., Wang, X., Schultz, A. J., Bakas, T. B., and Kanatzidis, M. G. 2002. *Inorg. Chem.* 41:6056–6061.
62. Condron, C. L., Martin, J., Nolas, G. S., Piccoli, P. M. B., Schultz, A. J., and Kauzlarich, S. M. 2006. *Inorg. Chem.* 45:9381–9386.
63. Bullock, R. M., Brammer, L., Schultz, A. J., Albinati, A., and Koetzle, T. F. 1992. *J. Am. Chem. Soc.* 114:5125–5130.

64. Brammer, L., Howard, J. A. K., Johnson, O., Koetzle, T. F., Spencer, T. F., and Stringer, A. M. 1991. *J. Chem. Soc. Chem. Commun.* 241–243.
65. Albinati, A., Bullock, R. M., Rappoli, B. J., and Koetzle, T. 1991. *Inorg. Chem.* 30:1414–1417.
66. Zheludev, A., Grand, A., Ressouche, E., Schweizer, J., Morin, B. G., Epstein, A. J., Dixon, D. A., and Miller, J. S. 1994. *J. Am. Chem. Soc.* 116:7243–7249.
67. Claiser, N., Souhassou, M., Lecomte, C., Gillon, B., Carbonera, C., Caneschi, A., Dei, A., Gatteschi, D., Bencini, A., Pontillon, Y., and Leliévre-Berna, E. 2005. *J. Phys. Chem. B* 109:2723–2732.
68. Chandler, G. S., Christos, G. A., Figgis, B. N., and Reynolds, P. A. 1992. *J. Chem. Soc. Faraday Trans.* 88:1961–1969.
69. ILL Annual Report 2005. Institut Laue Langevin, Grenoble, France, 2005. Online at http://www.ill.eu/quick-links/publications/annual-report/
70. Macchi, P., Schultz, A. J., Larsen, F. K., and Iversen, B. B. 2001. *J. Phys. Chem. A* 105:9231–9242.
71. Jones, G. L., Bakera, J., Chen, W. C., Colletta, B., Cowan, J. A., Dias, M. F., Gentile, T. R., Hoffmann, C., Koetzle, T., Lee, W. T., Littrell, K., Miller, M., Schultz, A., Snow, W. M., Tong, X., Yan, H., and Yue, A. 2005. *Physica B* 356:86–90.
72. Jones, G. L., Dias, F., Collett, B., Chen, W. C., Gentile, T., Piccoli, P. M. B., Miller, M. E., Schultz, A. J., Yan, H. Y., Tong, X., Snow, M., Lee, W. T., Hoffmann, C., and Thomison, J. 2006. *Physica B* 385–386:1131–1133.
73. Schleger, P., Puig-Molina, A., Ressouche, E., Rutty, O., and Schweizer, J. 1997. *Acta Crystallogr. A* 53:426–435.

4 Adventures of Quantum Chemistry in the Realm of Inorganic Chemistry

Constantinos A. Tsipis

CONTENTS

Recent improvements in the design of faster and more efficient algorithms have placed powerful computational quantum chemistry tools in the hands of all chemists. *Ab initio* and density functional calculations are routinely performed by nonspecialists in the field. These computational tools have nowadays become as useful to the bench chemist as spectrometers and vacuum lines. The impact of modern computational technology in the advancement of inorganic chemistry is outlined herein. Attempts have been made to present only the background information required to appreciate the general features of the computational quantum chemical techniques available and the types of chemical problems that can be reasonably solved in the growing field of quantum inorganic chemistry. It should be stressed that quantum chemical calculations assist experimental studies by accurately predicting chemical behavior. This is why a blend of theory and experiment must be effectively employed to solve a variety of difficult problems encountered in modern chemical research.

INTRODUCTION

The phenomenal increase in speed and computational power of computers has continued at an astonishing pace over the last decade. Chemistry, like many other disciplines, is being profoundly influenced by increased computing power. This has happened in part by enhancing many existing computational procedures, providing a new impetus to quantum mechanical and molecular simulations at the atomic level of detail. Computational chemistry plays an increasing role in chemical research. With significant strides in computer hardware and software, computational chemistry has achieved full partnership with theory and experiment as a tool for understanding and predicting the behavior of a broad range of chemical, physical, and biological phenomena. With massively parallel computers capable of peak performance of several teraflops already on the scene, and with the development of parallel software for efficient exploitation of these high-end computers, we can anticipate that computational chemistry will continue to change the scientific landscape in the twenty-first century. We are not far from the point where chemists can routinely design catalysts and other materials and predict biological activity or environmental fate from chemical structure.

Quantum chemistry, penetrating all chemistry, is the science that treats molecular behavior by one unified concept: the Schrödinger equation. Although there are specialists in the field, increasingly the computational quantum chemical techniques are being applied by chemists of many persuasions, who are experienced researchers, knowing chemistry, and now having computational tools available. Commercial programs incorporating the latest methods have become widely available and require little knowledge beyond the chemical formula to produce some results for a variety of properties. The ready availability and applicability of these programs provided a probe of structural, spectral, and reactivity characteristics frequently unavailable from experiment or, alternatively, facilitated interpretation of available experimental results. Noteworthy, the steadily increasing fraction of articles listed in the current chemical journals reference a computational chemistry software package.

Following the progress in the area of accurate quantum chemical treatment of systems containing transition metals in the last few years, a new exciting field of inorganic chemistry—that of quantum inorganic chemistry—has emerged. Computational methods for the analysis of transition metal containing molecular systems have undergone rapid progress in the recent past. Timely and important areas of computational chemistry concerning transition metal chemistry have been covered in recent special issues of *Chemical Reviews*[1] and *Coordination Chemistry Reviews*.[2] An earlier comprehensive review of the theoretical approaches to the description of the electronic structure and chemical reactivity of transition metal compounds up to 1990 has also been reported in *Coordination Chemistry Reviews*.[3] Moreover, Koga and Morokuma[4] reported an overview of theoretical studies of transition-metal-catalyzed reactions, while Gordon and Cundari surveyed the recent advances in the effective core potential (ECP) studies of transition metal bonding, structure, and reactivity.[5] Very recent reviews cover some more specific topics of computational chemistry related to quantum chemical studies of intermediates and reaction pathways in selected metalloenzymes and catalytic synthetic systems[6] and the electronic structures of metal sites in proteins and models.[7]

The aim of the present article is to familiarize experimental chemists with the latest computational techniques and how these techniques can be applied to solve a wide range of real-world problems encountered in the realm of inorganic chemistry and particularly in transition metal chemistry. In addition, this article is meant to be a tutorial on the intelligent use of quantum chemical methods for the determination of molecular structure, reactivity, and spectra of coordination and organometallic compounds.

APPLYING COMPUTATIONAL QUANTUM CHEMISTRY METHODS

The newcomer to the field of computational quantum chemistry faces three main problems:

1. **Learning the language.** The language of computational quantum chemistry is littered with acronyms. *What do these abbreviations stand for in terms of underlying assumptions and approximations?* If you want to use computational quantum chemistry methods, you need to decipher the acronyms.
2. **Technical problems.** *How does one actually run the program, and what does one look for in the output?* This point, related to both the hardware and software, needs to be solved "on location." As computer programs evolve they become easier to use. In addition, modern programs often communicate with the user in terms of a graphical interface, and many methods became essentially "black box" procedures. This effectively means that we no longer have to be highly trained theoreticians to run even quite sophisticated calculations. Molecular modeling software lists can easily be found on the Internet.[8,9]

3. **Quality assessment.** *How good is the result of the calculation?* The accuracy of the various components of the computational methodology is of crucial importance to computational chemistry if we want theory and experiment to become partners in the solution of chemical problems. Because of the ease by which calculations can be performed, this point has become the central theme in computational quantum chemistry. It is quite easy to run a series of calculations, which produce absolutely meaningless results, since the program cannot tell us whether the chosen method is valid for the problem we are studying. Therefore, quality assessment is an absolute requirement, but requires much more experience and insight than just running the program. Basic understanding of the theory behind the method and knowledge of the performance of the method for other systems are needed. In particular, when we want to make predictions, and not merely reproduce known results, we have to be able to judge the quality of the results obtained. Along this line, a benchmark level of confidence in terms of robustness of structural and energetic results needs to be established, but this is by far the most difficult task in computational chemistry.

ELECTRONIC STRUCTURE CALCULATION METHODS

Electronic structure calculation methods are well-defined mathematical procedures aiming to solve the electronic much-lauded Schrödinger wave equation in order to obtain the electronic wave function, $\Psi(\mathbf{r})$, which contains all the information for a particular state of a chemical system. A number of electronic structure calculation methods have been developed, each with advantages and disadvantages. These methods can be classified into the following three categories:

1. *Ab initio* methods
2. Semiempirical methods
3. Density functional methods

In this section we will briefly outline the basic principles and terminology of the three classes of electronic structure calculation methods, starting at first principles, for the beginner in the field to feel comfortable in using these methodologies.

BASIC PRINCIPLES AND TERMINOLOGY OF *AB INITIO* METHODS

Ab initio methods are electronic structure calculation methods used to model the electron density in an atom or molecule with respect to the average electron density, without using any adjustable or empirically derived parameters in calculating the molecular energy. The reference framework for the *ab initio* methods is the independent particle or molecular orbital (MO) model; the electrons are confined to certain regions of space. In practice, a number of approximations are made within the MO calculations that have a direct bearing on the reliability of the results obtained.

Born-Oppenheimer (BO) Approximation

The first approximation in calculating molecular wave functions is the Born-Oppenheimer (BO) "clamped nuclei" approximation. It allows the motions of nuclei and electrons to be considered separately. In this way, one obtains the time-independent nonrelativistic electronic Schrödinger equation, which describes the motion of electrons in the field of fixed nuclei:

$$H|\Psi_e(\mathbf{r};\mathbf{R})\rangle = E(\mathbf{R})|\Psi_e(\mathbf{r};\mathbf{R})\rangle$$

The electronic Hamiltonian, H, of a molecular system with N electrons and M nuclei contains one- and two-electron terms consisting of all of the kinetic and potential energy terms that act upon the electrons:

$$H = -\sum_{i=1}^{N}\frac{1}{2}\nabla_i^2 - \sum_{i=1}^{N}\sum_{A=1}^{M}\frac{Z_A}{r_{iA}} + \sum_{i=1}^{N}\sum_{j=1}^{N}\frac{1}{r_{ij}}$$

In the framework of the BO approximation the electronic wave function $\Psi_e(\mathbf{r};\mathbf{R})$ depends parametrically on the nuclear coordinates $\mathbf{R}$. By parametric dependence we mean that, for different arrangements of the nuclei, $\Psi_e(\mathbf{r};\mathbf{R})$ is a different function of the electronic coordinates. In the BO approximation, the energy $E(\mathbf{R})$ of the electronic Schrödinger equation provides a potential for the motion of the nuclei. In other words, the electrons are dragged along the nuclei, and the dressed nuclei (atoms in molecules) move in the field of the electronic distribution. These electronic energies' dependence on the positions of the nuclei causes them to be referred to as potential energy surfaces (PESs), which will be discussed later. Most computational chemical studies involve characterizing key features of the PES or integrating the trajectories on this surface.

Hartree-Fock (HF) Approximation

In the HF approximation the interelectronic repulsion term,

$$\sum_{i=1}^{N}\sum_{j=1}^{N}\frac{1}{r_{ij}}$$

of the electronic Hamiltonian is replaced with an effective potential $V(r)$, which represents the potential one electron "feels" when moving independently of the others in the field created by the fixed nuclei and the mean field of the other electrons. This substitution allows us to reduce the N-particle problem to a set of one-particle eigenvalue problems,

$$F(\mathbf{r}_i)\varphi_i(\mathbf{r}_i) = \varepsilon_i\varphi_i(\mathbf{r}_i)$$

and is the basis of molecular orbital theory. These nonlinear equations are referred to as Hartree-Fock or self-consistent field (SCF) equations. The term orbital (atomic, AO; molecular, MO) was created for the one-electron wave functions $\varphi_i(\mathbf{r}_i)$. An MO is an eigenfunction of an effective one-particle Hamiltonian, the so-called Hartree-Fock operator, F, and the electron in the MO is considered to have an orbital energy of ε_i.

Basis Set Approximation

In the basis set approximation, introduced by Roothaan in 1951, each MO $\varphi_i(\mathbf{r}_i)$ should be expanded algebraically as a finite set of basis functions χ_μ, which are normally centered on the atoms in the molecule. This gives,

$$\varphi_i(\mathbf{r}_i) = \sum_{\mu=1}^{N} c_{i\mu}\chi_\mu$$

The basis functions collectively are the *basis set.* The coefficients for a given MO can be thought of as arrays that, when squared and added together, will produce a density matrix **P** with elements given by

$$P_{\mu\nu} = 2\sum_{i=1}^{occ} c_{i\mu}c_{i\nu}$$

Here the sum runs over the occupied orbitals of the system. The density matrix and its associated orbital basis set are all that is needed to compute electronic properties of the molecular system.

Introducing the basis set approximation into the SCF equations, the eigenvalue problem is converted into a matrix problem,

$$\mathbf{FC} = \mathbf{SC}\varepsilon$$

which can be solved iteratively. From a practical standpoint, these simplifications allow an initial guess at the electron configuration of the chemical system to produce a guess at the effective potential, which can be used as an improvement to the guess of the orbitals, etc. Once the orbitals and the potential no longer change, the iterative calculation is complete; it is said to be *self-consistent*, which is why the HF approximation as practiced today is generally synonymous with SCF in the literature. In the HF approximation the energy of the molecular system is computed variationally with respect to the coefficients $c_{i\mu}$ and can be written as

$$E_{HF} = \sum_{\mu\nu} P_{\mu\nu}H_{\mu\nu} + \sum_{\mu\nu\lambda\sigma} P_{\mu\nu}P_{\lambda\sigma}\left(J_{\mu\nu\lambda\sigma} - K_{\mu\nu\lambda\sigma}\right) + V_{nucl}$$

Here the Hamiltonian matrix elements $H_{\mu\nu}$ are the terms referring to the kinetic energy and the potential energy of attraction between the electrons and the nuclei. The Coulomb integrals $J_{\mu\nu\sigma\lambda}$ are the terms referring to the potential energy of repulsion between pairs of electrons, while the exchange integrals $K_{\mu\nu\sigma\lambda}$ are the terms arising from the antisymmetry of the wave function with respect to exchange of any two electrons. Finally, the term V_{nuc} is the potential energy from the repulsion of pairs of nuclei.

The basis functions χ_{μ} can be thought of as atomic orbitals (*s*, *p*, *d*, etc.) from the rigorous solution of the Schrödinger equation for the hydrogen atom. For many-electron atoms, we don't know the actual mathematical functions for the atomic orbitals, so substitutes are used, usually either Slater-type orbitals (STOs) or Gaussian-type orbitals (GTOs). We don't concern ourselves with the exact form of STOs and GTOs. Suffice it to say that they are chosen to behave mathematically like the actual atomic orbitals of *s*-, *p*-, *d*-, *f*-type, etc. The use of GTOs is now commonplace. This is because it is very difficult to evaluate the necessary integrals over STOs when the orbitals in the integrand are centered on three or four different atoms. In practice, the basis functions are often fixed linear combinations of GTOs whose exponents have been judiciously chosen by minimizing atomic energies, matching STOs, or reproducing experimentally known molecular properties.

There is a bewildering number of basis sets tabulated and tested in the literature for almost every element of the periodic table, and still more are developed each year.[10,11] They are symbolized by acronyms, which are also used as keywords in the computer codes that perform quantum chemical calculations. Basis sets could be classified as follows:

Minimal basis sets
Split valence basis sets
Extended basis sets

Important additions to basis sets are polarization and diffuse basis functions. *Polarization* functions[12–14] are functions added in the basis set to distort the shape of the orbitals (in principle the atomic orbitals) due to their polarization introduced by the other nuclei in the molecular system. The distortion of an atomic orbital is accomplished by adding in basis functions of higher angular momentum quantum number. *Diffuse* functions are necessary to be added in the case of investigating *excited states* and *anionic species,* because in these cases the electronic density is more spread out over the molecule. To model this correctly, we have to use basis functions that themselves are more spread out. These functions are GTOs with small exponents.

For transition metal complexes, because of the large number of basis functions that must be included to model the nonvalence electrons in the metals, particularly those of the second and third rows, the inner shell electrons are replaced by a model potential called *effective core potential* (ECP).[5,15,16] The ECPs not only reduce the computational demands of calculations on transition metal compounds, but also allow the inclusion of relativistic effects for the heavier second- and third-row transition metals, in a straightforward manner. The relativistic effective core potentials

TABLE 4.1
Acronyms of the Most Commonly Used Basis Sets

Basis Set	Characteristics
STO-3G	A minimal basis set using 3 GTOs to approximate each STO.
3-21G	A split valence basis set. Inner shell basis functions made of 3 GTOs. Valence orbitals are split into 2 and 1 GTOs.
6-31G	A split valence basis set. Inner shell basis functions made of 6 GTOs. Valence orbitals are split into 3 and 1 GTOs.
6-31G(d) or 6-31G*	The 6-31G basis set with the addition of six *d*-type polarization functions to nonhydrogen atoms.
6-31G(d,p) or 6-31G**	The 6-31G(d) basis set with the addition of *p*-type polarization functions to hydrogen atoms.
6-31G+(d,p) or 6-31+G**	The 6-31G(d,p) basis set with the addition of *s*- and *p*-type diffuse functions to the atoms of the first and second rows.
6-31G++(d,p) or 6-31++G**	The 6-31G(d,p) basis set with the addition of a set of *s*- and *p*-type diffuse functions to the atoms of the first and second rows and a set of *s*-type functions to hydrogen.
6-311G	A split valence basis set. Inner shell basis functions made of 6 GTOs. Valence orbitals are triply split into 3, 1, and 1 GTOs.
D95	Dunning-Huzinaga full-double zeta.
CEP-121G	Stevens-Basch-Krauss ECP triple-split basis.
LanL2DZ	D95V on first row and Los Alamos ECP plus DZ on Na-Bi.
SDD	D95V up to Ar and Stuttgart-Dresden ECPs on the remainder of the periodic table.
Cc-pvdz, cc-pvtz, cc-pvqz, cc-pv5z, cc-pv6z	Dunning's correlation-consistent basis sets (double, triple, quadruple, quintuple-zeta, and sextuple-zeta, respectively).

(RECPs) are generated from the relativistic HF atomic core. The ECP schemes of LANL2DZ and SDD have performed extremely well with respect to the calculation of the equilibrium geometries of transition metal complexes and organometallics. In addition, extra basis sets of at least the 6-31G(d,p) quality must be used for the nonmetal and hydrogen atoms. Generally, the larger the basis set, the more accurate the calculation (within limits) and the more computer time that is required. A few commonly used basis sets are listed in Table 4.1.

CORRELATED OR POST-HF MODELS

Because of the central field imposed by the HF potential *V*(r), each electron is not explicitly aware of the others' presence. It is usually said that the HF models neglect *electron correlation*, because there is a finite probability that two electrons will occupy the same point in space. Electron correlation, the difference between the exact energy, E_{exact}, of a system and the HF limit energy, E_{HF}, which is obtained using an essentially complete basis set, is neglected in the HF approach. Correlation energy will always be negative because HF energy is an upper bound to the exact energy, E_{exact}:

$$E_{\text{correlation}} = E_{\text{exact}} - E_{\text{HF}} < 0$$

Higher levels of theory are necessary to resolve this issue by recovering the correlation energy. Most of the standard high-level techniques in quantum chemistry utilize the HF approximation as a starting point and then attempt to correlate electrons by more rigorous computational methods. A large number of such methods have been used to improve the HF method, which can be classified into two classes:

Variational methods
Perturbational methods

Variational Methods

A common method for incorporating correlation into HF calculations is to construct a modified wave function using the unoccupied (virtual) orbitals of the "reference" HF wave function. In this approach, which is termed the configuration interaction (CI) method, contributions from excited configurations (single excited, Ψ_i^a ; double excited, Ψ_{ij}^{ab} ; etc.), in which electrons are promoted from the occupied (filled) *i*, *j*, …, MOs to the virtual *a*, *b*, …, MOs, are mixed variationally with the ground-state wave function, Ψ_0, to give the following expression:

$$\Psi = c_0\Psi_0 + \sum_{ia} c_i^a \Psi_i^a + \sum_{ij} c_{ij}^{ab} \Psi_{iji}^{ab} + \ldots$$

If the set of the determinantal wave functions included in the CI calculation is complete, it is a full CI (FCI) and is both variational and size consistent. However, an FCI calculation is extremely expensive and impractical for all but the tiniest of molecular systems. Therefore, the CI expansion should be limited in one way or another. More realistically, the set of determinantal wave functions could include only the double (D) excited configurations (denoted CID) or the single (S) and double (D) excited configurations (denoted CISD). It is also typical to augment the CISD calculations with triple (T) and/or quadruple (Q) excitations, with the most popular variant being the quadratic configuration interaction denoted as QCISD. In general, CI is not the practical method of choice for calculation of correlation energy, because FCI is not possible, convergence of the CI expansion is slow, and the integral transformation is time-consuming. However, an advantage of the CI method is that it is variational, so the calculated energy is always greater than the exact energy.

Other approaches aim to optimize not only the mixing coefficients of the various configurations, but also the coefficients of the basis functions in the molecular orbitals. The latter are frozen at the HF values in the CI methods described above. This more complex approach is called multiconfiguration self-consistent field (MCSCF). It can give quite good results with a modest number of configurations. The complete active space self-consistent field (CASSCF) method is an example of this approach. Using CASSCF, the references are selected by choosing an "active space" of several chemically important orbitals and performing an FCI in the span of the active space. The MCSCF method requires considerable care in the selection of the basis set and especially the active space, and should not be considered for routine use. MCSCF methods are essential for the study of processes in which transitions between potential energy surfaces occur, such as in photochemical reactions.[10,17–22]

Perturbational Methods

There are two important perturbative post-HF methods: many-body perturbation theory (MBPT) and coupled cluster (CC).

In perturbation theory we define a perturbed Hamiltonian $H(\lambda)$:

$$H(\lambda) = F_0 + \lambda[H(\lambda) - F_0)$$

where F_0 is the HF operator with an appropriate eigenfunction Ψ_0 if $\lambda = 0$, and the exact (FCI) eigenfunction Ψ is obtained if $\lambda = 1$, while the energy $\mathrm{E}(\lambda)$ of the perturbed system is expanded in MacLaurin series:

$$\mathrm{E}(\lambda) = E_0 + \lambda \mathrm{E}_1 + \lambda^2 \mathrm{E}_2 + \lambda^3 \mathrm{E}_3 + \ldots$$

including corrections of first, E_1, second, E_2, third, E_3, order, etc.

A particularly successful application to molecules and the correlation problem goes back to Møller and Plesset in 1933.[23] This is now the MP*n* method. The first important correction is the second-order term, E_2, and this leads to MP2. MP2 is relatively economic to evaluate and gives a reasonable proportion of the correlation energy. Higher-order terms become more and more expensive. MP3 is commonly used, but does not seem to give much improvement over MP2. An enormous practical advantage is that MP2 is fast (of the same order of magnitude as SCF), while it is rather reliable in its behavior and size consistent. A disadvantage is that it is not variational, so the estimate of the correlation energy can be too large. In practice, MP2 must be used with a reasonable basis set (6-31G* or better).

In the coupled cluster method,[22,24–26] which is by far the most popular high-level *ab initio* quantum chemical method today, the electron correlation effects are treated in a different way. The CC method allows the mixing in of the higher configurations by constructing an exponential operator e^{T} that acts on the HF ground state to produce configurations from only a certain class of excitations:

$$\Psi = e^{\mathrm{T}}\Psi_0$$

T is an excitation operator:

$$T = T_1 + T_2 + \ldots + T_{\mathrm{n}}$$

The excitations are usually limited to single, double, and triple substitutions. Since carrying out the exponential as a sum will generate many new configurations that are sums of products of the configurations, this technique is known as the coupled cluster singles, doubles, and triples (CCSDT) theory. Triple excitations can also be included in an approximate but less time-consuming way, leading to the CCSD(T) method. Armed with this powerful technique, something like 97% of the correlation energy can be extracted from just the first three terms in *T*. Unfortunately, these most sophisticated correlation methods at the present time are extremely costly in terms of time and resources.

BASIC PRINCIPLES AND TERMINOLOGY OF SEMIEMPIRICAL METHODS

The common *ab initio* molecular orbital techniques can only be applied with difficulty to transition metal systems. They are limited in their practical applicability because of their heavy demands of CPU time and storage space on disk or in the computer memory. At the HF level the problem is seen to be in the large number of two-electron integrals that need to be evaluated. Without special tricks this is proportional to the fourth power of the number of basis functions. The first strategy used to reduce computational effort is to consider only valence electrons in the quantum mechanical treatment. The core electrons are accounted for in a core-core repulsion function, together with the nuclear repulsion energy. The next step is to replace many of the remaining integrals by parameters, which can either have fixed values or depend on the distance between the atoms on which the basis functions are located. At this stage empirical parameters can be introduced, which can be derived from measured properties of atoms or diatomic molecules. In the modern semiempirical methods, however, the parameters are mostly devoid of this physical significance: They are just optimized to give the best fit of the computed molecular properties to experimental data. Different semiempirical methods differ in the details of the approximations (e.g., the core-core repulsion functions) and in particular in the values of the parameters.

Nowadays, semiempirical MO theory is usually taken to mean the modern variants of neglect of differential diatomic overlap (NDDO)[27] theory. New versions of the NDDO methods have recently been developed that include *d*-orbitals for second-row and heavier elements: modified neglect of diatomic overlap (MNDO/d),[28] extended Austin model 1 (AM1),[29] parametric method (PM3),[30] and PM3(tm).[31] Commercial programs like MOPAC embody a suite of methods from which a knowledgeable modeler can make an optimum selection. Recently, a slightly extended and reparameterized version of PM3 termed PM5 has been made available in MOPAC.[32]

BASIC PRINCIPLES AND TERMINOLOGY OF DENSITY FUNCTIONAL METHODS

Given the computational problems associated with the application of *ab initio* molecular orbital-based methods to problems of structure and reactivity in coordination and organometallic compounds, most theoretical studies in this area employ density functional theory (DFT) methods.[33–35] In a nutshell, DFT is based on the fact that all molecular electronic properties can be calculated if the electron density $\rho(\mathbf{r})$ is known. Thus, the molecular properties are functionals of $\rho(\mathbf{r})$ because the electron density itself is a function of the spatial coordinates **r**. In the framework of DFT the ground-state electronic energy of an N electron system can be expressed by

$$E_{DFT} = \sum_{\mu\nu} P_{\mu\nu} H_{\mu\nu} + \sum_{\mu\nu\lambda\sigma} P_{\mu\nu} P_{\lambda\sigma} J_{\mu\nu\lambda\sigma} + E_X(\rho) + E_C(\rho) + V_{nuc}$$

where $E_X(\rho)$ and $E_C(\rho)$ are two empirically derived functions that replace the exchange matrix and are referred to as the exchange and correlation functionals, respectively. A starting point based upon a SCF procedure is developed that ultimately determines the ground-state charge and spin densities, through solution of the single-particle Kohn-Sham equations.[36] DFT provided a sound basis for the development of computational strategies for obtaining information about the structure, energetics, and properties of molecules at much lower costs than traditional *ab initio* wave function techniques. An excellent publication by Koch and Holthausen in 2000, titled *A Chemist's Guide to Density Functional Theory*,[33] offers an overview of the performance of DFT in the computation of a variety of molecular properties, thus providing a guide for practicing, not necessarily quantum, chemists. Moreover, a recent survey of chemically relevant concepts and principles extracted from DFT, the so-called conceptual DFT by Geerlings et al., is very informative.[37]

Despite its simple origins, DFT works very well in most cases. For about the same cost of doing an HF calculation, DFT includes a significant fraction of the electron correlation. Note that DFT is not an HF method, nor is it (strictly speaking) a post–Hartree-Fock method. The wave function is constructed in a different way, and the resulting orbitals are often referred to as Kohn-Sham (KS) orbitals. The KS orbitals appear to be as robust as HF orbitals for qualitative interpretation and rationalization of molecular properties. The chief difference between the SCF and KS-DFT approaches lies in the exchange correlation (XC) density functionals. If the exact form for the XC functionals were known, the KS-DFT approach would give the exact energy. There is, however, a number of approximate functionals available, which can be classified as:

- Functionals based on the local density approximation (LDA):[38] For these functionals the energy depends only on the charge density $\rho(\mathbf{r})$.
- Functionals based on the generalized gradient approximation (GGA):[39–41] These functionals use not only the value of the electron density, $\rho(\mathbf{r})$, but also its gradient, $\nabla\rho(\mathbf{r})$.
- Meta-GGA functionals: For these functionals the energies depend also on the Laplacian of the density $\nabla^2\rho(\mathbf{r})$ or the orbital kinetic energy.
- Hybrid density functionals combining GGA functionals with a parameterized proportion of the exchange energy calculated by HF theory.[42]

Some of the most popular functionals you may come across are:

BP86: Developed by Becke and Perdew in 1986
B3PW91: A three-parameter functional expression including the PW91 correlation functional
BLYP: Developed by Becke, Lee, Yang, and Parr
B3LYP: A modification of BLYP in which a three-parameter functional developed by Axel Becke is used
MPW1K: Modified Perdew-Wang one-parameter model for kinetics developed by Donald G. Truhlar et al.[43]

The choice of the functional is the only limitation of the DFT method. At the present time, there is no systematic way of choosing the functional, and the most popular ones in the literature have been derived by careful comparison with experiment. It should be emphasized that great care has to be exercised in choosing the DFT functional to properly calculate spin states in transition metal complexes.

QUALITY AND RELIABILITY OF QUANTUM CHEMICAL RESULTS

The combination of a quantum chemical computational method and a basis set defines a model chemistry. A model chemistry should be uniformly applicable and tested on as many systems as possible to assess its performance. The strategic use of model chemistries that can be practically applied for inorganic molecules with present-day hardware and software, based on the computational method and the basis set, is compiled in Table 4.2.

The various correlation methods are displayed vertically in order of increasing sophistication from top to bottom. Basis sets are displayed horizontally, becoming more flexible from left to right. At the bottom of Table 4.2, full configuration interaction (FCI) represents complete solution within the finite space defined by the basis set. At the far right, the results of applying a complete basis set are found (in principle but not in practice). At the bottom right, application of a complete basis set with FCI corresponds to full solution of the time-independent nonrelativistic Schrödinger equation. Each empty box represents a well-defined size-consistent theoretical model. Clearly, we may test each level to find how far we have to proceed from the top left to the bottom right for acceptable agreement between theory and experiment. In practice, full models usually have to make some compromises to achieve a wide range of applicability. If the prediction of energies is most important, a common practice is to carry out a geometry optimization at some lower level of theory (model 1) and then make a final, more expensive computation at a higher level (model 2). A useful notation for this type of composite model is model 2// model 1.

The reliability of chemical properties calculated with the most rigorous quantum chemical computational methods, listed in order of difficulty, is as follows:

- Molecular structure (±1%)
- Reaction enthalpies (±1 kcal/mol)
- Reaction free energies (±2 kcal/mol)
- Vibrational frequencies (±5%)

TABLE 4.2
Basis Set versus Quality of Theory in *Ab Initio* Calculations, Illustrating Their Strategic Use in Transition Metal Chemistry

	Basis Set								
Method	Minimal	DZ	6-31G**	DZP	TZP	PVTZ	QZP	....	Complete Basis
HF	Geometry								
DFT	Energy								
MP2									
CISD									
MP3									
CCD			Energy Correlation						
CCSD									
MP4									
MCSCF									
QCISD(T)									
CCSD(T)									
MR-CI									
. . .									
FULL CI									TRUE

CAPABILITIES OF COMPUTATIONAL QUANTUM CHEMISTRY METHODS

Using computational quantum chemical methods, the following operations may be performed:

- Geometry optimization
- Single-point energy calculation
- Predicting barriers and reaction paths
- Calculation of wave functions and detailed descriptions of molecular orbitals
- Calculation of atomic charges, dipole moments, multipole moments, electrostatic potentials, polarizabilities, etc.
- Calculation of vibrational frequencies, infrared (IR), and Raman intensities
- Calculation of nuclear magnetic resonance (NMR) chemical shifts
- Calculation of ionization energies and electron affinities
- Time-dependent calculations
- Inclusion of the electrostatic effects on solvation

GEOMETRY OPTIMIZATION

Geometry optimization is the computational procedure of finding the equilibrium geometry of a molecule, e.g., the geometry of the lowest possible energy. The procedure calculates the wave function and energy at a starting geometry, which can usually be input in the form of a Z-matrix or in cartesian coordinates, and then proceeds to move to a new geometry that will give a lower energy. This is then repeated until the lowest-energy geometry close to the starting point is obtained. The optimization procedure calculates the forces on the atoms by evaluating the gradient of the energy with respect to atomic coordinates analytically or, in some cases, numerically. The mathematical and computational machinery for structure optimization is based on various algorithms, the complexity of which depends on the desired accuracy of the electronic level. Sophisticated algorithms (such as those based on the steepest descent, Newton-Raphson, simplex, Fletcher-Powell, and combined methods) are used to select a new geometry at each step, which gives rapid convergence to the geometry with the lowest energy.

It is important to recognize that the optimization procedure will not necessarily find the geometry of lowest energy, the so-called global minimum (point **a**) in the potential energy surface (Figure 4.1).

It may find a local minimum (point **e**). It is the starting geometry that determines which of the minima would be obtained by optimization. Thus, if the starting geometry was at point **b** of the PES, the global minimum is obtained, while if it was at point **d**, the local minimum is obtained. Besides the location of minima (global or local) on the PES, the optimization process could also locate some other stationary points corresponding to transition states or other maxima (saddle points). We can sort out the types of stationary points by carrying out frequency calculations. With the development of common efficient analytical energy gradients, $g_i = \partial V(x)/\partial x_i$, and second derivatives, $G_{ij} = \partial^2 V(x)/\partial x_i \partial x_j$ (Hessian matrix), methods,

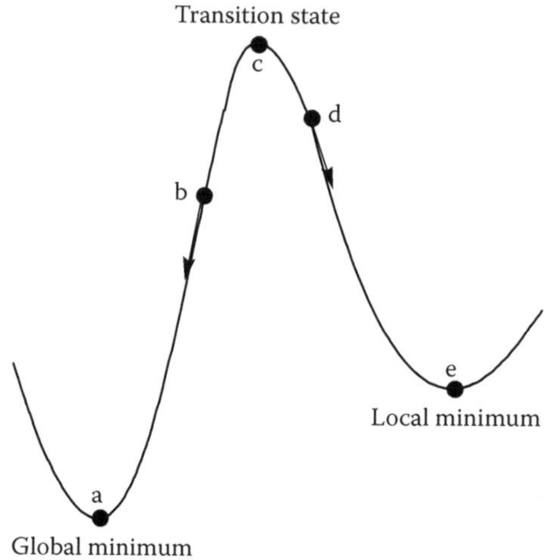

FIGURE 4.1 Part of a potential energy surface (PES) in a 2D space.

complete optimization of geometric parameters for minima and saddle points has become more common.

If the geometry obtained from an optimization run is a local or indeed the global minimum, all the frequencies (e.g., the eigenvalues of the Hessian matrix) will be real and positive. If we have a transition state (TS) or any stationary point other than a minimum, some of the frequencies will be complex (negative eigenvalues of the Hessian) and are often called imaginary frequencies. Transition structures (point **c**) usually connecting two stationary structures (points **a** and **e**) will have one imaginary frequency. In summary, the nature of stationary points on the PES can be characterized by the number of imaginary frequencies (NImag): NImag = 0 for a minima, NImag = 1 for a TS (first-order saddle point), NImag = n for a maximum (n-order saddle point). Note that frequency calculations should only be carried out at the geometry obtained from an optimization run and with the same basis set and method.

Currently, the most common geometry optimizations for transition metal coordination compounds and organometallics are performed at the HF, MP2, and DFT levels of theory. In particular, the new generation of gradient-corrected DFT methods, such as the B3LYP, BP86, and BW91 variants, are efficient and accurate computational methods for these studies. The single-reference HF and MP2 methods could not yield reliable results for the geometry optimization of first-row transition metal compounds, since the compactness of the 3*d* orbitals leads to the presence of strong near-degeneracy effects, thus leading to multireference character in the state of interest. It is always a good idea to do a geometry optimization with a small basis set and a poor method before moving to the basis set and method of choice for the molecular system of interest in relation to the molecular size and the computational resources available. An acceptable chemically accurate calculation of the geometrical parameters should give for the bond lengths and bond angles values within 0.01–0.02 Å and 1–2° of experiment, respectively. Notice that the only way to obtain molecular structures that converge toward the correct result is to use increasingly larger basis sets and more complete treatment of electron correlation. As a general rule, increasing the electron correlation increases bond length (and as a consequence, reduces bond angles), while basis set enlargement exhibits the opposite behavior.

SINGLE-POINT ENERGY CALCULATIONS

This procedure simply calculates the energy, wave function, and other requested properties at a single fixed geometry. It is frequently carried out after a geometry optimization, but with a larger basis set or a more superior method than is possible with the basis set and method used to optimize the geometry. Thus, for a very large system, such as transition metal compounds, the geometry may be optimized at the B3LYP level with a small basis set, but energy is calculated with the MP2, MCSCF, MCPF, or CCSD(T) method and a larger basis set. In general terms, DFT methods give a much better and more reliable description of the geometries and relative energies than HF or MP2 methods, except for some weak bonding interactions. Hydrogen-bonding and proton-transfer reactions are generally not treated well by DFT.[44,45] It should be stressed that DFT methods (such as B3LYP and B3P86) became the dominant computational tool for treating the transition metal compounds. Only CC methods, which are very costly, would appear to equal or exceed the accuracy of the best DFT methods.

PREDICTING BARRIERS AND REACTION PATHS

An important use of electronic structure calculations is the determination of barrier heights for chemical reactions, because such barriers play a crucial role in determining the rates of the reactions. The essential features of a chemical reaction mechanism are contained in the minimum energy path(s)—the path(s) of steepest descent connecting reactants and products via transition states. A schematic of such a minimum energy path (also referred to as an intrinsic reaction coordinate (IRC)) is shown in Figure 4.2. The reaction coordinate represents a composite change in all geometric parameters (angles and bond lengths) as the reaction proceeds. Short of determining an entire reaction coordinate, there are a number of structures and their energies

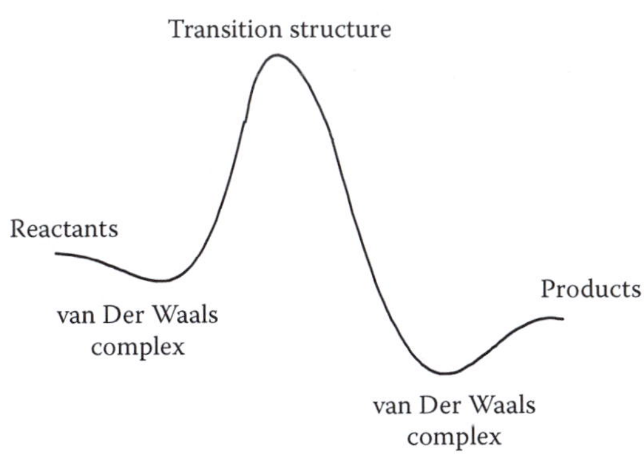

FIGURE 4.2 Hypothetical reaction coordinate.

that are important to defining a reaction mechanism. For the simplest single-step reaction, there would be five of these structures, shown in Figure 4.2:

1. The reactants
2. The precursor van der Waals complex between the reactants
3. The transition structure
4. The precursor van der Waals complex between the products
5. The products

Reactants and products correspond to energy minima, whereas transition states linking products to reactants usually correspond to first-order saddle points on the energy surface (although unusual symmetries can produce higher-order transition states, including those of the monkey-saddle type). Thus, the location of stationary points (particularly minima and saddle points) on potential energy surfaces represents an important and challenging problem in computational chemistry.

The PES's topology governs how much of a hill or barrier there is to each elementary reaction step and whether the process is thermodynamically favorable. This is referred to as the *energetic reaction profile*. The recent survey of some practical methods in use to explore PESs for chemical reactions by Schlegel is very informative.[46]

As a general rule of thumb, transition structures are more difficult to describe than equilibrium geometries. The location of saddle points on the PES is more difficult than finding local minima. However, a number of clever algorithms have been developed in the past decade. Generally, if a program is given a molecular structure and told to find a TS, it will first compute the Hessian matrix. The nuclei are then moved in a manner that increases the energy in directions corresponding to negative eigenvalues of the Hessian and decreases energy where there are positive eigenvalues. This is a quasi-Newton technique, which implicitly assumes that the PES has a quadratic shape. Thus, the optimization will only be able to make the correct geometry if the starting geometry is sufficiently close to the TS geometry to make this a valid assumption. One excellent technique is to start with a geometry with bond lengths of the bonds being formed or broken intermediate to their bonding and van der Waals lengths.

Since TS structure calculations are so sensitive to the starting geometry, a number of techniques for finding reasonable starting geometries have been proposed. One very useful technique is to start from the reactant and product structures, which are more easily obtained than TS structures. In this category belong the quadratic synchronous transit methods (QST2 and QST3). In QST2 the software package will require the user to provide as input the structures of reactants and products, while in QST3 a possible transition structure is also required. The QST techniques fail for multistep reactions, but they can be used individually for each step.

TS structures can also be obtained by following the reaction path from the equilibrium geometry to the TS. This technique is known as eigenvalue following (EF) because the user specifies which vibrational mode should lead to a reaction given sufficient kinetic energy.

Another way to reliably find a TS is based on the scan of the PES through a series of calculations representing a grid of points on the PES. The saddle point can then be found by inspection or, more accurately, by using mathematical techniques to interpolate between the grid points. However, this is a very time-consuming method and is only used when the research requires obtaining a PES for reasons other than finding the TS.

Nonetheless, the aforementioned saddle point searches sometimes fail to converge, or they converge to critical points that are minima, and thereby more robust algorithms are still needed.

Once the TS has been obtained, it may be useful to consider the exact path that led back to the reactants and forward to the products using the IRC method. The IRC provides a nice rubric for thinking about the course of a chemical reaction and, combined with variational state theory, allows the calculation of the reaction rates as well.

MOLECULAR ORBITALS AND ELECTRON DENSITY

Molecular orbitals (MOs) are not real physical quantities. Orbitals are a mathematical convenience that help us think about bonding and reactivity, but they are not physical observables. Many quantum chemistry programs have graphical interfaces that allow the schematic representation of MOs. Among the various MOs of a molecule, the highest occupied (HOMO) and lowest unoccupied (LUMO) are of particular interest for the qualitative prediction of the chemical reactivity and selectivity. Most important is the electron density related to the probability that an electron will occupy a precise point of space, which is a measurable, real physical quantity. Notice that any property is a result of the electron density. A surface of constant electron density (isosurface) serves several functions, depending on the value of the density at the surface. Surfaces of very high electron density identify atomic positions (the basis of the x-ray diffraction experiment). Surfaces corresponding to somewhat lower electron density can be used to "assign" bonds in molecules where there may be more than one reasonable alternative. Perhaps the most important use of electron density surfaces is to depict overall molecular size and shape, the same information that is provided by a space-filling model. For instance, the 0.001 electron density isosurface is often used to show the spatial extent of a molecule. It surrounds that part of space that has the largest probability of containing the electrons. Moreover, electron density isosurfaces indicate the preferred positions for the nucleophilic and electrophilic attack of the molecule of interest.

ATOMIC CHARGES, DIPOLE MOMENTS, AND MULTIPOLE MOMENTS

One of the most generally used concepts in chemistry is the atomic charge, which has the following intuitive meaning: When two noninteracting atoms **A** and **B** (**1**) form a chemical bond (**2**), the atomic charge of **A** resulting from the formation of the chemical bond is the amount of electron density gained from (**2a**) or lost to (**2b**) atom **B**:

$\mathbf{A}^0\ \ \mathbf{B}^0$	$\mathbf{A}^{\delta+}\text{–}\mathbf{B}^{\delta-}$	$\mathbf{A}^{\delta-}\text{–}\mathbf{B}^{\delta+}$
1	**2a**	**2b**

Several methods for the quantification of the atomic charges have been developed over the years, trying to solve the problem of how can one partition the electron density, which does not by definition belong to any atom, over the different atoms. Different ways of partitioning lead to different numbers for the atomic charges, and therefore possibly to different ways of interpreting the nature of the chemical bond.

The oldest and also best known definition of the atomic charge is the Mulliken population analysis.[47] This method uses the basis functions χ_μ, in terms of which the molecular wave function $\varphi_i(\mathbf{r})$ is expressed. The total number of electrons N of the molecule is then given by

$$N = n_i \sum_{i=1}^{occ} \langle i|i\rangle = \sum_{A,B} \sum_{\mu\in A} \sum_{\nu\in B} \sum_{i=1}^{occ} n_i c_{\mu i} c_{\nu i} \langle \mu|\nu\rangle$$

$$= \sum_{A,B} \sum_{\mu\in A} \sum_{\nu\in B} \sum_{i=1}^{occ} n_i P_{\mu\nu} S_{\mu\nu}$$

with orbital occupations n_i. The diagonal elements of $P_{\mu\mu}S_{\mu\mu}$ represent the net Mulliken population that basis functions acquire in the molecule. The gross Mulliken population Q_μ for all basis functions is obtained by assigning half of each total overlap population $(P_{\mu\nu}S_{\mu\nu} + P_{\nu\mu}S_{\nu\mu})$ to this basis function:

$$Q_\mu = P_{\mu\mu}S_{\mu\mu} + \sum_{\nu\neq\mu} \frac{1}{2}\left(P_{\mu\nu}S_{\mu\nu} + P_{\nu\mu}S_{\nu\mu}\right)$$

The Mulliken charge of atom A is obtained by summing up the gross population Q_μ for all basis functions χ_μ centered on that atom and subtracting them from the corresponding charge Z_A, as shown in

$$Q_A^{Mulliken} = Z_A - \sum_{\mu\in A} Q_\mu$$

The Mulliken charges based on the wave function representation as a linear combination of basis functions are heavily basis-set dependent and in principle do not converge with increasing basis-set size. Moreover, the half-to-half partitioning of the Mulliken population analysis yields totally unphysical charges. Attempts to circumvent these problems lead to the development of a population scheme, the natural population analysis (NPA).[48] The NPA is performed on the optimized polyatomic wave function employing explicitly orthogonal (natural) atomic orbitals, and thus solves the overlap problem. Natural bond orbital (NBO) analysis transforms the input basis set to various localized basis sets (natural atomic orbitals, NAOs; hybrid orbitals, NHOs; bond orbitals, NBOs; and localized molecular orbitals, NLMOs):

$$\text{Input basis set} \rightarrow \text{NAOs} \rightarrow \text{NHOs} \rightarrow \text{NBOs} \rightarrow \text{NLMOs}$$

The localized sets may be subsequently transformed to delocalized natural orbitals (NOs) or canonical molecular orbitals (CMOs). Each step of the above sequence involves an orthonormal set that spans the full space of the input basis set and can be used to give an exact representation of the calculated wave function and the expectation values of selected properties of the system.

More realistic atomic charges are obtained by population analysis schemes that are based on the electron density. In this category belong the methods developed by Bader,[49] Hirshfeld,[50] Politzer and coworkers,[51] Van Alsenoy and coworkers,[52] and the Vorodoi deformation density (VDD) suggested recently.[53] An assessment of the Mulliken, Bader, Hirshfeld, Weinhold, and VDD methods for charge analysis reported very recently is very informative.[53]

Within the BO approximation the electric dipole moment μ is calculated according to

$$\mu = e\sum_{A=1}^{M} Z_A \mathbf{R}_A - e\int \psi_i \sum_{i=1}^{N} r_i \psi_i d\tau$$

$$= e\sum_{A=1}^{M} Z_A \mathbf{R}_A - e\int P_i(\mathbf{r})\mathbf{r} d\tau$$

The first and second terms of the above equation represent the nuclear and electronic contributions to the dipole moment of the molecule, respectively.

Multipole moments (quadrupole, octapole, exapole, and dodecapole) can also be calculated by computational quantum chemical techniques, and the entire set of the electric moments is required to completely and exactly describe the distribution of charge in a molecule.

ELECTROSTATIC POTENTIALS

The molecular electrostatic potential (MEP) is the energy that a positive charge (an electrophile) "feels" at any location in a molecule. The MEP is defined according to

$$MEP = \sum_{A=1}^{M} \frac{Z_A}{|\mathbf{r} - \mathbf{R}_A|} - \int \frac{d\mathbf{r}'\rho(\mathbf{r})}{|\mathbf{r}' - \mathbf{r}|}$$

where the first and second terms represent the nuclear and electronic contributions to MEP, respectively.

The electrostatic potential is a physical property of a molecule related to how a molecule is first seen or felt by another approaching species, and thereby has proven to be particularly useful in rationalizing the interactions between molecules and molecular recognition processes. This is because electrostatic forces are primarily responsible for long-range interactions between molecules. The electrostatic potential varies through space, and it can be visualized in the same way as the electron density. A portion of a molecule that has a negative electrostatic potential surface will be susceptible to electrophilic attack—the more negative, the better. As noncovalent interactions between molecules often occur at separations where the van der Waals radii of the atoms are just touching, it is often most useful to examine the MEP in this region. For this reason, the MEP is often calculated at the molecular surface and is visualized from color mapping the value of the electrostatic potential onto an electron density surface that corresponds to a conventional space-filling model. The resulting model simultaneously displays molecular size and shape and electrostatic potential value.

VIBRATIONAL FREQUENCIES

Frequency runs are performed for two reasons, either to actually predict the frequencies and the IR and Raman intensities or to confirm whether a stationary point found corresponds to a local (or global) minimum or to a saddle point. The calculation of the vibrational frequencies offers several advantages. First, the calculation leads not only to the position of the peaks, but also to the exact vibrational motion (normal vibrational mode) of the molecule to which the peak corresponds. Programs that offer graphical display of these motions are widely available, thus adding further in the visualization and understanding of the origin of the effect we often see in the laboratory.

It should be noted that most quantum chemical studies of vibrational frequencies are carried out within the double harmonic approximation (i.e., using only second derivatives for the force constants and the derivative of the dipole moment for the intensities). Because calculations produce harmonic vibrational frequencies that do not include anharmonic effects—the so-called mechanical and electrical anharmonicities—we must be careful when comparing experimental and theoretical results. As computational results of vibrational frequencies at various levels of theory have systematic errors, one can scale the computed values by an empirical factor characteristic for each level of theory.[54,55]

NMR CHEMICAL SHIFTS

Nuclear magnetic resonance (NMR) spectra can also be computed using *ab initio* and density functional theories. However, the main problem in all calculations of magnetic properties (i.e., NMR chemical shifts and magnetizabilities) using finite basis sets is gauge invariance. This simply means that the results depend on the chosen gauge origin, e.g., the chosen origin of the coordinate system (e.g., the calculated NMR chemical shifts will change if the molecule is "translated" along an axis). This clearly unphysical behavior can be avoided by assigning a local gauge origin to each basis function, which is known as gauge-including atomic orbitals (GIAOs).[56–60] Other approaches to the gauge origin problem are the individual gauge for localized orbitals (IGLO) method of Kutzelnigg,[61] the local orbitals–local origins (LORG) method of Hansen and Bouman,[62] and the continuous set of gauge transformations (CSGT) method.[63–65] The GIAO approach is the most elegant solution to the gauge invariance problem and, in contrast to the IGLO method, is easily extended to correlated approaches.

The final results of an NMR calculation at any level of theory are the corresponding absolute isotropic magnetic shielding tensor elements, σ^{iso} (in ppm), as well as the anisotropic shielding, σ^{aniso} (in ppm), of all nuclei in the molecule. In order to compare with experimental results, one needs to carry out equivalent NMR calculations on the reference compound used and take the difference between the two calculations:

$$\delta = \sigma_{ref} - \sigma$$

With respect to basis sets, the following recommendations can be made. In the case of ^{11}B and ^{13}C nuclei, a basis set of DZ quality is in most cases sufficient for relative shifts. Larger basis sets (i.e., TZ or even QZ) are in most cases not needed for the accurate prediction of chemical shifts. On the other side, ^{15}N, ^{17}O, and ^{19}F NMR chemical shift calculations require larger basis sets of at least TZ plus polarization quality (TZP). For other nuclei, not very much can be said at the moment, and the user is strongly urged to carefully check the basis set dependence to ensure reliable theoretical results. However, limited experience suggests that for second-row elements, quite large basis sets are needed for accurate calculations.

In recent years the viability of DFT methods to the prediction of metal and ligand NMR shielding parameters in transition metal complexes has been highlighted in a steadily increasing number of studies and is now summarized in a series of recent review articles.[2c,66–68] The nuclear magnetic shielding tensor elements of a number of transition metal complexes ($[Co(CN)_6]^{3-}$, $[Co(NH_3)_6]^{3+}$, $Cr(CO)_6$, $[CrO_4]^{2-}$, $Fe(CO)_5$, $Fe(C_5H_5)_2$, $[MnO_4]^-$, $[Mn(CO)_6]^+$, $VOCl_3$, VF_5) were successfully predicted by DFT calculations within the LORG framework.[69] Moreover, the *ab initio*/GIAO methodology has been successfully applied to the theoretical calculation of complexation-induced chemical shifts ($\Delta\delta$) of zinc, ruthenium, rhodium, and tin-porphyrin complexes, where the ^1H NMR resonance signals experience large ring-current-induced upfield shifts.[70]

IONIZATION ENERGIES AND ELECTRON AFFINITIES

According to Koopman's theorem, the energy of an electron in an orbital is often equated with the energy required to remove the electron to give the corresponding ion. In this respect, the energy of the HOMO is interpreted as the negative of an ionization potential, I_p, while that of the LUMO is interpreted as the negative of an electron affinity, E_a:

$$\varepsilon_{HOMO} = -I_p$$

$$\varepsilon_{LUMO} = -E_a$$

However, when applying Koopman's theorem, two important caveats must be remembered. The first is that the orbitals in the ionized state M^+ are assumed to be the same as in the state M: They are frozen. In other words, the MOs for M^+ are not allowed to change (i.e., there is no relaxation). The second caveat is that there is no difference in the correlation energies of M and M^+. Correlation effects will tend to increase the ionization potential, while admitting orbital relaxation will tend to diminish the ionization potential. In many cases these two factors fortuitously cancel for valence electrons, with Koopman's theorem offering a reasonably good measure of the actual I_p.

For a molecule, if the energy difference is computed at the geometry of M, we obtain the so-called *vertical* ionization potentials (VIPs). However, if we allow M^+ to relax to its optimum geometry before evaluating its energy, we get *adiabatic* ionization potentials (AIPs). Obviously, the I_p and E_a computed in the framework of Koopman's theorem are also vertical.

Considering that the numbers of electrons change during the ionization process, correlation effects are of particular importance for a proper description of the electron detachment processes. Therefore, highly correlated methods, such as CCSD(T) or modern DFT methods combined with a relatively large basis set, should be applied for the calculation of adiabatic I_p and E_a. In particular, for E_a diffuse functions should be added to the basis set to properly describe the anion M^-. For the $M + e \rightarrow M^-$ process there are actually three different quantities that can be computed: the vertical electron detachment energy (VEDE; energy to eject the electron at the geometry of the anion); the adiabatic electron affinity (AEA; when the neutral M is allowed to relax to its optimum geometry); and the vertical electron affinity (VEA; where the geometry of the neutral is assumed for the anion).

Ionization potentials, electron affinities, and proton affinities are reproduced fairly well within gradient-corrected DFT. However, more reliable results are obtained via propagator methods, such as the outer valence Green's function (OVGF) method.[71]

TIME-DEPENDENT CALCULATIONS

Time-dependent response theory (TDRT) is a powerful tool for the computation of electronic transitions (electronic spectra) and dynamic polarizabilities and hyperpolarizabilities.

Electronic spectra are more difficult to model. The problem originates from the fact that to obtain the electron density and energy in all *ab initio* calculations, the energy is minimized with respect to the density. If an excited-state energy and density are sought in this manner, there must be something in the theory to prevent the solution from collapsing to the ground state. A reliable quantum chemical treatment of electronic excitation in atoms and molecules requires, in general, a proper inclusion of static and dynamic effects of electron correlation. This makes it necessary to carry out extended multireference CI (MR-CI) calculations or complete active space plus perturbation theory in second order (CAS-PT2) to reach an accuracy of about 0.1 eV in excitation energies.

To study the electronic spectra of transition metal coordination compounds, four types of quantum chemical computational techniques can be applied: (1) the variational SCF, CI, and MCSCF approaches; (2) the time-dependent density functional theory (TD-DFT) approaches;[72–74] (3) the coupled cluster approaches, such as the equation of motion CCSD (EOM-CCSD);[75] and (4) the single-state (SS) or multistate (MS) second-order perturbational approaches, such as the SS-CASP2 and MS-CASP2.[76] These methods are present in most of the quantum program packages available. An excellent review of the recent applications of quantum chemical computational techniques in the electronic spectroscopy and photoreactivity of transition metal coordination compounds by Daniel[2d] is very informative.

A single excitation interaction (CIS) calculation[77] is the most common procedure to get excited-state energies. This has the advantage of being easier to use, but can only be used to treat excited states that are largely single excitations. A CIS calculation is not extremely accurate. However, it is able to compute many excited-state energies easily. An essential feature of these calculations is that they report oscillator strengths for the transitions predicted. This is to help in comparing to actual spectra, since there are many excited states that can be calculated for a molecule but that have zero or near-zero intensities, and therefore are not observed in the normal optical experiment.

INORGANIC CHEMISTRY BY ELECTRONIC STRUCTURE CALCULATION METHODS

In the past few years, several excellent reviews have appeared that describe the application of electronic structure calculation methods to a variety of difficult problems encountered in transition metal chemistry.[1–7] Therefore, here we will focus on a few selected case applications of electronic structure calculation methods to transition metal and organometallic chemistry, aiming to stimulate the interest of inorganic chemists willing to learn new science and to adapt to a somewhat different style than what they are accustomed to, welcoming a combination of theory, computation, and experiment toward solution of difficult chemical problems of interest. Such a synergistic process allows an evolution of solutions that can progressively address more of the complexity of the realistic problem and incorporate new physical data as they become available.

EXPLORING BONDING AND NONBONDING INTERMETALLIC M···M INTERACTIONS

Many important classes of transition metal coordination compounds and metalloenzymes exhibit bonding and nonbonding intermetallic interactions. A detailed understanding of the nature of these interactions is of particular importance to understand the structure and reactivity of these molecular systems. The structural variability in dimetal systems that contain the simple diamond-shaped $M_2(\mu\text{-}X)_2$ core structure has attracted considerable interest from both the experimental and theoretical points of view. Of particular interest is a new class of bis(μ-oxo) dicopper and diiron systems comprising iron and copper centers in nonheme, multimetal enzymes, such as monooxygenases, fatty acid desaturase, and tyrosinase, which function in the activation of dioxygen to catalyze a diverse array of organic transformations. Aspects of the chemistry of copper and iron compounds containing the $M_2(\mu\text{-}O)_2$ core have been recently reviewed.[78]

The structural behavior of the $M_2(\mu\text{-}X)_2$ core has been analyzed theoretically by Alvarez et al.[79] and Mealli and coworkers.[80] Following the arguments of Alvarez, the formation or breaking of *trans*-annular bonds in these systems is related to the number of framework electrons, e.g., the number of electrons involved in bonding between the metals and the bridging ligands. It was found that the allocation of the M-M-based valence electrons into molecular orbitals of σ, π, δ, δ^*, π^*, and σ^* symmetries provides a wealth of information regarding the chemistry that these systems will possess.

A number of theoretical studies have led to a detailed picture of the bonding within the $Cu_2(\mu\text{-}O)_2$ core. The diamagnetic nature of complexes involving this core is consistent with the d^8 electronic configuration of Cu(III) and the tetragonal nature of the coordination sphere. The d^8-d^8 $[LCu(III)(\mu\text{-}O)_2Cu(III)L]^{2+}$ (where L is a tridentate ligand) compounds having two bridging bonds, and a sufficient number of electrons to fill up all six σ, σ^*, π, π^*, δ, and δ^* Cu···Cu orbitals do not involve intermetallic Cu···Cu bonds.

A similar bonding pattern has been established through quantum chemical calculations at the DFT level of theory, for an important class of phosphido-bridged $\mu\text{-}PR_2$ compounds exhibiting a variety of M···M′ (M = M′ = Pt or Pd; M = Pt, M′ = Pd) interactions.[81,82] The ability of such DFT methods (B3LYP functional, LANL2DZ, 6-31G basis set) to reproduce experimental geometry and identify the M···M′ bonding interactions and its variation on oxidation in this class of complexes was demonstrated.

The d^8-d^8 phosphido-bridged diplatinum $[(C_6F_5)_2Pt(\mu\text{-}PPh_2)_2Pt[(C_6F_5)_2]^{2-}$ and trinuclear $[(C_6F_5)_2M(\mu\text{-}PPh_2)_2M'(\mu\text{-}PPh_2)_2M''(C_6F_5)_2]^{2-}$ (M, M′, M″ = Pd(II), Pt(II) compounds having sufficient number of electrons to fill up all six MOs) do not exhibit any intermetallic M···M′ interaction. The calculations[81,82] indicated that oxidation of these complexes should occur on the metal centers, since the highest occupied orbitals are mainly metal-based orbitals that are close in energy. In the oxidized d^7-d^7 phosphido-bridged $[(C_6F_5)_2Pt(\mu\text{-}PPh_2)_2Pt[(C_6F_5)_2]$ and $[(C_6F_5)_2M(\mu\text{-}PPh_2)_2M'(\mu\text{-}PPh_2)_2M''(C_6F_5)_2]$ species having only ten electrons to occupy the six MM′ orbitals, two bridging bonds result in a formal single Pt-Pt bond. The MOs contributing to the formation of the Pt(III)-Pt(III) bond in the oxidized species are depicted schematically in Scheme 4.1.

When the bridging ligand is hydride, as in the case of the catalytically active d^8-d^8 $[(R_3Si)(PCy_3)Pt(\mu\text{-}H)_2Pt(R_3Si)(PCy_3)]$ hydrido-bridged diplatinum compound, the Pt-Pt interaction acquires a triple bond character (there are σ, π, and δ MOs). This is substantiated by the respective molecular orbital interactions (Scheme 4.2), which correspond to σ- (HOMO-9), π- (HOMO-7), and δ- (HOMO-5) bonding interactions.

DFT calculations at the B3LYP/LANL2DZ level successfully reproduced the geometrical parameters of the hydrido-bridged diplatinum complexes (Figure 4.3), providing for the first time the position of the bridging hydride ligands, which otherwise could not be located by the x-ray crystallographic study.[83] The optimized geometries are of C_{2h} symmetry involving asymmetric $Pt(\mu\text{-}H)_2Pt$ bridges; the Pt-H and H···Pt bond distances differ by about 0.3 Å.

Employing gradient-corrected levels of DFT, ^{1}H, ^{31}P, ^{28}Si, and ^{195}Pt chemical shifts were calculated at the GIAO B3LYP/LANL2DZ level of theory using the B3LYP/LANL2DZ equilibrium geometries. The chemical shift of the bridging hydride ligands was predicted to occur around 13.5 ppm, as expected for the resonances attributable to a platinum hydride found typically in the range τ = 13–27 ppm. The reliability of the computed ^{1}H NMR chemical shifts was assessed by the

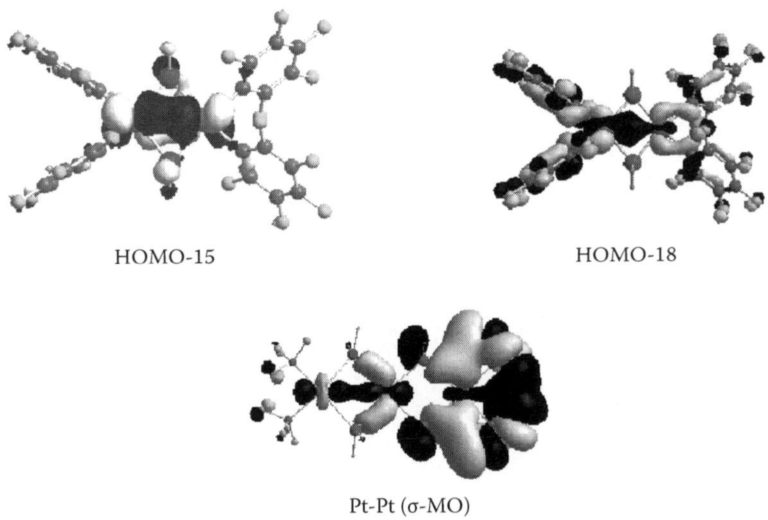

SCHEME 4.1 (See color insert following page 86) The MOs contributing to the formation of the Pt(III)-Pt(III) bonds in $[(C_6F_5)_2Pt(\mu\text{-}PH_2)_2Pt[(C_6F_5)_2]$ and $[(CF_3)_2Pt(\mu\text{-}PH_2)_2Pt(\mu\text{-}PH_2)_2Pt(CF_3)_2]$ complexes.

calculation of the ^{1}H NMR chemical shifts of complex $[Pt_2(H)_2(\mu\text{-}H)(\mu\text{-}dpm)_2]^+$ (dpm = $H_2PCH_2PH_2$), involving both symmetrical bridging and terminal hydride ligands at the same level of theory. It was found that the bridging hydride resonances are at τ 15.7 ppm, a value very close to the experimental one of $\tau = 15.85$ ppm of the real $[Pt_2(H)_2(\mu\text{-}H)(\mu\text{-}dppm)_2]^+$ (dppm = $Ph_2PCH_2PPh_2$) complex.[84]

The single two-electron reduction for the Fe-Fe-bonded dinuclear complexes $Fe_2(CO)_6(\mu_2\text{-}PR_2)_2$ (R = Me or CF_3), with the diamond-shaped $Fe_2(\mu\text{-}PH_2)_2$ core structure, was studied by DFT methods.[85] These complexes, being also d^7-d^7 phosphido-bridged dimetal complexes, upon reduction undergo the simple structural change related to the cleavage of the intermetallic Fe-Fe bond. The experimentally observed cleavage of the Fe-Fe bond upon addition of electrons was reproduced in all calculations. This is because the added electrons occupy the LUMO of the complexes, which are Fe-Fe σ-antibonding. It was found that inclusion of solvation and/or ion pair effects in the computational model is crucial to correctly reproduce the experimentally observed energetic profile that favors the disproportionation reaction of the

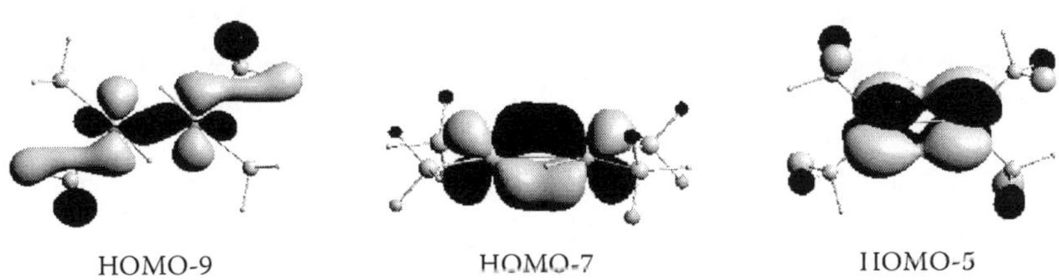

SCHEME 4.2 (See color insert) The most relevant molecular orbital interactions describing the Pt-Pt interactions in the model complex $[(H_3Si)(PH_3)Pt(\mu\text{-}H)_2Pt(H_3Si)(PH_3)]$.

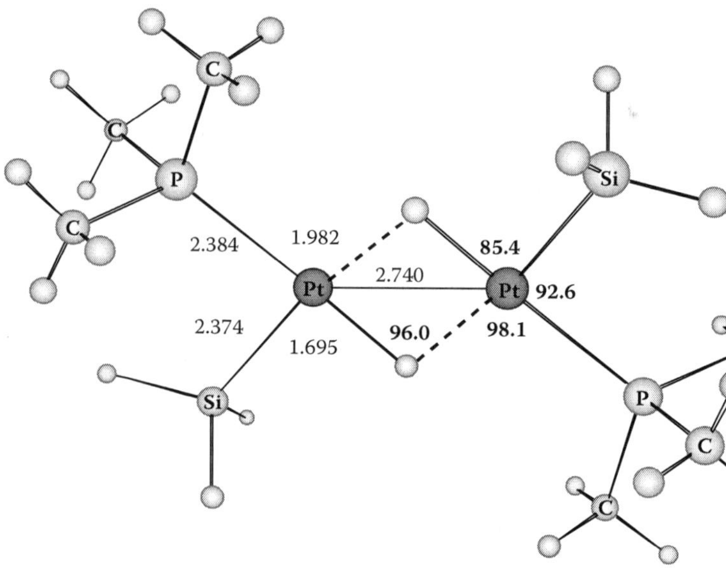

FIGURE 4.3 Equilibrium geometry of a model complex [{Pt(SiH_3)(μ-H)(PMe_3)}$_2$] computed at the B3LYP/LANL2DZ level of theory.

singly reduced complexes. The model that includes both a countercation and solvation is energetically and chemically the most realistic. Moreover, the self-consistent reaction field (SCRF) method to treat solvation effects, such as COSMO, was found to be well suited for modeling the changes in electronic structure arising from solvation.

Ab initio calculations at the HF level using relatively small basis sets and ECPs for Pt, P, and Cl atoms have been carried out for a series of model Pt(II) dimers with bridging S^{2-} and RS^- ligands, involving the $Pt_2(\mu_2\text{-}S)_2$ core, in order to elucidate the determinants of the structure, planar or hinged.[86] It was found that electronic rather than steric effects govern the geometry of the $Pt_2(\mu_2\text{-}S)_2$ core; the decrease of the through-ring antibonding interaction between the in-plane sulfur *p* orbitals with folding appears to be the determinant for hinging.

DFT calculations at the B3LYP/SDD level of theory also threw light on the bonding mechanism in a series of halo-bridged Cu(I) dimers, with a diamondlike structure of D_{2h} symmetry.[87] It was predicted that the Cu-X bond involves both σ- and π-dative bonding components. Most important is the presence of π-type MOs delocalized over the entire four-membered $Cu(\mu\text{-}X)_2Cu$ ring, which supports a ring current and could probably account for the nearly equivalent Cu-X bonds in the rhombus. Moreover, all $[Cu(\mu\text{-}X)(PH_3)_2]_2$ dimers exhibit a σ-type MO corresponding to weak Cu···Cu interactions supporting through-ring intermetallic interactions, which seems to be responsible for the stabilization of the otherwise unstable antiaromatic $Cu(\mu\text{-}X)_2Cu$ ring (the metallacycle ring having a total of eight framework electrons does not conform to the Hückel rule of aromaticity). The intermetallic Cu···Cu distance is tuned by the identity of the halide ligand, X, following paradoxically the trend:

$$Cu(\mu-I)Cu < Cu(\mu-Br)Cu < Cu(\mu-Cl)Cu$$

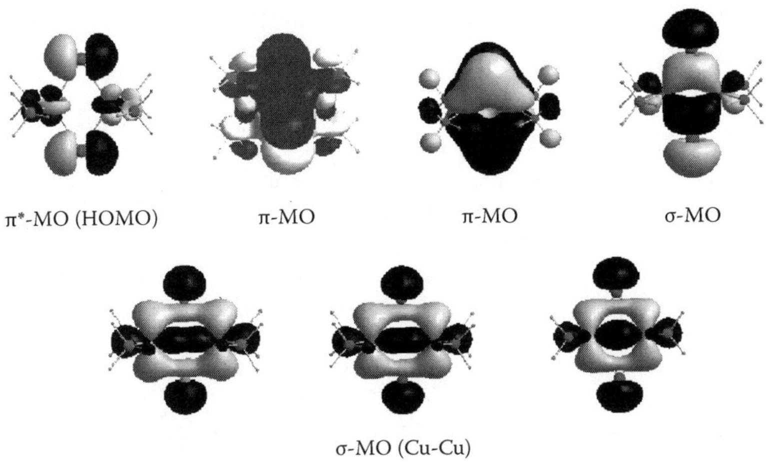

SCHEME 4.3 (See color insert) The most important molecular orbitals of the halo-bridged copper(I) dimers.

The shortest Cu···Cu distance was observed for the best electron-donating bridged iodide ligand. An electron density topological analysis assigned a bond order for the Cu···Cu interaction in the chloro-derivative of 0.2, which increases to 0.6 in the hydrido-bridged dimer with a much shorter Cu···Cu separation distance. All halo-bridged copper(I) dimers exhibit a similar pattern of molecular orbital level diagram, with a successive lowering of the respective eigenvalues, ongoing from the chloro- to iodo-derivatives. This is consistent with the higher stability of the iodo complex in the series. The most important molecular orbitals of the halo-bridged copper dimers are collected in Scheme 4.3.

The nature of the very short bond (around 260 pm) between platinum and thallium in the experimentally known $[(NC)_5Pt\text{-}Tl(CN)_n]^{n-}$ (n = 0–3) compound has been investigated[88] by electronic structure calculation methods at RHF, DFT, and MP2 levels using model compounds $[H_5Pt\text{-}TlH_n]^{n-}$. The simplest picture of the short Pt-Tl bonds corresponds to an approximate $\sigma^2\pi^4$ triple bond comprising two parts, σ donation from Tl to the empty coordination site of L_5Pt^- and π donation from the off-axis Pt-L bonds to the $6p\pi$ orbital of Tl. This bonding character is consistent with the observed oxidation states between Pt^{IV}-Tl^{I} and Pt^{II}-Tl^{III}. Moreover, relativistic DFT study of $[(NC)_5Pt\text{-}Tl(CN)]^-$ demonstrated that the remarkable experimentally observed spin-spin coupling pattern, e.g., ^2J(Tl-C) >> ^1J(Tl-C) and J(Pt-Tl) ~ 57 kHz, is semiquantitatively reproduced if both relativistic effects and solvation are taken into account. Solvent effects are very substantial and shift the Pt-Tl coupling by more than 100%. Relativistic increase of s-orbital density at the heavy nuclei, charge donation by the solvent, and the specific features of the multicenter C-Pt-Tl-C bond are responsible for the observed coupling pattern.[89]

The mechanism of metallophilic interactions, e.g., the attractions between formally closed-shell metal d^{10} centers (archetypically a pair of Au(I) cations), has been exhaustively investigated by Pyykkö and coworkers using the HF and MP2 *ab initio* methods. The theoretical treatment of these strong closed-shell interactions in inorganic chemistry has also been reviewed by Pyykkö.[90] Energy partitioning schemes have been used to investigate the metallophilic attraction between Au(I), Ag(I), and Cu(I) centers in model dimers of $[X\text{-}M\text{-}PH_3]_2$ (X = H, Cl) type.[91] The main part of the attraction involves pair correlation between one M(d^{10}) entity and non-M(d^{10}) localized orbitals of the partner monomer. It was also predicted that at r_e(M-M) dispersive and nondispersive components are about equally important for all systems considered. The contribution, due to n*d*-n*d* interaction, falls drastically along the series Au > Ag > Cu.

The aurophilicity phenomenon has been the subject of many theoretical treatments. From these studies, it was concluded that aurophilicity is a genuine correlation dispersion effect, enhanced by induction and, in particular, relativistic corrections, which may all be quite strong in heavy atomic systems. The coexistence of Au(I)-Au(I) contacts and hydrogen bonding in the model systems $[H_2P(OH)AuCl]_2$ and $[H_2P(OH)AuH_2(O)]_2$ studied at the MP2 level illustrated that the two interactions are comparable.[92] The aurophilic interactions between gold atoms in the different oxidation states Au(I)-Au(I), Au(I)-Au(III), and Au(III)-Au(III) have also been investigated at the MP2 level of theory, and the dispersion, induction, and electrostatic multipole components to the aurophilic attraction were analyzed using intermolecular models.[93]

More recently[94] the electronic and geometric structures of $Cl\text{-}M\text{-}PH_3$ and $[Cl\text{-}M\text{-}PH_3]_2$ (M = Cu, Ag, Au) have been studied computationally using post-HF *ab initio* and DFT methods. Including electron correlation through the QCISD, CCSD, and CCSD(T) calculations changes the trend in metallophilic interaction; the interaction decreases along the series Ag → Au → Cu. Moreover, the overall performance of different quantum chemical approaches (HF, MP2, and five different DFT methods) concerning the Au-Au distances and bond dissociation energies for fifteen molecules containing the Au(I) species has very recently been investigated by Wang and Schwarz.[95] It was found that simple local spin density functionals (LSDF) of the Slater (or Slater plus Vosko) type yield rather reasonable results, while common gradient corrected DFs are not recommended, nor are the large-core pseudopotentials for Au. As far as the aurophilic bonding mechanism is concerned, both one-electron (i.e., electrostatic, polarization, charge transfer, and orbital interactions) and two-electron (i.e., correlation, dispersion) effects contribute significantly to the Au(I)-Au(I) interactions.

Finally, heteronuclear metal-metal contacts between gold(I) and group 11, 12, and 13 elements have been investigated both experimentally and theoretically, and their chemistry has been reviewed by Bardaji and Laguna.[96] The metallophilic interactions in these gold-metal contacts are predicted to be of intermediate strength relative to the homonuclear metal-metal interactions. In general, metallophilic interactions are found in bridged or unbridged pairs, oligomers, chains, and sheets, involving, among others, main group elements (such as Se and Te), d^{10} ions (such as Cu(I), Ag(I), and Au(I)), s^2 ions (such as Tl(I) and In(I)), and d^8 ions (such as Ir(I)). A recent survey of such interactions by Gade[97] is very informative. Of particular importance are the extremely strong $6s^2$-$6s^2$ intermetallic interactions in $[AuBa]^-$ and $[AuHg]^-$ diatomics studied recently by Wesendrup and Schwerdtfeger.[98]

EXPLORING THE MECHANISM OF CATALYTIC GAS PHASE REACTIONS INVOLVING TRANSITION METALS

In recent years gas phase reactions between transition metal atoms or ions and the inherently stable thermodynamically small molecules, such as CO_2, CH_4, NH_3, H_2O, and H_2, have attracted considerable interest.[5,99–113] This interest is primarily due to the fact that these reactions may serve as models for understanding the reactivity pattern of transition-metal-based catalysts in the condensed phase. Moreover, the complexes formed may serve as potential precursors for the challenging task of activation of these molecules and provide simple model systems to probe fundamental bonding interactions.[114] Transition metal ions are particularly suitable for this purpose because of their rich gas phase chemistry.[115] Only a few representative examples of catalytic gas phase reactions involving transition metals will be discussed herein, for the inorganic chemist to get insight into how electronic structure calculation methods help in understanding the corresponding reaction mechanisms.

The coordination of carbon dioxide to first-transition-row metal cations and the insertion reaction of the metal into one C=O bond of carbon dioxide have been studied theoretically at the B3LYP level.[110b] Binding energies have also been determined at the CCSD(T) level using large basis sets. The most favorable coordination of CO_2 is the linear end-on M^+-**O**CO one, due to the electrostatic nature of the bonding. The ground-state and binding energies are mainly determined by several mechanisms for reducing metal-ligand repulsion. For the early transition metals (Sc^+, Ti^+, and V^+), the insertion reaction is exothermic and the OM^+CO structure is more stable than the linear M^+-**O**CO isomer, because of the very strong MO^+ bond that is formed. More recently,[110c] the electronic and geometrical structures of 3*d*-metal neutral, MCO, anionic $[MCO]^-$, and cationic $[MCO]^+$ monocarbonyls (M = Sc to Cu) were computed using DFT with generalized gradient approximation for the exchange correlation potential. The calculated adiabatic electron affinities and ionization potentials are in good agreement with experiment. The reaction $[MCO]^{0,-,+} + CO \rightarrow [MC]^{0,-,+} + CO_2$ was found to be significantly less endothermic than the 130.3 kcal/mol found for gas phase CO. Among the neutrals, this reaction is endothermic by 30.0 kcal/mol for Mn, while Fe is found to be the second best atom, with the reaction being endothermic by 36.7 kcal/mol.

The gas phase reaction of the nickel atom with a CO_2 molecule was investigated at the B3LYP and CCSD(T) levels of theory.[109c] The lowest-energy path proceeds through the $^3A''$ ONiCO intermediate and yields $^3\Sigma^-$ NiO and CO. An electron transfer from the Ni atom to the CO_2 molecule initiates the metal insertion reaction, and then the insertion and oxygen-abstraction steps take place in a concerted fashion along with the charge-transfer processes. The insertion reaction is direct and needs to overcome an energy barrier of 34.6 kcal/mol. The $Ni + CO_2 \rightarrow NiO + CO$ reaction was found to be endothermic by 37.4 kcal/mol, in good agreement with experiment (36.6 kcal/mol).

The reactions of the ground-state $V(s^2d^3)$ atoms with the isovalent CO_2, CS_2, and OCS molecules have been studied at the B3LYP level.[108a] The comparative study of the reactions,

$$V(s^2d^3) + XCY \rightarrow V(XCY) \rightarrow TS \rightarrow XVCY \rightarrow VX + CY$$

with XCY = CO_2, CS_2, and OCS, showed that insertion reaction proceeds on a single potential energy surface ($^4A''$) along one intrinsic reaction coordinate, the CX (X = O, S) bond length. Comparison of relative energies illustrated that insertion products are always more stable than the related coordination species, with the following increasing orders of stability: V(OCS) < V(CO_2), V(CS_2), and OVCS < OVCO < SVCS, SVCO. Only the V + CO_2 → VO + CO and V + OCS → VS + CO reactions are exothermic, with a low or no activation barrier for the insertion reaction.

The $^2A'$ and $^2A''$ potential energy surfaces for the Sc + CO_2 → ScO + CO reaction were explored by B3LYP and CCSD(T) methods independently by Pápai et al.[109b] and Hwang and Mebel[108] in order to gain insight into the mechanism of the reaction. Both $^2A'$ and $^2A''$ state Sc + CO_2 reactions lead to spontaneous metal insertion into CO_2. However, only the $^2A'$ state insertion complex can readily dissociate into ScO + CO. The entrance channel of the reaction corresponds to the η^2-**C,O** coordination of the CO_2 molecule.

Very recently[111b] the mechanism of the CO_2-to-CO reduction by ground-state Fe atoms has been investigated in the framework of electronic structure calculations at the B3LYP level of theory, using the 6-31G(d) and 6-311+G(3df) basis sets. The Fe + CO_2 → FeO + CO reaction predicted to be endothermic by 23.24 kcal/mol could follow two possible alternative pathways, depending on the nature of the entrance channel involving formation of either a Fe(η^2-**OCO**) or a Fe(η^3-**OCO**) intermediate. The Fe(η^2-**OCO**) intermediate was found to be weakly bound with respect to Fe(^5D) and CO_2 dissociation products, but strongly bound relative to the separated Fe^+(^6D) and $[CO_2]^-$ anion; the computed Fe-CO_2 bond dissociation energy is predicted to be 207.3 kcal/mol. On the other hand, the Fe(η^3-**OCO**) intermediate is unbound with respect to Fe(^5D) and CO_2 dissociation products by 8.3 kcal/mol, but also strongly bound relative to the separated Fe^+(^6D) and $[CO_2]^-$ anion; the computed Fe-CO_2 bond dissociation energy is predicted to be 198.3 kcal/mol. Both reaction pathways involve an intramolecular insertion reaction of the Fe atom into the O-C bond of the Fe(η^2-**OCO**) or Fe(η^3-**OCO**) intermediates, yielding the isomeric OFe(η^1-**CO**) or OFe(η^1-**OC**) insertion products, respectively, with a relatively low activation barrier.

The mechanism of the reaction YS^+ + CO_2 → YO^+ + COS in the gas phase has been studied at the B3LYP level of theory by Xie.[112] The reaction proceeds via two four-center transition states with a cyclic intermediate complex. The reaction mechanism includes seven stationary points (reactants, three intermediate complexes, two transition states, and products) on the reaction potential surface. The activation barriers of the two transition states are –8.3 and 2.1 kcal/mol, respectively, indicating that the second reaction step should be the rate-determining reaction step.

The catalytic activity of bare cationic transition metal monoxides (MO^+) toward methane is a key to the mechanistic aspects in the direct methane hydroxylation. The reaction pathway and energetics for methane-to-methanol conversion by first-row transition metal oxide cations has thoroughly been investigated by Yoshizawa et al.[106] using the B3LYP density functional approach. Both high-spin and low-spin

PESs were characterized at the B3LYP/6-311G(d,p) level of theory. The methane activation process proceeds via two transition states following the course: MO^+ + $CH_4 \rightarrow OM^+(CH_4) \rightarrow$ [TS1] $\rightarrow HO - M^+ - CH_3 \rightarrow$ [TS2] $\rightarrow M^+(CH_3OH) \rightarrow M^+$ + CH_3OH (M = Sc, Ti, V, Cr, Mn, Fe, Co, Ni, and Cu). A crossing between high-spin and low-spin PESs occurs once near the exit channel for ScO^+, TiO^+, VO^+, CrO^+, and MnO^+, but it occurs twice in the entrance and exit channels for FeO^+, CoO^+, and NiO^+. In most cases the high-spin and low-spin PESs have crossing points, in the vicinity of which spin inversion can take place. This nonadiabatic electronic process is a manifestation of the so-called *two-state reactivity* (TSR) concept, which is of particular interest in organometallic chemistry and oxidation catalysis.[99f] The TRS phenomenon involving participation of spin inversion in the rate-determining step can dramatically affect reaction mechanisms, rate constants, branching ratios, and temperature behaviors of organometallic transformations. Some recent case studies involving TRS, namely, the $V^+ + CS_2$ system, the hydroxylation of methane by FeO^+, and the β-hydrogen transfer in the $[FeC_5H_5]^+$ cation, are presented and discussed by Schröder and Schwarz.[99f]

Density functional B3LYP and CCSD(T) calculations have also been employed to investigate potential energy surfaces for the reactions of neutral scandium, nickel, and palladium oxides with methane. The results show that NiO and PdO are reactive toward methane and can form molecular complexes with CH_4 bound by 9–10 kcal/mol without a barrier. At elevated temperatures, the dominant reaction channel is direct abstraction of a hydrogen atom by the oxides from CH_4 with a barrier of about 16 kcal/mol, leading to MOH (M = Ni, Pd) and free methyl radical. On the other hand, scandium oxide is not reactive with respect to methane at low and ambient temperatures. At elevated temperatures (when a barrier of about 22 kcal/mol can be overcome), the reaction can produce the CH_3ScOH molecule, but the latter is not likely to decompose to the methyl radical and ScOH or Sc + CH_3OH because of the high endothermicity of these processes.[108b] The minimal energy reaction pathway for methane conversion to methanol using nickel and palladium oxides involves both triplet and singlet PESs, which is another example of the TRS concept. The same holds true for the insertion reaction of Zn, Cd, and Hg atoms with methane and silane studied at the BPW91 density functional and MP2 levels of theory.[116]

The TSR paradigm is also characteristic for the dehydrogenation reactions of H_2O, NH_3, and CH_4 molecules by $Mn^+(^7S, ^5S)$ and Fe^+ $(^6D, ^4F)$ cations investigated at the B3LYP level using a DZVP basis set optimized *ad hoc* for the employed functional.[113] In all cases, the low-spin ion-dipole complex, which is the most stable species on the respective PESs, is initially formed. In the second step, a hydrogen shift process leads to the formation of the insertion products, which are more stable in a low-spin state. The low- and high-spin PESs have crossing points, in correspondence of which spin inversion takes place. The topological analysis of the *electron localization function* (ELF) has been used to characterize the nature of the bonds for all of the minima and transition states along the paths. Depending on the presence of lone pairs in the ligand, two different bonding mechanisms have been identified. This has been achieved by analyzing DFT calculations (B3LYP approach) using the bonding evolution theory. The different domains of structural stability occurring along the reaction path have been identified as well as the bifurcation catastrophes responsible for the changes in the topology of the systems. The analysis provides a chemical description of the reaction mechanism in terms of agostic bond formation and breaking. After the formation of the first reaction intermediate, all three reactions are equivalent from a bonding evolution viewpoint, since the presence of a trisynaptic basin, which corresponds to the condensation of two covalent bonds into a three-center bond, is verified in all three cases.

The reactions of first-row transition metal cations with water have also been investigated in detail.[100] Both the low- and high-spin PESs have been characterized at the B3LYP/DZVP level of theory. Energy differences between key low- and high-spin species and total reaction energies for the possible products have been predicted at even higher levels of theory. The expected trend for the $M(OH_2)^+$ ion-molecule dissociation energies has been well described through the row. Whereas the only exothermic products of the $M^+ + H_2O$ reaction for M = Sc to V were $MO^+ + H_2$ with exothermicity decreasing from Sc to V; the aforementioned reactions were endothermic for M = Cr to Cu, with endothermicity increasing through the series. As in the $Fe^+ + H_2O$ system, less difference between the high- and low-spin structures was observed for the late transition metals than in the early transition metal systems. The reason for this is that the high-spin Fe^+ to Cu^+ cations have at least one set of paired electrons, while the Sc^+ to Mn^+ high-spin cations do not. Both high- and low-spin PESs cross once in the entrance channel for Sc^+ to Mn^+. Two crossings were observed, at the entrance and exit channels, on the Fe^+ PESs. Finally, the surfaces of Co^+ to Cu^+ demonstrate one crossing near the exit channel.

The mechanism of H_2 oxidation by FeO^+ has been investigated by density functional calculations using the B3LYP, BP86, and FT97 functionals and an extended basis set.[105b] Three mechanisms were considered: addition-elimination, rebound, and oxene-insertion. The addition-elimination and rebound mechanisms are competitive, exhibiting TRS with a crossing between sextet and quartet states. TSR provides a low-energy path for bond activation and is predicted to be the dominant pathway at room temperature. Both TSR mechanisms are concerted: The addition-elimination mechanism involves 2 + 2 addition in the bond activation step, while the rebound mechanism is effectively concerted involving the H-abstraction followed by a barrierless rebound of the H-radical.

Finally, the interaction of first-row transition metal dications (Mn^{2+} to Cu^{2+}) with a single water molecule has been studied at the B3LYP and CCSD(T) levels of theory using triple-zeta basis sets.[107] The results obtained provided strong evidence for the possibility to observe the $[M(OH_2)]^{2+}$ complexes as long-lived species in the gas phase, with abundance increasing toward the left-hand side in the periodic table. In particular, for $[Cu(OH_2)]^{2+}$ and $[Cu(OH_2)_2]^{2+}$ complexes, although the dissociation products $Cu^+ + H_2O^+$ from $[Cu(OH_2)]^{2+}$ are lower in energy than $Cu^{2+} + H_2O$, an avoided crossing of the $Cu^{2+}\cdots OH_2$ and $Cu^+\cdots OH_2^+$ PESs provides a minimum in the adiabatic PES, which predicts that $[Cu(OH_2)]^{2+}$ should be a stable gas phase ion. The activation barrier for the decomposition of $[Cu(OH_2)]^{2+}$ to $Cu^+ + H_2O^+$ is about 27 kJ/mol, whereas the barrier height for the decomposition of $[Cu(OH_2)_2]^{2+}$ to $[Cu(OH_2)]^+ + H_2O^+$ is much larger (167 kJ/mol). Based on these results, El-Nahas et al.[117] predicted the existence of $[Cu(OH_2)_2]^{2+}$ in the gas phase, under suitable conditions, and probably the much less stable $[Cu(OH_2)]^{2+}$. Actually, recent experimental results[118] vindicated the theoretical predictions that the cupric ion solvated by one and two water molecules should be a stable species and therefore exist in the gas phase.

EXPLORING THE CATALYTIC CYCLE OF SYNTHESIS REACTIONS CATALYZED BY TRANSITION-METAL-CONTAINING CATALYSTS

In the last few years several excellent reviews have appeared that describe the application of electronic structure calculation methods to reactions of transition metal compounds. Niu and Hall[1a] have reviewed the theoretical studies on reactions of transition metal complexes, while Torrent et al.[1d] gave an extensive overview of theoretical studies of some transition-metal-mediated reactions of industrial and synthetic importance. In a more recent review, Harvey et al.[2n] surveyed a number of recent studies concerning spin-forbidden reactions that occur in coordination chemistry. These reactions concern ligand dissociation/association, oxidative addition, and migratory insertion reactions, which play an important role in homogeneous catalytic processes. For spin-forbidden reactions the *minimum energy crossing points* (MECPs) between states of different spin can be located for large, realistic transition-metal-containing systems yielding important new insight into their mechanism. Moreover, recent books have also been devoted to computational organometallic chemistry[119] and the theoretical aspects of homogeneous catalysis.[120]

It is our wish to comment here on the more recent developments in the area, presenting a few case examples of homogeneous catalytic processes investigated by electronic structure calculation methods. In general terms, computational modeling of catalytic processes requires a high-level quantum chemical treatment. However, the downside is that quantum chemical methods require tremendous computational resources. Due to the expense of quantum chemical calculations, the detailed study of a catalytic cycle often involves a stripped-down model of the chemical system. This methodology is followed when we want to treat relatively big-sized molecular systems at a high level of theory. The use of models resulting in substitution of peripheral components by simpler substituents and removal of the solvent molecules does not alter the description of the core region of the compounds and is ultimately the most efficient and productive route to modeling the electronic structure and related properties of relatively big-sized transition metal coordination compounds. It should be emphasized that great care has to be exercised in choosing the simpler substituents. These substituents might exhibit electronic properties similar to those they are going to substitute. This is because in some cases the peripheral components act only as spectators, but in others, they can substantially influence the PES. Thus, quantum chemical models that neglect the surrounding molecular environment might lead to limited or even erroneous conclusions. Another popular approach to simulate complex molecular systems, such as the transition-metal-catalyzed synthesis or enzyme reactions at the quantum chemical level, is the combined quantum mechanics and molecular mechanics (QM/MM) method.[121–123] In this hybrid method, part of the molecular potential, such as the catalyst's or metalloenzyme's active site, is determined by QM calculation, while the remainder of the molecular

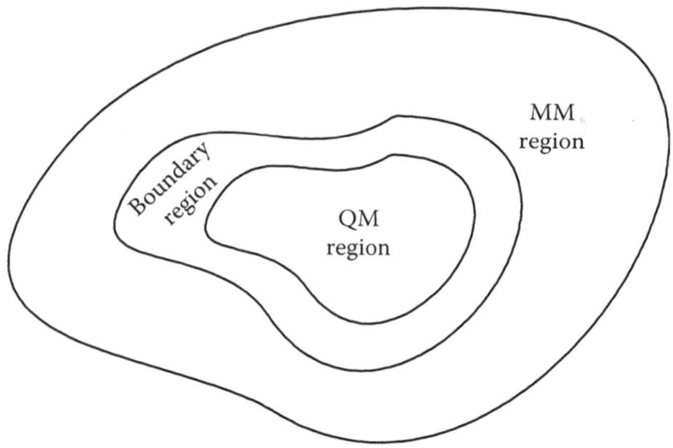

SCHEME 4.4 Schematic representation of the setup of a QM/MM calculation. A small region containing the reacting atoms is treated by a quantum mechanical method (e.g., molecular orbital theory), and its surroundings are represented more simply by molecular mechanics.

potential is determined using a much faster molecular mechanics (MM) calculation. The QM/MM method's promise is that it allows for simulation of bond breaking and formation at the active site, while still taking into account the role of the extended system in an efficient and computationally tractable manner.

The QM/MM method's key feature is that a QM calculation is performed on a truncated QM model of the active site with the large ligands removed and replaced by capping atoms. Then, an MM calculation is performed on the remainder of the system. The division of the full system into two subsystems (Scheme 4.4) may require cutting a covalent bond. Different techniques have been developed to treat this problem, fulfilling the valence of the quantum atom placed in the boundary with hydrogen atoms or frozen orbitals.

The QM (e.g., reacting) region feels the influence of the molecular mechanics environment; for example, the atomic charges of the MM atoms affect the QM atoms. By combining the accuracy of a QM method (for a small reacting system) with the speed of an MM method (for its surroundings), reactions in large molecules and in solution can be studied. In this way, the wave function of the quantum subsystem, and thus any related property, can be obtained under the influence of the environment. In the QM/MM method, the molecular system is described by a mixed Hamiltonian:

$$H = H_{\mathrm{QM}} + H_{\mathrm{MM}} + H_{\mathrm{QM/MM}}$$

and the total energy of the system can be written as

$$E = E(\mathrm{QM}) + E(\mathrm{QM/MM}) + E(\mathrm{MM}) + E(\mathrm{boundary})$$

The energy of the QM system, E(QM), and the energy of the MM system, E(MM), are calculated exactly as they would be in a standard calculation at those levels. E(QM/MM) is the interaction energy between the QM and MM regions (e.g., due to the interaction of the MM atomic charges with the QM system), while E(boundary)

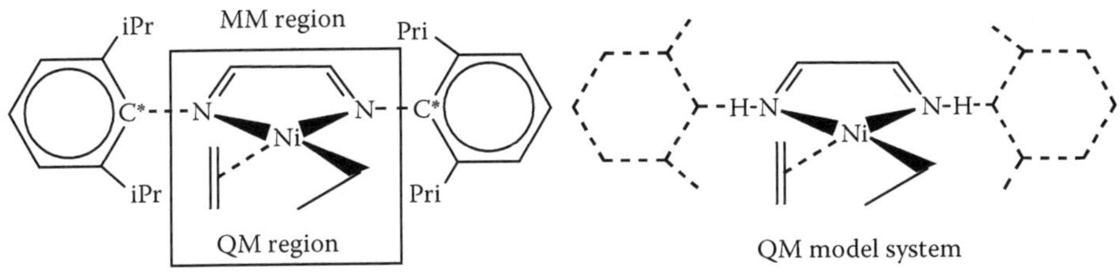

FIGURE 4.4 An example of QM/MM partitioning in a Ni-diimine olefin polymerization catalyst.

is the energy due to any boundary restraints applied to the outer edge of the MM region to maintain its structure.

A characteristic example of the successful application of the hybrid QM/MM method for a more realistic computational simulation of complex systems concerns the Brookhart-type Ni-diimine and group 4 diamide olefin polymerization catalysts.[124] The structures of the real system and the corresponding model QM system for which the electronic structure calculation is performed are shown in Figure 4.4. The hydrogen atoms cap the electronic system and are termed dummy atoms. The dummy atoms correspond to "link" atoms in the real system, which are labeled C.

More recently, Tobisch and Ziegler reported on a comprehensive theoretical investigation of the influence of ligand L on the regulation of the selectivity for the $[Ni^{0}L]$-catalyzed cyclodimerization of 1,3-butadiene based on DFT and a combined DFT/MM (QM/MM) method.[125] The role of electronic and steric effects has been elucidated for all crucial elementary steps of the entire catalytic cycle. Moreover, both the thermodynamic and kinetic aspects of the regulation of the product selectivity have been fully understood.

DFT has also been used to study the reaction mechanism and structure-reactivity relationships in the stereospecific 1,4-polymerization of butadiene catalyzed by neutral allylnickel(II) halide $[Ni(C_3H_5)X]_2$ ($X^- = Cl^-, Br^-, I^-$) complexes[126] using stripped-down models of the catalytic system for the computational effort to be kept moderate. All crucial steps of the catalytic cycle, such as monomer π complex formation, symmetrical and unsymmetrical splitting of dimeric π complexes, and *anti-syn* isomerization, have been scrutinized.

The *regio-* and *stereo*-selectivity in palladium-catalyzed electrophilic substitution of aryl chlorides with aromatic and heteroaromatic aldehydes has recently been studied at the B3PW91/DZ+P level of theory.[127] The two most important factors controlling the selectivity were found to be the location of the phenyl functionality in the η^1-moiety of the bis(allyl)palladium intermediate and the relative configuration of the phenyl substituents in the cyclic six-membered transition state of the reaction.

A few theoretical studies[128–132] have recently been devoted to the study of olefin metathesis reactions catalyzed by the Grubbs-type ruthenium carbene $[(PR_3)(L)Cl_2Ru{=}CH_2]$ (L = PR_3 or imidazol-2-ylidene, NHC) complexes.[128] Both the associative and dissociative mechanisms of olefin metathesis reaction catalyzed by the model $[(PH_3)(L)Cl_2Ru{=}CH_2]$ (L = PH_3 or NHC) complexes have been thoroughly investigated by Vyboishchikov and coworkers[129] using DFT techniques with the BP86 and B3LYP functionals. These functionals, combined with medium-sized basis sets, provided realistic geometries, relative energies, and vibrational frequencies. Moreover, for further validation, additional CCSD(T) calculations were performed on the BP86 geometries. The dissociative mechanism with olefin coordination in a *trans*-position with respect to the ancillary ligand L is favored over all other alternatives. The π-bonded olefin and the metallacycle species were identified as intermediates, while the rate-determining step was found to be the ring closure-opening processes of these species in terms of the calculated ΔG^o_{298} values. However, the calculated ΔH^o_{298} values for phosphine dissociation from the precatalyst are in opposite trend with experimental results, but conforming to the strength of the *trans* influence of the ancillary ligand L. At the same time, Cavallo[130] reported a DFT study of the first-generation real Grubbs-type catalysts $[(PCy_3)(Cl)_2(L)Ru{=}CHPh]$, with L = PCy_3 or NHC, along with some of Herrmann's[131] cationic Ru-based catalysts with chelating bisphospine ligands, following a dissociative mechanism with *trans*-olefin coordination to the active catalyst. The author found that the binding energy of PCy_3 ligand to the Ru central atom of the precatalyst is larger when the *trans* ancillary ligand L is NHC rather than PCy_3. These results are in line with the experimental data reported by Grubbs as well as with the DFT results reported by Weskamp and coworkers.[131] More recently, Bernardi et al.[132] investigated the mechanism of the metathesis reaction catalyzed by Grubbs-type catalysts at the B3LYP level of theory using the DZVP basis set. Among the three different paths considered for olefin metathesis reaction with the model catalysts $[Cl_2(L)_2Ru{=}CH_2]$ (L = PH_3 or PPh_3), the following two seem to be the most probable: the mechanism conforming to the proposed mechanism by Grubbs[128] and the mechanism supporting the results reported by Herrmann et al.[131]

Most recently, the important synthesis reaction, the copper-catalyzed cyclopropanation reaction, has been studied theoretically.[133–135] Bühl et al.[133] performed DFT calculations at the BP86/AE1 level of theory with augmented Wachters' basis on Cu and 6-31G(d) basis on all other elements using model catalytic systems. For the big-sized models the D95V and Stuttgart-Dresden (SDD) relativistic effective core potentials were used. The calculations illustrated that the mechanism of the Cu-catalyzed olefin cyclopropanation involves, as key intermediates, Cu(carbene) species that have been confirmed either free or in the form of complexes with the reactant olefin.

The exploration of the PES of a medium-sized reaction model of copper-catalyzed cyclopropanation reactions at the B3LYP/6-311+G(2d,p)//B3LYP/6-31G(d) level of DFT by Fraile et al.[134] elucidated the key steps of the catalytic cycle. It was found that the cyclopropanation step takes place through a direct carbene insertion of the copper-carbene species to yield a catalyst-product complex, which can finally regenerate the starting complex. In a more recent paper[135] the authors investigated the role of a counterion in Cu-catalyzed enentioselective cyclopropanation reactions at the same level of theory using medium-sized reaction models. The presence of a coordinated counteranion was found to significantly affect the geometries of the reaction intermediates and transition structures, but not the overall reaction mechanism or the key reaction pathway. The rate-determining step was found to be the nitrogen extraction from a catalyst-diazoester complex to generate a copper-carbene intermediate.

The mechanism of the copper-catalyzed cyclopropanation reaction has also been studied by Rasmussen et al.[136] at various levels of density functional and *ab initio* theory using a variety of basis sets. The calculations were performed on small model systems using quantum chemical computational techniques of high level, or on real systems using the hybrid QM/MM computational approach. Solvation corrections were evaluated by using B3LYP/LACVP together with the BP-SCRF solvation model. The selectivity-determining step in the copper-catalyzed cyclopropanation proceeds by a concerted but very asynchronous addition of a metallacarbene to the alkene. Ligand-substrate interactions influencing the enantio- and diastereo-selectivity have been identified, and the preferred orientation of the alkene substrate during the addition is suggested.

SUMMARY AND OUTLOOK

In this work, we have provided a guided tour of the use of quantum chemical computational techniques in solving problems that inorganic chemists face, enabling them to acquire enough information to form a good concept of what the work is about. Theoretical studies have significantly improved our understanding of the metal-ligand bonding modes and shone light on the details of the inorganic reaction mechanisms for a number of industrially and synthetically important reactions. The recent advances in the field of quantum inorganic chemistry hold great promise for the future discovery of new series of novel transition metal coordination compounds with particular properties and functionality, guiding experimentalists to synthesize them. A general recipe for a successful application of computational quantum chemistry is as follows:

1. Choose the appropriate computational method. DFT is the widely accepted computational technique for transition metal coordination chemistry.
2. Choose with care the basis set of the highest possible quality for the system of interest. For heavy metals use relativistic ECPs, such as LANL2DZ or SDD, while for the rest of the atoms use at least the 6-31G(d) basis set. For anionic species, diffuse functions must be added as well.
3. Perform single-point calculations at a higher level of theory, such as CCSD(T) with a larger basis set, on the optimized geometry, when it is computationally possible, with the computational resources available to get more realistic energetic data.
4. Construct with care a model with computationally convenient size to represent the real big-sized systems or use the hybrid QM/MM method. By combining the accuracy of a QM method (for a small reacting system) with the speed of an MM method (for its surroundings), reactions in large molecules and in solution can be studied.
5. Include solvation in studies of reaction mechanisms involving charged species to correctly predict the energetic reaction profile.
6. Determine the structures and energies of reactants, transition states, intermediates, and products when reaction mechanisms are studied.

7. Investigate the factors controlling the reactivity and selectivity of the particular reaction and create rules for the design of new compounds.
8. Compute spectra, such as IR, NMR, Raman, UV-VIS, PES, etc., which can help in identifying new compounds and predict their reactivity.
9. Apply charge density and energy partitioning schemes to get insight into the bonding mechanism and the reactivity of the molecular system of interest.

In spite of the progress carried out in the field of quantum inorganic chemistry, a long path is still waiting to be walked. The scope of applications of electronic structure calculation methods is still far from mature. We have to look for new methodological developments, as well as for more careful assessment of applicability. There is an ongoing effort to improve DFT methods through a systematic improvement of the functionals based on a pure theoretical basis aiming to find the exact functional. Ultimately, highly parallel machines may very well be the most effective way to solve many of the computational bottlenecks. Combining large numbers of powerful computers, an efficient parallel algorithm may be able to extend some of the higher-level electronic structure calculation methods into the metalloenzyme and metal cluster regime. A daunting challenge for the future is to accurately model chemical reactions in phases and at the active site of metalloenzymes. An ability to do so would be of great importance in designing new biological catalysts as well as fully understanding the chemical mechanism of those that already exist. This would be of significant technological as well as scientific importance. One could imagine that many new molecules could be made, and made much more efficiently, by such catalysts. Recently, there has been much interest in using DFT for predicting the correct spin ground state of high spin species very abundant in the realm of inorganic chemistry. Moreover, the scope of applications of DFT grows rapidly with calculations of new molecular properties being added to actively developed software. Finally, effective core potential methods coupled with parallel supercomputing are expected to constitute an exciting approach toward the goal of developing methods for addressing the chemistry of all heavy elements of the periodic table.

REFERENCES

1. (a) Niu, S., and M. B. Hall. 2000. *Chem. Rev.*, 100, 353; (b) Loew, G. H., and D. L. Harris. 2000. *Chem. Rev.*, 100, 407; (c) Siegbahn, P. E. M., and M. R. A. Blomberg. 2000. *Chem. Rev.*, 100, 421; (d) Torrent, M., M. Solà, and G. Frenking. 2000. *Chem. Rev.*, 100, 439; (e) Rohmer, M.-M., M. Bènard, and J.-M. Poblet. 2000. *Chem. Rev.*, 100, 495; (f) Dedieu, A. 2000. *Chem. Rev.*, 100, 543; (g) Maseras, A., A. Lledós, E. Clot, and O. Eisenstein. 2000. *Chem. Rev.*, 100, 601; (h) Alonso, J. A. 2000. *Chem. Rev.*, 100, 637; (i) Harrison, J. F. 2000. *Chem. Rev.*, 100, 679; (k) Frenking, G., and N. Fröhlich. 2000. *Chem. Rev.*, 100, 717; (l) Hush, N. S., and J. R. Reimers. 2000. *Chem. Rev.*, 100, 775; (m) Ceulemans, A., L. F. Chibotaru, G. A. Heylen, K. Pierloot, and L. G. Vanquickenborne. 2000. *Chem. Rev.*, 100, 787; (n) Cundari, T. R. 2000. *Chem. Rev.*, 100, 807.

2. (a) Comba, P., and R. Remaenyi. 2003. *Coord. Chem. Rev.*, 238–239, 9; (b) Ellis, D. E., and O. Warschkow. 2003. *Coord. Chem. Rev.*, 238–239, 31; (c) Autschbach, J., and T. Ziegler. 2003. *Coord. Chem. Rev.*, 238–239, 83; (d) Daniel, C. 2003. *Coord. Chem. Rev.*, 238–239, 143; (e) Newton, M. D. 2003. *Coord. Chem. Rev.*, 238–239, 167; (f) Ciofini, I., and C. A. Daul. 2003. *Coord. Chem. Rev.*, 238–239, 187; (g) Lovell, T., F. Himo, W.-G. Han, and L. Noodleman. 2003. *Coord. Chem. Rev.*, 238–239, 211; (h) Erras-Hanauer, H., T. Clark, and R. van Eldik. 2003. *Coord. Chem. Rev.*, 238–239, 233; (i) Georgakaki, I. P., L. M. Thomson, E. J. Lyon, M. B. Hall, and M. Y. Darensbourg. 2003. *Coord. Chem. Rev.*, 238–239, 255; (j) Friesner, R. A., M.-H. Baik, B. F. Gherman, V. Guallar, M. Wirstam, R. B. Murphy, and S. J. Lippard. 2003. *Coord. Chem. Rev.*, 238–239, 267; (k) Boone, A. J., C. H. Chang, S. N. Greene, T. Herz, and N. G. J. Richards. 2003. *Coord. Chem. Rev.*, 238–239, 291; (l) Webster, C., and M. B. Hall. 2003. *Coord. Chem. Rev.*, 238–239, 315; (m) Renhold, J., A. Barthel, and C. Mealli. 2003. *Coord. Chem. Rev.*, 238–239, 333; (n) Harvey, J. N., R. Poli, and K. M. Smith. 2003. *Coord. Chem. Rev.*, 238–239, 347; (o) Sapunov, V. N., R. Schmid, K. Kirchner, and H. Nagashima. 2003. *Coord. Chem. Rev.*, 238–239, 363; (p) Macchi, P., and A. Sironi. 2003. *Coord. Chem. Rev.*, 238–239, 383.
3. Tsipis, C. A. 1991. *Coord. Chem. Rev.*, 108, 163.
4. Koga, K., and K. Morokuma. 1991. *Chem. Rev.*, 91, 823.
5. Gordon, M. S., and T. R. Cundari. 1996. *Coord. Chem. Rev.*, 147, 87.
6. Noodleman, L., T. Lovell, W.-G. Han, J. Li, and F. Himo. 2004. *Chem. Rev.*, 104, 459.
7. Solomon, E. I., R. K. Szilagyi, S. DeBeer George, and L. Basumallick. 2004. *Chem. Rev.*, 104, 419.
8. http://www.chem.swin.edu.au/chem_ref.html#Software
9. http://cmm.info.nih.gov/modeling/universal_software.html
10. Cramer, C. J. 2002. *Essentials of Computational Chemistry*. Chichester: Wiley.
11. Feller, D., and E. R. Davidson. 1990. In *Reviews in Computational Chemistry*, ed. K. B. Lipkowitz and D. B. Boyd. Vol. 1, 1. New York: VCH.
12. Krishnan, R., M. J. Frisch, and J. A. Pople. 1980. *J. Chem. Phys.*, 72, 4244.
13. Wilson, A. K., T. van Mounk, and T. H. Dunning. 1996. *J. Mol. Struct.*, 388, 339.
14. Hay, P. J. 1977. *J. Chem. Phys.*, 66, 4377.
15. Stevens, W. J., M. Krauss, H. Basch, and P. G. Jasien. 1992. *Can. J. Chem.*, 70, 612.
16. Frenking, G., I. Antes, M. Bohme, S. Dapprich, A. W. Ehlers, V. Jonas, A. Neuhaus, M. Otto, R. Stegmann, A. Veldkamp, and S. F. Vyboischclukov. 1996. In *Reviews in Computational Chemistry*, ed. K. B. Lipkowitz and D. B. Boyd. Vol. 8, 63. New York: VCH.
17. Schmidt, M. W., and M. S. Gordon. 1998. *Annu. Rev. Phys. Chem.*, 49, 233.
18. Gilson, H. S. R., and M. Krauss. 1998. *J. Phys. Chem. A*, 102, 6525.
19. Roos, B. O., K. Anderson, M. P. Fülscher, P. A. Malmqyist, L. Serrano-Andres, K. Pierloot, and M. Merchan. 1996. *Adv. Chem. Phys.*, 93, 219.
20. Palmer, I. J., I. N. Ragazos, F. Bernardi, M. Olivucci, and M. A. Robb. 1993. *J. Am. Chem. Soc.*, 115, 673.
21. Olivucci, M., I. N. Ragazos, F. Bernardi, and M. A. Robb. 1993. *J. Am. Chem. Soc.*, 115, 3710.
22. Bartlett, R. J., and J. F. Stanton. 1994. In *Reviews in Computational Chemistry*, ed. K. B. Lipkowitz and D. B. Boyd, 65. Vol. 5. New York: VCH.
23. Møller, C., and M. S. Plesset. 1934. *Phys. Rev.*, 46, 618.
24. Čížek, J. 1966. *J. Chem. Phys.*, 45, 4526.
25. Čížek, J. 1969. *Adv. Chem. Phys.*, 14, 35.
26. Čížek, J., and J. Paldus. 1971. *Int. J. Quantum. Chem.*, 5, 359.
27. Pople, J. A., D. P. Santry, and G. A. Segal. 1965. *J. Chem. Phys.*, 43, 129.

28. (a) Dewar, M. J. S., and W. Thiel. 1977. *J. Am. Chem. Soc.*, 99, 4899; (b) Thiel, W. 1998. In *Encyclopedia of Computational Chemistry,* ed. P. V. R. Schleyer, N. L. Allinger, T. Clark, J. Gasteiger, P. A. Kollman, H. F. Schaefer III, and P. R. Schreiner, 1599. Vol 3. Chichester: Wiley.
29. (a) Dewar, M. J. S., E. G. Zoebisch, E. F. Healy, and J. P. Stewart. 1985. *J. Am. Chem. Soc.*, 107, 3902; (b) Holder, A. J. 1998. In *Encyclopedia of Computational Chemistry,* ed. Schleyer, N. L. Allinger, T. Clark, J. Gasteiger, P. A. Kollman, H. F. Schaefer III, and P. R. Schreiner, 8. Vol. 1. Chichester: Wiley.
30. (a) Stewart, J. J. P. 1989. *J. Comput. Chem.*, 10, 209; (b) Stewart, J. J. P. 1998. In *Encyclopedia of Computational Chemistry,* ed. Schleyer, N. L. Allinger, T. Clark, J. Gasteiger, P. A. Kollman, H. F. Schaefer III, and P. R. Schreiner, 2080. Vol. 3. Chichester: Wiley.
31. Hehre, W. J., and J. Yu. 1995. *Book of Abstracts*, Part 1. 210th ACS National Meeting, Chicago, August 20–24.
32. http://www.chachesoftware.co./mopac/index.shtml
33. Koch, W., and M. C. Holthausen. 2000. *A Chemist's Guide to Density Functional Theory*. Weinheim: Wiley-VCH.
34. Bartolotti, L. J., and K. Flurchick. 1995. In *Reviews in Computational Chemistry*, ed. K. B. Lipkowitz and D. B. Boyd. Vol. 7, 187. New York: VCH.
35. Parr R. G., and W. Yang. 1989. *Density-Functional Theory of Atoms and Molecules*. Oxford: Oxford University Press.
36. Ellis, D. E. 1995. *Density Functional Theory of Molecules, Clusters, and Solids*. Dordrecht: Kluwer Academic Publishers.
37. Geerlings, P., F. De Proft, and W. Langenaeker. 2003. *Chem. Rev.*, 103, 1793.
38. Vosko, S. H., L. Wilk, and M. Nusair. 1980. *Can. J. Phys.*, 58, 1200.
39. Perdew, J. P., and Y. Wang. 1992. *Phys. Rev. B*, 45, 13244.
40. Perdew, J. P., and Y. Wang. 1986. *Phys. Rev. B*, 33, 8800.
41. Perdew, J. P. 1986. *Phys. Rev. B*, 33, 8822.
42. (a) Becke, A. D. 1993. *J. Chem. Phys.*, 98, 5648; (b) Becke, A. D. 1988. *Phys. Rev. B*, 38, 3098; (c) Lee, C., W. Yang, and R. G. Parr. 1988. *Phys. Rev. B*, 37, 785; (d) Curtiss, L. A., K. Raghavachari, P. C. Redfern, and J. A. Pople. 1997. *J. Chem. Phys.*, 106, 1063.
43. (a) Lynch, B. J., P. L. Fast, M. Harris, and D. G. Truhlar. 2000. *J. Phys. Chem. A*, 104, 4811; (b) Lynch, B. J., and D. G. Truhlar. 2001. *J. Phys. Chem. A,* 105, 2936.
44. Shadhukhan, S., D. Munoz, C. Adamo, and G. E. Scuseria. 1999. *Chem. Phys. Lett.*, 306, 83.
45. Pavese, M., S. Chawla, D. Lu, J. Lobaugh, and G. A. Voth. 1997. *J. Chem. Phys.*, 107, 7428.
46. Schlegel, H. B. 2003. *J. Comput. Chem.*, 24, 1514.
47. (a) Mulliken, R. S. 1955. *J. Chem. Phys.*, 23, 1833; (b) Mulliken, R. S. 1955. *J. Chem. Phys.,* 23, 1841; (c) Mulliken, R. S. 1955. *J. Chem. Phys.*, 23, 2338; (d) Mulliken, R. S. 1955. *J. Chem. Phys.*, 23, 2343.
48. (a) Reed, A. F., R. B. Weinstock, and F. A. Weinhold. 1985. *J. Chem. Phys.*, 83, 735; (b) Reed, A. F., L. A. Curtiss, and F. A. Weinhold. 1988. *Chem. Rev.*, 88, 899.
49. Bader, R. F. W. 1990. *Atoms in Molecules. A Quantum Theory*. Oxford: Clarendon Press.
50. Hirshfeld, F. L. 1977. *Theoret. Chim. Acta*, 44, 129.
51. (a) Politzer, P., and R. R. Harris. 1970. *J. Am. Chem. Soc.*, 92, 6451; (b) Politzer, P., and E. W. Stout, Jr. 1971. *Chem. Phys. Lett.*, 8, 519; (c) Politzer, P. 1971. *Theoret. Chim. Acta*, 23, 203.
52. Rousseau, B., A. Peeters, and C. Van Alsenoy. 2001. *J. Mol. Struct.*, 538, 235.
53. Fonseca, C., G. J.-W. Handgraaf, E. J. Baerends, and F. M. Bickelhaupt. 2004. *J. Comput. Chem.*, 25, 189.
54. Scott, A. P., and L. Radom. 1996. *J. Phys. Chem.*, 100, 16502.

55. Halls, M. D., J. Velkovski, and H. B. Schlegel. 2001. *Theor. Chem. Accounts*, 105, 413.
56. London, F. 1937. *J. Phys. Radium.*, 8, 397.
57. McWeeny, R. 1962. *Phys. Rev.*, 126, 1028.
58. Ditchfield, R. 1974. *Mol. Phys.*, 27, 789.
59. Dodds, J. L., R. McWeeny, and A. J. Sadlej. 1980. *Mol. Phys.*, 41, 1419.
60. Wolinski, K., J. F. Hilton, and P. Pulay. 1990. *J. Am. Chem. Soc.*, 112, 8251.
61. Kutzelnigg, W. 1980. *Isr. J. Chem.*, 19, 193.
62. Hansen, A. E., and T. D. Bouman. 1985. *J. Chem. Phys.*, 82, 5035.
63. Keith, T. A., and R. F. W. Bader. 1992. *Chem. Phys. Lett.*, 194, 1.
64. Keith, T. A., and R. F. W. Bader. 1993. *Chem. Phys. Lett.*, 210, 223.
65. Cheeseman, J. R., M. J. Frisch, G. W. Trucks, and T. A. Keith. 1996. *J. Chem. Phys.*, 104, 5497.
66. Bühl, M., M. Kaupp, O. L. Malkina, and V. G. Malkin. 1998. *J. Comput. Chem.*, 20, 91.
67. Schreckenbach, G., and T. Ziegler. 1998. *Theor. Chem. Accounts*, 99, 71.
68. Kaupp, M., O. L. Malkina, V. G. Malkin, and P. Pyykkö. 1998. *Chem. Eur. J.*, 4, 118.
69. Wilson, P. J., R. D. Amos, and N. C. Handy. 2000. *Phys. Chem. Chem. Phys.*, 2, 187.
70. Gomila, R. M., D. Quinonero, C. Garau, A. Frontera, P. Ballester, A. Costa, and P. M. Deyà. 2002. *Chem. Phys. Lett.*, 36, 72.
71. (a) Ortiz, J. V. 1988. *J. Chem. Phys.*, 89, 6348; (b) von Niessen, W., J. Schirmer, and L. S. Cederbaum. 1984. *Comp. Phys. Rep.*, 1, 57; (c) Zakrzewski, V. G., and J. V. Ortiz. 1995. *Int. J. Quant. Chem.*, 53, 583; (d) Ortiz, J. V. 1998. *Int. J. Quant. Chem. Symp.*, 22, 431; (e) Ortiz, J. V. 1999. *Int. J. Quant. Chem. Symp.*, 23, 321; (f) Ortiz, J. V., V. G. Zakrzewski, and O. Dolgounircheva. 1997. In *Conceptual Perspectives in Quantum Chemistry*, ed. J.-L. Calais and E. Kryachko, 465. Dordrecht: Kluwer Academic.
72. Jamorski, C., M. E. Casida, and D. R. Salahub. 1996. *J. Chem. Phys.*, 104, 5134.
73. Casida, M. E. 1996. *Recent Advances in Density Functional Methods*. Singapore: World Scientific.
74. Casida, M. E. 1996. In *Recent Developments and Applications of Modern DFT*, ed. J.M.S. Ed, 391–434. Amsterdam: Elsevier Science.
75. (a) Comeau, D. C., and R. J. Bartlett. 1993. *Chem. Phys. Lett.*, 207, 414; (b) Geertsen, J., M. Rittby, and R. J. Bartlett. 1989. *Chem. Phys. Lett.*, 164, 57; (c) Sekino, H., and R. J. Bartlett. 1984. *Int. J. Quantum Chem. Symp.*, 18, 255; (d) Stanton, J. F., and R. J. Bartlett. 1993. *J. Chem. Phys.*, 98, 7029; (e) Stanton, J. F., and R. J. Bartlett. 1993. *J. Chem. Phys.*, 98, 9335.
76. (a) Andersson, K., P.-A. Malmqvist, B. O. Roos, A. J. Sadlej, and K. Wolinski. 1990. *J. Phys. Chem.*, 94, 5483; (b) Andersson, K., P. A. Malmqvist, and B. O. Roos. 1992. *J. Chem. Phys.*, 96, 1218; (c) Andersson, K., B. O. Roos, P.-A. Malmqvist, and P.-O. Widmark. 1994. *Chem. Phys. Lett.*, 230, 391/434; (d) Finley, J., P.-A. Malmqvist, B. O. Roos, and L. Serrano-Andrés. 1998. *Chem. Phys. Lett.*, 288, 299.
77. Foresman, J. B., M. Head-Gordon, J. A. Pople, and M. J. Frisch. 1992. *J. Phys. Chem.*, 96, 135.
78. Que, L., Jr., and W. B. Tolman. 2002. *Angew. Chem. Int. Ed.*, 41, 1114.
79. Alvarez, S., A. A. Palacios, and G. Aullón. 1999. *Coord. Chem. Rev.*, 185–186, 431, and references cited therein.
80. (a) Mealli, C., and D. M. Proserpio. 1990. *J. Am. Chem. Soc.*, 112, 5484; (b) Mealli, C., and A. Orlandini. 1999. In *Metal Clusters in Chemistry*, ed. P. Braunstein, L. A. Oro, and P. R. Raithby, 143–62. Vol. 1. Weinheim: Wiley-VCH.
81. Alonso, E., J. M. Casas, F. A. Cotton, X. Feng, J. Forniés, C. Fortuño, and M. Tomás. 1999. *Inorg. Chem.*, 38, 5034.
82. Alonso, E., J. M. Casas, J. Forniés, C. Fortuño, A. Martin, A. Guy Orpen, C. A. Tsipis, and A. C. Tsipis. 2001. *Organometallics*, 20, 5571.

83. Tsipis, C. A., A. C. Tsipis, and C. E. Kefalidis. 2004. *Fundamental World of Quantum Chemistry*. Vol. 3. Dordrecht, The Netherlands: Kluwer Academic Publishers.
84. Brown, M. P., R. J Puddephatt, M. Rashidi, and K. R. Seddon. 1978. *J. Chem. Soc. Dalton. Trans.*, 516.
85. Baik M.-H., T. Ziegler, and C. K. Schauer. 2000. *J. Am. Chem. Soc.*, 122, 9143.
86. Captevila, M., W. Clegg, P. González-Duarte, A. Jarid, and A. Lledós. 1996. *Inorg. Chem.*, 35, 490.
87. Aslanidis, P., P. J. Cox, S. Divanidis, and A. C. Tsipis. 2002. *Inorg. Chem.*, 41, 6875.
88. Pyykkö P., and M. Patzschke. 2003. *Faraday Discuss.*, 124, 41.
89. Autschbach, J., and T. Ziegler. 2001. *J. Am. Chem. Soc.*, 123, 5320.
90. (a) Pyykkö, P. 1997. *Chem. Rev.*, 97, 597; (b) Pyykkö, P., and T. Tamm. 1998. *Organometallics*, 17, 4842.
91. Magnko, L., M. Schweizer, G. Rauhut, M. Schütz, H. Stoll, and H.-J. Werner. 2000. *Phys. Chem. Chem. Phys.*, 4, 1006.
92. Mendizabal, F., P. Pyykkö, and N. Runeberg. 2003. *Chem. Phys. Lett.*, 370, 733.
93. Mendizabal, F., and P. Pyykkö. 2004. *Phys. Chem. Chem. Phys.*, 6, 900.
94. O'Grady, E., and N. Kaltsoyannis. 2004. *Phys. Chem. Chem. Phys.*, 6, 680.
95. Wang, S.-G., and W. H. E. Schwarz. 2004. *J. Am. Chem. Soc.*, 126, 1266.
96. Bardaji, M., and A. Laguna. 2003. *Eur. J. Inorg. Chem.*, 3069.
97. Gade, L. H. 2001. *Angew. Chem. Int. Ed.*, 40, 3573.
98. Wesendrup, P., and P. Schwerdtfeger. 2000. *Angew. Chem. Int. Ed.*, 39, 907.
99. (a) Schröder, D., A. Fiedler, M. F. Ryan, and H. Schwarz. 1995. *J. Phys. Chem.*, 98, 1994; (b) Schröder, D., and H. Schwarz. 1995. *Angew. Chem. Int. Ed.*, 34, 1973; (c) Holthausen, M. C., A. Fiedler, H. Schwarz, and W. Koch. 1996. *J. Phys. Chem.*, 100, 6236; (d) Dieterle, M., J. N. Harvey, C. Heinemann, J. Schwarz, D. Schröder, and H. Schwarz. 1997. *Chem. Phys. Lett.*, 277, 399; (e) Bronstrup, M., D. Schröder, and H. Schwarz. 1999. *Chem. Eur. J.*, 5, 1176; (f) Schröder, D., and H. Schwarz. 2000. *Acc. Chem. Res.*, 33, 139.
100. (a) Irigoras, A., J. E. Fowler, and J. M. Ugalde. 1998. *J. Phys. Chem. A.*, 102, 293; (b) Irigoras, A., J. E. Fowler, and J. M. Ugalde. 1999. *J. Am. Chem. Soc.*, 121, 574; (c) Irigoras, A., J. E. Fowler, and J. M. Ugalde. 1999. *J. Am. Chem. Soc.*, 121, 8549; (d) Irigoras, A., J. E. Fowler, and J. M. Ugalde. 2000. *J. Am. Chem. Soc.*, 122, 114.
101. Clemmer, D. E., and P. B. Armentrout. 1994. *J. Phys. Chem.*, 98, 68.
102. Hall, C., and R. N. Perutz. 1996. *Chem. Rev.*, 96, 3125.
103. Nakao, Y., T. Taketsugu, and K. Hirao. 1999. *J. Chem. Phys.*, 110, 10863.
104. Kooi, S. E., and A. W. Castleman. 1999. *Chem. Phys. Lett.*, 315, 49.
105. (a) Danovich, D., and S. Shaik. 1997. *J. Am. Chem. Soc.*, 119, 1773; (b) Filatov, M., and S. Shaik. 1998. *J. Phys. Chem. A*, 102, 3835.
106. (a) Yoshizawa, K., Y. Shiota, and T. Yamade. 1998. *J. Am. Chem. Soc.*, 120, 564; (a) Shiota, Y., and K. Y. Yoshizawa. 2000. *J. Am. Chem. Soc.*, 122, 12317.
107. El-Nahas, A. M. 2001. *Chem. Phys. Lett.*, 345, 325.
108. (a) Hwang, D.-Y., and A. M. Mebel. 2002. *Chem. Phys. Lett.*, 357, 51; (b) Hwang, D.-Y., and A. M. Mebel. 2002. *J. Phys. Chem. A.*, 106, 12072.
109. (a) Pápai, I., Y. Hannachi, S. Gwizdala, and J. Mascetti. 2002. *J. Phys. Chem. A.*, 106, 4181; (b) Pápai, I., G. Schubert, Y. Hannachi, and J. Mascetti. 2002. *J. Phys. Chem. A.*, 106, 9551; (c) Hannachi, Y., J. Mascetti, A. Stirling, and I. Pápai. 2003. *J. Phys. Chem. A.*, 107, 6708.
110. (a) Souter, P. F., and L. Andrews. 1997. *J. Am. Chem. Soc.*, 119, 7350; (b) Sodupe, M., V. Branchadell, M. Rosi, and C. W. Bauschlicher Jr. 1997. *J. Phys. Chem. A.*, 101, 7854; (c) Gutsev, G. L., L. Andrews, and C. W. Bauschlicher Jr. 2003. *Chem. Phys.*, 290, 47.
111. (a) Karipidis, P., A. C. Tsipis, and C. A. Tsipis. 2003. *Collect. Czech. Chem. Commun.*, 68, 423; (b) Pantazis, D. A., A. C. Tsipis, and C. A. Tsipis. 2003. *Collect. Czech. Chem. Commun.*, 69, 13.
112. Xie, X.-G. 2004. *Chem. Phys.*, 299, 33.

113. (a) Michelini, M. del C., E. Sicilia, N. Russo, M. E. Alikhani, and B. Silvi. 2003. *J. Phys. Chem. A*, 107, 4862; (b) Chiodo, S., O. Kondakova, M. del C. Michelini, N. Russo, E. Sicilia, A. Irigoras, and J. M. Ugalde. 2004. *J. Phys. Chem. A*, 108, 1069.
114. Dincan, M. A. 2000. *Int. J. Mass Spectrom.*, 200, 545, and references therein.
115. Freiser, B. S. 1996. *Organometallic Ion Chemistry*. Dordrecht: Kluwer.
116. Alikhani, M. E., 1999. *Chem. Phys. Lett.*, 313, 608.
117. El-Nahas, A. M., N. Tajima, and K. Hirao. 2000. *Chem. Phys. Lett.*, 318, 333.
118. Stone, J. A., and D. Vukomanovich. 2001. *Chem. Phys. Lett.*, 346, 419.
119. Cundari, T. R. 2001. *Computational Organometallic Chemistry*. New York: Marcel Decker.
120. van Leeuwen, P. W. N. M., J. H. van Lenthe, and K. Morokuma. 1995. *Theoretical Aspects of Homogeneous Catalysis: Applications of Ab Initio Molecular Orbital Theory*. Kluwer, Dordrecht, The Netherlands.
121. Field, M. J., P. A. Bash, and M. Karplus. 1990. *J. Comp. Chem.*, 11, 700.
122. Maseras, F., and K. Morokuma. 1995. *J. Comp. Chem.*, 16, 1170.
123. (a) Freindorf, M., and J. Gao. 1996. *J. Comp. Chem.*, 17, 386; (b) Gao, J. 1996. *Acc. Chem. Res.*, 29, 298; (c) Gao, J. 1996. *Rev. Comp. Chem.*, 7, 119; (d) Gao, J., P. Amara, C. Alhambra, and M. J. Field. 1998. *J. Phys. Chem. A*, 102, 4714; (e) Reuter, N., A. Dejaegere, B. Maigret, and M. Karplus. 2000. *J. Phys. Chem. A*, 104, 1720.
124. (a) Deng, L., T. K. Woo, L. Cavallo, P. M. Margl, and T. Ziegler. 1997. *J. Am. Chem. Soc.*, 119, 6177; (b) Deng, L., T. Ziegler, T. K. Woo, P. Margl, and L. Fan. 1998. *Organometallics*, 17, 3240; (c) Woo, T. K., P. M. Margl, L. Deng, L. Cavallo, and T. Ziegler. 1999. *Catalysis Today*, 50, 479; (d) Woo, T. K., S. Patchkovskii, and T. Ziegler. 2000. *Comput. Sci. Eng.*, 28.
125. (a) Tobisch, S., and T. Ziegler. 2002. *J. Am. Chem. Soc.*, 124, 4881; (b) Tobisch, S., and T. Ziegler. 2002. *J. Am. Chem. Soc.*, 124, 13290.
126. Tobisch, S., and R. Taube. 2001. *Chem. Eur. J.*, 7, 3681.
127. Wallner, O. A., and K. J. Szabó. 2003. *Chem. Eur. J.*, 9, 4025.
128. (a) Sanford, M. S., M. Ulman, and R. H. Grubbs. 2001. *J. Am. Chem. Soc.*, 123, 749; (b) Sanford, M. S., J. A. Love, and R. H. Grubbs. 2001. *J. Am. Chem. Soc.*, 123, 6543.
129. Vyboishchikov, S. F., M. Bühl, and W. Thiel. 2002. *Chem. Eur. J.*, 8, 3962.
130. Cavallo, L. 2002. *J. Am. Chem. Soc.*, 124, 8965.
131. Weskamp, T., F. J. Kohl, W. Hierlinger, D. Gleich, and W. A. Herrmann. 1999. *Angew. Chemie. Int. Ed.*, 38, 2416.
132. Bernardi, F., A. Botoni, and G. P. Miscione. 2003. *Organometallics*, 22, 940.
133. Bühl, M., F. Terstegen, F. Löffler, B. Meynhardt, S. Kierse, M. Müller, C. Näther, and U. Lüning. 2001. *Eur. J. Org. Chem.*, 2151.
134. Fraile, J. M., J. I. García, V. Martínez-Merino, J. A. Mayoral, and L. Salvatella. 2001. *J. Am. Chem. Soc.*, 123, 7616.
135. Fraile, J. M., J. I. García, M. J. Gil, V. Martínez-Merino, J. A. Mayoral, and L. Salvatella. 2004. *Chem. Eur. J.*, 10, 758.
136. Rasmussen, T., J. F. Jensen, N. Èstergaard, D. Tanner, T. Ziegler, and P.-O. Norrby. 2002. *Chem. Eur. J.*, 8, 177.

5 NMR Techniques for Investigating the Supramolecular Structure of Coordination Compounds in Solution

Gianluca Ciancaleoni, Cristiano Zuccaccia, Daniele Zuccaccia, and Alceo Macchioni

CONTENTS

INTRODUCTION

Weak interactions[1] are central to chemistry. They strongly affect the conformation and aggregation of biomolecules, the packing of solid materials, and the performances and stereo-differentiation of homogeneous catalysts.

In coordination chemistry weak interactions are crucial substantially because their energy can be comparable with that of some coordinative bonds (for example, X-H agostic bonds) and because the bonds involving a metal are usually easily polarizable. As a consequence, in some cases new types of weak interactions like dihydrogen bond,[2] i.e., the hydrogen bond between X-H (X = O, N, etc.) and M-H (M = transition metal), have been discovered.

The methodologies for investigating weak interactions in the solid state are well developed, and they mainly rely on x-ray diffraction studies.[3] In solution, some information can be obtained by "classical" techniques, but detailed insights on the presence and action of weak interactions are hardly obtainable. This is particularly true if the structure of the noncovalent supramolecular assemblies (supramolecular structure), i.e., the relative position of the interacting moieties and average size, has to be determined. Information about the supramolecular structure can be obtained by means of advanced nuclear magnetic resonance (NMR) spectroscopy.[4] Intermolecular nuclear Overhauser effect (NOE) NMR experiments,[5] based on the detection of dipolar (through the space) coupling between nuclei, are ideal for elucidating the relative orientation of the moieties. Diffusion NMR experiments[6] allow the molecular size of supramolecular adducts to be estimated. In fact, the average hydrodynamic radius of the supramolecule can be evaluated from the translational self-diffusion coefficient measured, taking advantage of the Stokes-Einstein equation.

This chapter focuses just on these two classes of NMR experiments. The essential theoretical aspects of NOE ("NOE NMR" and "NOE Dependence on the Internuclear Distance" sections) and diffusion NMR ("Diffusion NMR" and "From D_t to the Hydrodynamic Dimensions" sections) spectroscopies are recalled trying to maintain the mathematical formalism at minimum. Exchange NMR experiments are also briefly mentioned in the "Exchange Spectroscopy" section. Soon after having introduced a new theoretical concept, some selected examples on its application are reported in order to immediately furnish the reader with its practical importance. Particularly, examples on the NOE determination of the relative orientation(s) within supramolecular adducts are described in the "Examples on the Elucidation of the Supramolecular Structure by NOE NMR" section. The "Examples on the Estimation of Internuclear Distances" and "Examples of Supramolecular Host-Guest Exchange Processes" sections deal with examples on the quantitative determination of intermolecular distances and supramolecular exchange processes, respectively. Finally, applications of diffusion NMR spectroscopy are illustrated in the "Examples of Applications of Diffusion NMR Techniques" section, starting from simple cases for which the standard Stokes-Einstein equation can be used to elaborate the diffusion data, and passing to more complicated cases that necessitate taking into account the size and shape of supramolecular adducts.

Efforts have been made to select examples that are, as much as possible, informative and explicative of the theoretical aspects and, at the same time, cover different fields of supramolecular chemistry of transition metal complexes.

NOE NMR

The nuclear Overhauser effect (NOE) is a consequence of the dipole-dipole interaction between nuclei, and it critically depends on the competition between multiple- and zero-quantum relaxation mechanisms.[7] For the simplest system consisting of two nonscalarly coupled spins, **I** (interesting) and **S** (saturated), separated by a constant distance (r_{IS}), the steady-state NOE is defined as the fractional enhancement of the signal of spin **I** when the resonance of spin **S** is selectively saturated:

$$NOE_I\{S\} = \frac{I - I^0}{I^0} \tag{5.1}$$

where I^0 is the equilibrium intensity of I. From a phenomenological point of view, a variation of the intensity of the resonance due to spin **I** is observed (Figure 5.1a). This fractional variation of intensity (NOE) can be visualized in 1D difference spectra or in 2D spectra as cross-peaks (out-of-diagonal), as shown in Figure 5.1b and 1c, respectively.

As will be better illustrated in the following, NOEs strongly depend on the internuclear distance and are usually classified as strong (r_{IS} < 2.5 Å), medium (2.5 < r_{IS} < 3.3), or weak (r_{IS} > 3.3); the maximum distance at which NOE can be observed is ca. 5 Å. The detection of intermolecular NOEs in supramolecular adducts is usually sufficient to disclose the relative orientation(s) of the constituting units based on a simple *triangulation reasoning*. This procedure asks for suitable "reporter" nuclei on the units of the supramolecular adducts. Before entering a little more into the details of NOE theory ("NOE Dependence on the Internuclear Distance" section), let us consider some examples concerning the determination of the supramolecular structure of ion pairs and host-guest adducts.

Examples on the Elucidation of the Supramolecular Structure by NOE NMR

Relative Cation-Anion Orientation(s) in Transition-Metal-Complex Ion Pairs

NOE NMR techniques have been utilized to determine the ion pair structure in solution of organic salts[8] and transition metal salts.[9] For example, let us consider the $^{19}F,^{1}H$-HOESY spectrum of *trans*-[Ru(PMe_3)$_2$(CO)(COMe)(pz_2-CH_2)]PF_6 (**1**, pz = pyrazolyl ring) recorded in CD_2Cl_2 (Figure 5.2).

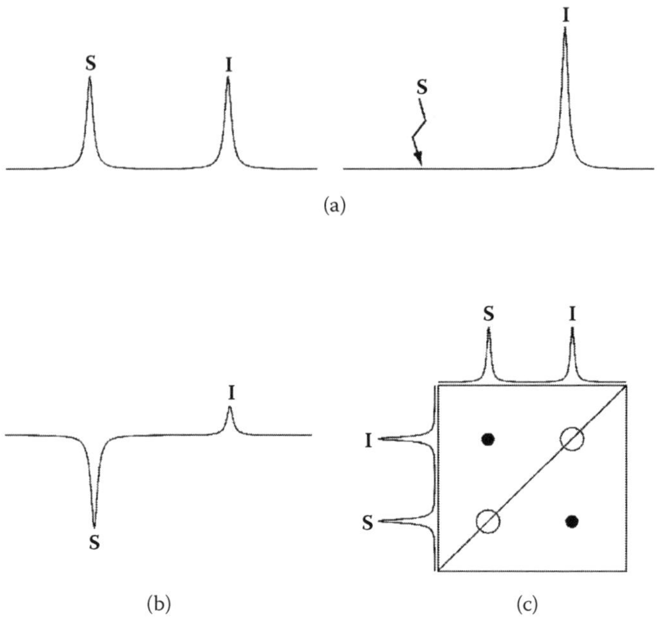

FIGURE 5.1 (a) NMR spectra of the **IS** spin system before (left) and after (right) the saturation of **S**. NOE manifests as a net intensity variation of **I** resonance in the 1D difference spectrum (b) or an out-of-diagonal cross-peak in the 2D NOESY spectrum (c).

It shows selective interionic NOEs between the fluorine resonances of PF_6^- counterion and the CH_2 and 5 and 5′ protons of the bis-pyrazolyl-methane ligand and, with smaller intensity, protons belonging to the phospine ligands (Figure 5.2). No interaction is observed with the remaining resonances of the organic ligand (H3, H3′, H4, and H4′) and with the acetyl group. It can be concluded that the anion selectively locates close to the pyrazolyl ligand, as depicted in Figure 5.2. According to quantomechanical calculations, the main factor determining the observed interionic structure is the noncentrosymmetric distribution of electron density around the metal center with a partial accumulation of the positive charge on the periphery of the pyrazolyl ligand.[11] The same relative anion-cation orientation is observed with different anions, such as BPh_4^- and BF_4^-. In the case of BPh_4^-, there is a further stabilization caused by lipophilic π-π stacking interactions between the phenyl groups of the anion and the pyrazolyl rings. These interactions lead to an energy gain of about 4 kJ/mol[12] and compensate for the reduction of Coulomb interactions due to an average increase of the anion-cation distance going from small to large anions such as BPh_4^-.[13]

Elucidating the interionic structure of ion pairs may be extremely important to rationalize the effects of counterion on the performances of ionic transition metal catalysts.[14] In order to illustrate this, let us consider the palladium catalysts for CO/*p*-methylstyrene copolymerization, shown in Scheme 5.1.[15] Their catalytic activity is markedly influenced by the coordination ability of the counterion only in complexes bearing *ortho*-unsubstituted ligands (**2a,b**): When the most coordinating $CF_3SO_3^-$ anion is used, the productivity is reduced by about three times with respect to that

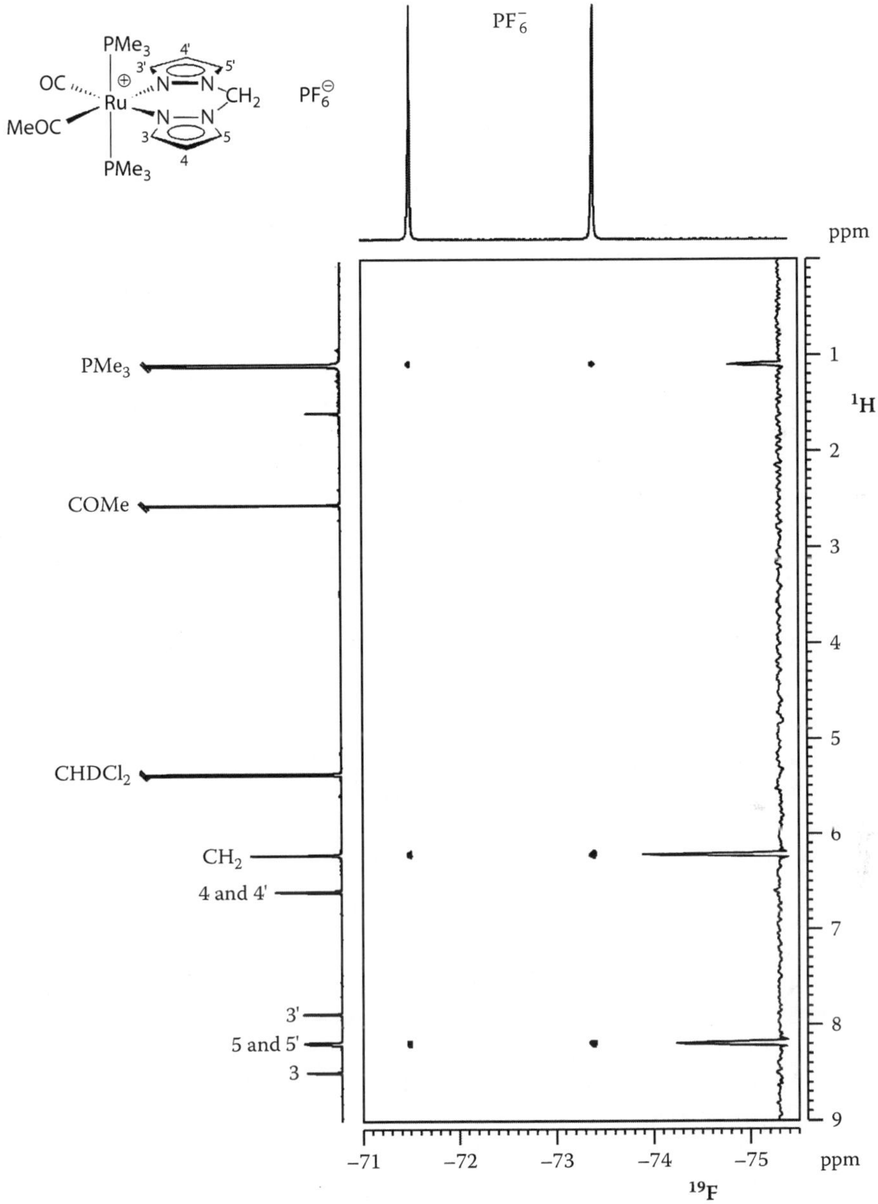

FIGURE 5.2 ^{19}F,^{1}H-HOESY spectrum (376.65 MHz, 298 K, CD_2Cl_2) of complex **1** showing selective interactions of PF_6^- with CH_2, H5, H5′, and PMe_3. The vertical trace on the right is relative to one component of the fluorine doublet. (Adapted from Zuccaccia et al. *J. Am. Chem. Soc.*, 123, 11020–11028, 2001.)

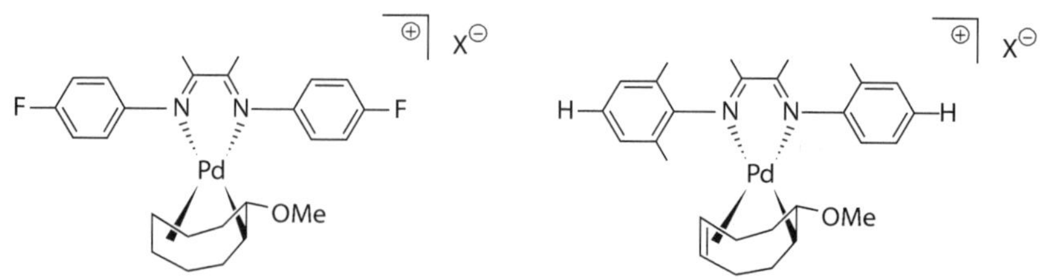

X = PF_6 (**2a**), CF_3SO_3 (**2b**)

X = PF_6 (**3a**), CF_3SO_3 (**3b**), BArF (**3c**)
BArF = $B(3,5\text{-}(CF_3)_2C_6H_3)_4$

SCHEME 5.1

with the PF_6^- anion. On the contrary, when aryl diimines bear methyl groups in *ortho* positions (complexes **3a–c**), the catalytic activity is almost independent of the nature of the counterion. These experimental findings nicely correlate with the average interionic structure observed in solution. In fact, $^{19}F,^{1}H$-HOESY experiments show that the anion interacts with most of the proton resonances of the cation in **2a** and **2b**, indicating that the latter can closely approach the metal center (Figure 5.3a). Thus a more coordinating anion can better compete with the substrate and strongly depress catalyst activity. On the contrary, $^{19}F,^{1}H$-HOESY experiments of complexes **3a–c** show interionic NOEs only between the anion and the methyl groups belonging to the N-N ligand (Figure 5.3b). This indicates that the steric hindrance provided by the methyl groups relegates the anion in the backbone of the diimine ligand, where there is a partial accumulation of the positive charge. Since the accessibility of the anion to the metal center is strongly reduced, its nature is less critical in determining the catalytic activity.

A confinement of the counterion in the periphery of square planar complexes can also be achieved by means of exclusively electronic effects. This has been found in the platinum (II) complex $[PtMe(dpa)(Me_2SO)]PF_6$ [dpa = bis(2-pyridyl)amine] (Figure 5.4).[16] Different from the previous case, the apical positions are free of steric hindrance; nevertheless, the anion does not locate in the apical coordination sites of the metal. In fact, the $^{19}F,^{1}H$-HOESY NMR spectrum shows selective interionic contacts between the fluorine atoms of the counterion and the bridging N-H, H3, and H3′ protons of the nitrogen ligand and a weak interaction with the dimethyl sulfoxide protons. This means that the preferential position of the counterion is close to the amine N-H proton of the nonplanar six-membered ring formed by the dpa coordinated to the metal. The N–H bridging proton has a strong electropositive character and gives a robust hydrogen bonding interaction with the counteranion, resulting in an electronic protection of the metal center. The average interionic solution structure is similar to that observed in the solid state for the analogous complex containing the $CF_3SO_3^-$ anion, where the presence of a strong hydrogen-bonding interaction between the amine hydrogen and an oxygen atom of $CF_3SO_3^-$ is clearly detected.

Host-Guest Adducts

While the formation of host-guest supramolecular adducts can be evaluated by classical NMR studies,[17] NOE investigations allow the relative orientation of the host

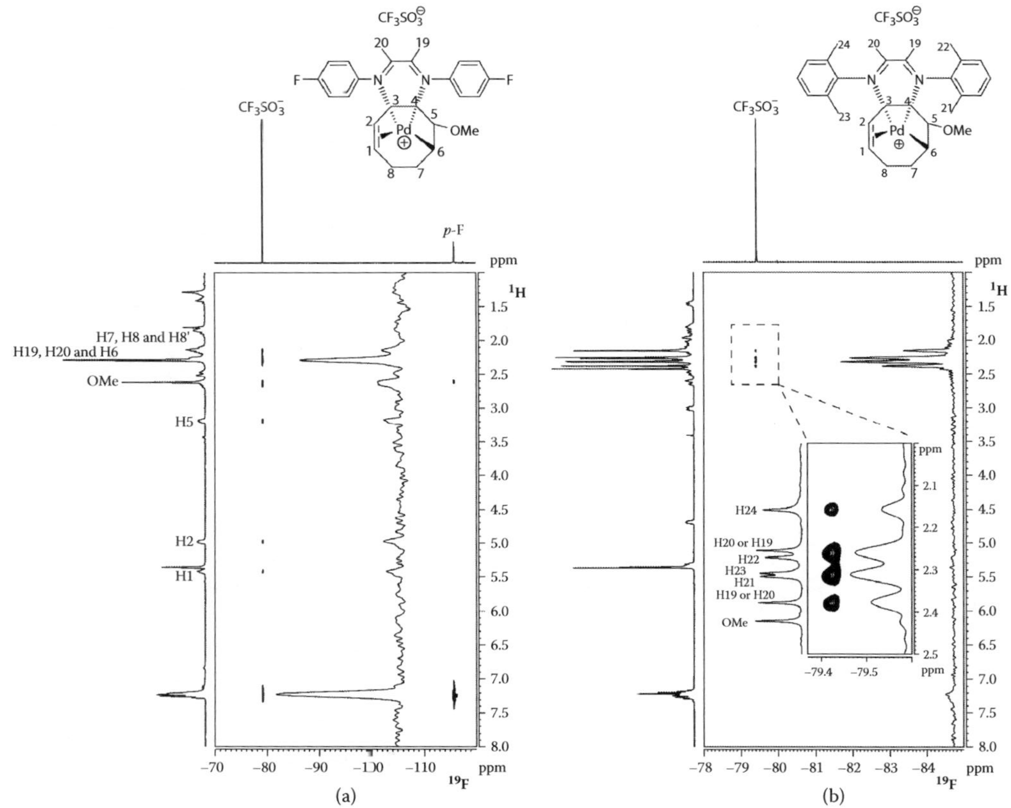

FIGURE 5.3 $^{19}F,^{1}H$-HOESY spectra (376.65 MHz, 273 K, CD_2Cl_2) of complexes **2b** (a) and **3b** (b). The vertical traces on the right of each spectrum are relative to the fluorine resonance of $CF_3SO_3^-$. (Adapted from Binotti et al., *Chem. Eur. J.*, 13, 1570–1582, 2007.)

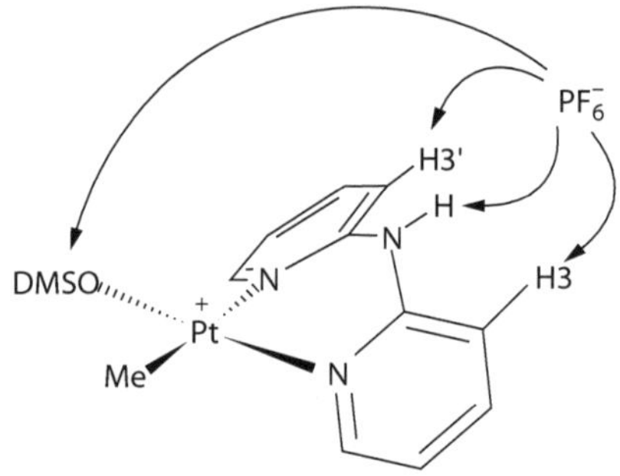

FIGURE 5.4 Sketch of the relative anion-cation orientation.

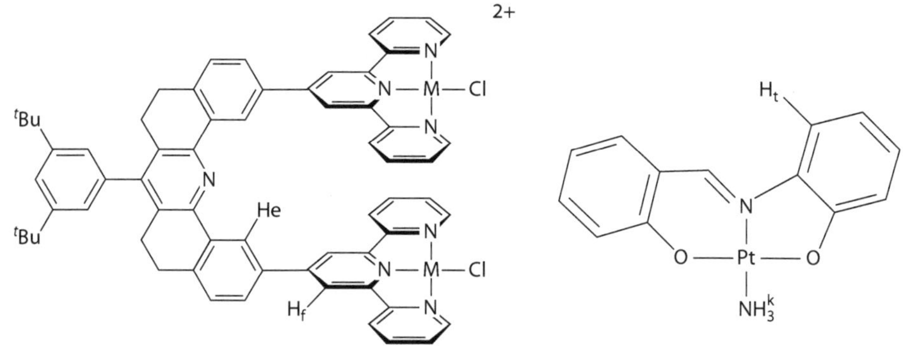

M = Pd (**4**), Pt (**5**)

6

SCHEME 5.2

and the guest within the supramolecular structure to be directly determined. For instance, let us consider the specific interaction between the molecular "cleft" **4** (or **5**) and the guest **6** shown in Scheme 5.2.[18] Association constants of 12,000 ± 2,900 M^{-1} and 51,000 ± 8,400 M^{-1} are determined for the adducts **4/6** and **5/6**, respectively, by ^{1}H-NMR titration studies.[19] These values are much larger than the first association constant of **4** with 9-methylanthracene (600 M^{-1}), suggesting an important role of the metal-metal bonding. The ^{1}H-NOESY NMR experiment gives a unique piece of information about the structure of the host-guest complex. First, it shows that **6** intercalates between the two terpyridyl units as found in the solid state by single-crystal x-ray diffraction. In addition, H_e/H_k and H_e/H_t NOEs are observed in the case of **4/6**, whereas only H_e/H_t NOE is observed in the case of **5/6**. This means that **6** can assume two different orientations with respect to the cleft **4**, but only one with the cleft **5** (Figure 5.5).

Another interesting example concerns the anionic organometallic hosts deriving from the supramolecular self-assemblies of four metal units (M = Ga^{III}, Al^{III}, In^{III}, Fe^{III}, Ti^{IV}, or Ge^{IV}) with six organic ligands [L = *N,N'-bis*(2,3-dihydroxybenzoyl)-1,5-diaminonaphthalene]. This affords well-defined tetrahedrons with the ligands spanning each edge and the metal ions occupying the vertices (Scheme 5.3).[20] These

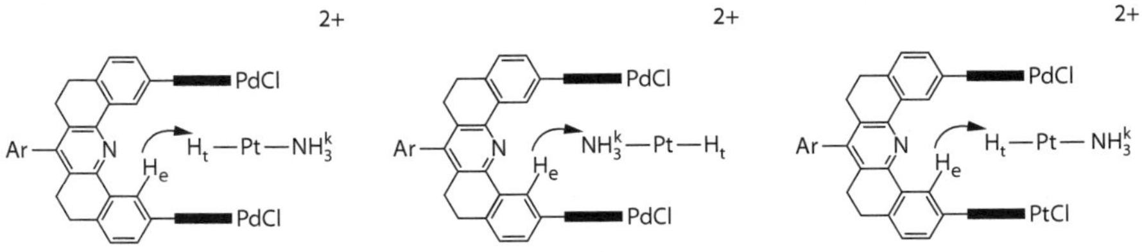

FIGURE 5.5 Different **4/6** and **5/6** host-guest orientations.

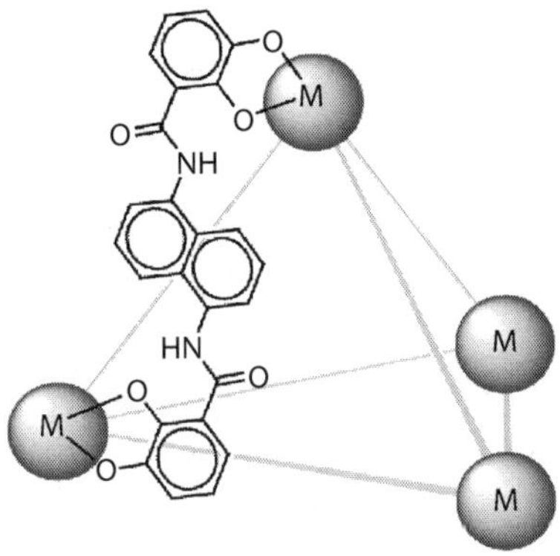

SCHEME 5.3

structures are highly soluble in water due to the high ionic character; nevertheless, they retain an internal hydrophobic pocket that can host a variety of cationic organic or organometallic guests. Independent of the nature of the guest [organic[21] as *N,N,N′,N′*-tetraethylethylenediamine-H^+ or organometallic[22] as $CpRu(\eta^6\text{-}C_6H_6)^+$], selective NOEs are observed between protons belonging to the guest and naphthalene protons of the host (Figure 5.6).

On the contrary, the guests and the cathecol protons of the host do not show any dipolar interaction, indicating that the naphthalene rings prevalently constitute the interior of the cavity, while the hydrogen atoms of the catechol rings are likely directed toward the exterior of the host. This conclusion is further supported by the supramolecular structure of a system containing both an encapsulated NEt_4^+ cation and an external organometallic cation [$Cp^*(PMe_3)Ir(Me)(PTA)^+$; PTA = 1,3,5- triaza-7-phosphaadamantane], the latter being too large to enter the host cavity.[23] While the strongest NOEs were observed between the organic cation and the naphthalene resonances of the host (Figure 5.7), smaller and selective interactions are detected between the exterior iridium complex and the catecholate protons of the ligands.

The last example of this section concerns the determination of the supramolecular structure of the host-guest assembly shown in Figure 5.8.[24] Quite surprisingly, the cationic tetrahedral coordination cage host (**7**$^{24+}$), having an $M_{12}L_6$ stoichiometry [M and L denote (en)Pd^{2+} coordination block (***Pd***) and 1,4-bis(3,5-pyrimidyl) benzene (**8**), respectively], accommodates the NBu_4^+ cationic guest (**9**$^+$) despite the 24+ charge of the host framework. The ^{1}H-NOESY NMR spectrum (Figure 5.8) of the inclusion complex shows a clear dipolar correlation between the C^fH_2 and C^gH_3

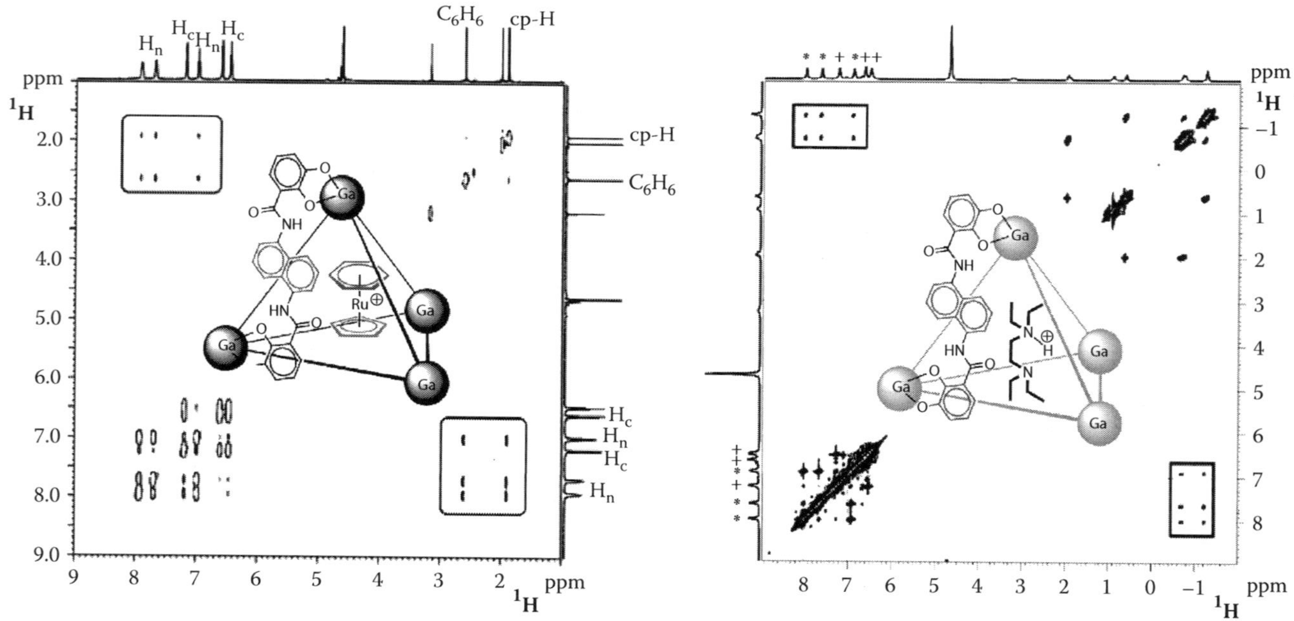

FIGURE 5.6 (See color insert following page 86) ^{1}H-NOESY spectra of the supramolecular adducts formed by the multianionic Ga_4L_6 host and organometallic (left) and organic (right) guests. HDO cross-peaks are deleted for clarity in both spectra. (Adapted from Pluth et al., *J. Am. Chem. Soc.*, 129, 11459–11467, 2007; Fiedler et al., *Inorg. Chem.*, 43, 846–848, 2004.)

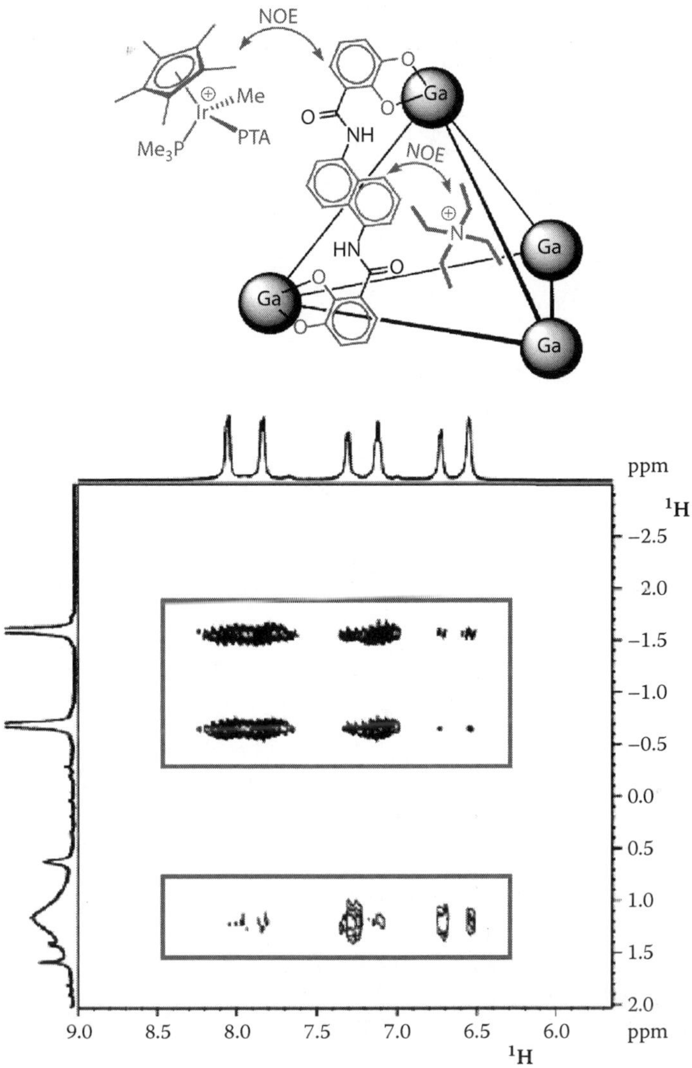

FIGURE 5.7 (See color insert) A section of the [1]H-NOESY spectrum of the anionic inclusion complex ($Ga_4L_6 \subset NEt_4^+$) ion paired with an external $Cp^*(PMe_3)Ir(Me)(PTA)^+$ organometallic cation. The red box highlights correlations between encapsulated NEt_4^+ and the three symmetry equivalent Ga_4L_6 host naphthyl protons. The blue box highlights correlations between broad exterior Cp* peaks of the organometallic cation at 1 ppm and the three symmetry equivalent Ga_4L_6 host catecholate protons. (Adapted from Leung et al., *J. Am. Chem. Soc.*, 128, 9781–9797, 2006.)

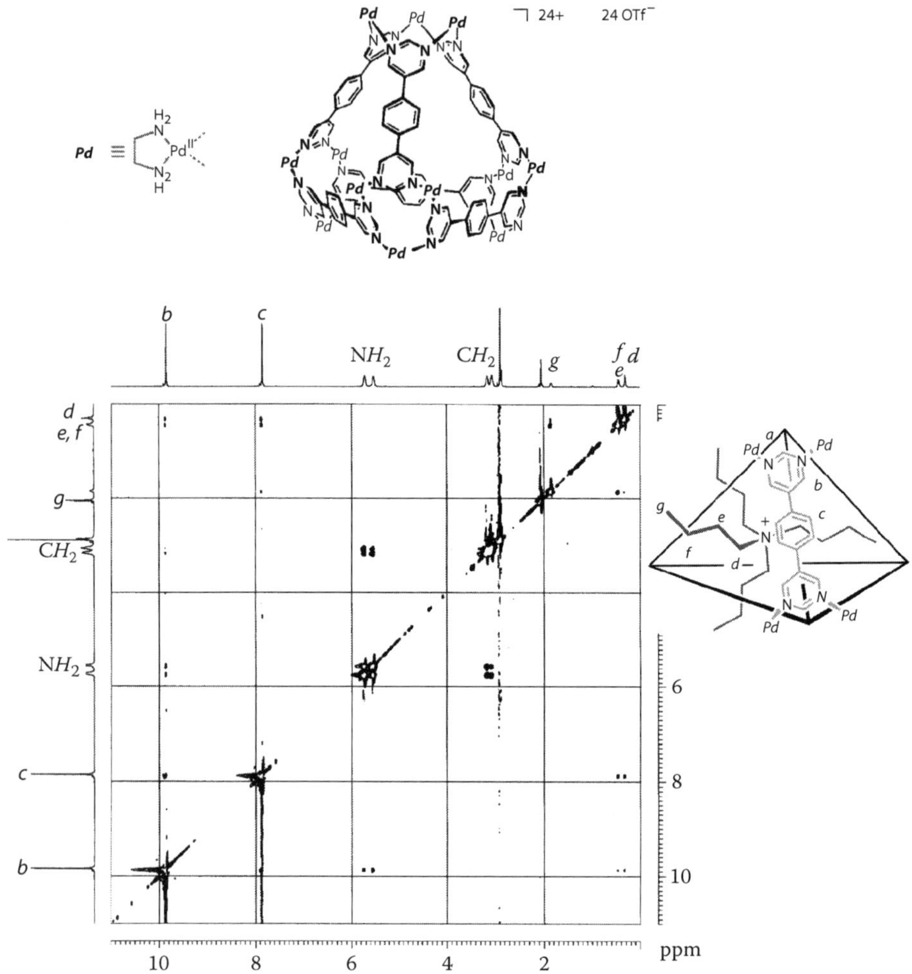

FIGURE 5.8 ^{1}H-NOESY NMR spectrum of the inclusion complex $[\mathbf{7} \subset \mathbf{9}]^{25+}$ showing the dipolar correlations between the protons belonging to the butyl groups and the phenylene protons of the molecular cage. (Adapted from Bourgeois et al., *J. Am. Chem. Soc.*, 125, 9260–9261, 2003.)

of the butyl group and the phenylene protons of **8**, indicating that the ammonium nitrogen is localized in the center of the tetrahedron $\mathbf{7}^{24+}$, with alkyl chains pointing out through the four triangular windows of the cage. X-ray investigations in the solid state explain the unusual inclusion of the cationic guest into a multicationic host: In fact, an interstitial layer formed by four BF_4^- anions is sandwiched between the host and the guest, resulting in a sort of onionlike supramolecular structure. Interestingly, the recognition pattern is very sensitive to the nature of the cation and anion: In the absence of BF_4^- the NBu_4^+ cation is not encapsulated by the host.

NOE Dependence on the Internuclear Distance

The explicit expression of the steady-state NOE for the simplest system consisting of two nonscalarly coupled spins, **I** and **S**, separated by a constant distance (r_{IS}) and tumbling isotropically in solution, is

$$NOE_I\{S\} = \frac{I - I^0}{I^0} = \frac{\gamma_S}{\gamma_I}\frac{W_{2IS} - W_{0IS}}{W_{2IS} + W_{1I} + W_{0IS}} = \frac{\gamma_S}{\gamma_I}\frac{\sigma_{IS}}{\rho_{IS}} \tag{5.2}$$

where γ_S and γ_I are the gyromagnetic ratios of **I** and **S**, respectively, while W_{0IS}, W_{1I}, and W_{2IS} are the transition probability of zero-, single-, and double-quantum pathways, respectively.[25] The term ($W_{2IS} - W_{0IS}$) represents the rate at which the NOE is transferred from **S** to **I** and is indicated by the symbol σ_{IS} (*cross-relaxation* rate constant). The quantity ($W_{2IS} + 2W_{1I} + W_{0IS}$) represents the *dipolar longitudinal relaxation* rate constant and is indicated by ρ_{IS}.[5] In the case of isotropic molecular tumbling, σ_{IS} and ρ_{IS} can be expressed as follows:

$$\sigma_{IS} = \left(\frac{\mu_0}{4\pi}\right)^2 \frac{\hbar^2\gamma_I^2\gamma_S^2}{10}\left(\frac{6\tau_C}{1+(\omega_I+\omega_S)^2\tau_C^2} - \frac{\tau_C}{1+(\omega_I-\omega_S)^2\tau_C^2}\right) r_{IS}^{-6} \tag{5.3}$$

$$\rho_{IS} = \left(\frac{\mu_0}{4\pi}\right)^2 \frac{\hbar^2\gamma_I^2\gamma_S^2}{10}\left(\frac{6\tau_C}{1+(\omega_I+\omega_S)^2\tau_C^2} + \frac{3\tau_C}{1+\omega_I^2\tau_C^2} - \frac{\tau_C}{1+(\omega_I-\omega_S)^2\tau_C^2}\right) r_{IS}^{-6} \tag{5.4}$$

where μ_0 is the permeability constant in a vacuum, ħ is the Planck's constant divided by 2π, τ_c is the rotational correlation time, and ω_I and ω_S are the resonance frequencies of **I** and **S** nuclei, respectively. Since both σ_{IS} and ρ_{IS} depend on the six inverse powers of the internuclear distance, the measurement of the steady-state NOE in this two-spin system cannot be directly related to the internuclear distances.[26] On the contrary, the measurement of the rate at which the NOE is transferred from **S** to **I** (kinetic of NOE buildup), in our ideal two-spin system, allows determining σ_{IS} and, consequently, can give direct information about the internuclear distance between **I** and **S** (r_{IS}).

The time course of the transient NOE experiments can be expressed by the following equation:

$$NOE_I\{S\}(\tau_m) = e^{-(R-\sigma_{IS})\tau_m}(1 - e^{-2\sigma_{IS}\tau_m}) \tag{5.5}$$

where R represents the total longitudinal relaxation rate constants of both **I** and **S** spins, assumed to be equal, and τ_m is the mixing time. The first derivative with respect to the mixing time, at time zero, indicates that the initial buildup rate of the NOE is only proportional to σ_{IS}:

$$\left.\frac{d(NOE_I\{S\})}{d\tau_m}\right|_{\tau_m=0} = 2\sigma_{IS} \tag{5.6}$$

It is extremely important to outline that even in a multispin system all enhancements behave as they were in a two-spin system approximation at the early stage of the NOE buildup. σ_{IS} can be *quantitatively* derived (1) by limiting the NOE studies to the linear buildup (Equation 5.6) of the **I** enhancement for short mixing times ($\tau_m \rightarrow 0$) or (2) by measuring the complete kinetics of NOE buildup and fitting the experimental data with Equation 5.5. Once σ_{IS} is known, the r_{IS} distance can be determined by Equation 5.3 after having evaluated the rotational correlation time (τ_c), the other variable on which σ depends. On the other hand, by comparing σ_{IS} with that relative to two nuclei (**A** and **B**), whose internuclear distance is known (r_{AB} = calibration or reference distance[27]), an estimation of r_{IS} can be obtained by Equation 5.7, assuming that the proportionality constant between σ and r^{-6} is the same for the two couples of nuclei (**IS** and **AB**):

$$\frac{\sigma_{IS}}{\sigma_{AB}} = \left(\frac{r_{IS}}{r_{AB}}\right)^{-6} \tag{5.7}$$

For couples of nuclei having the same nature ($\gamma_I\gamma_S = \gamma_A\gamma_B$), the latter condition is satisfied when **I-S** and **A-B** vectors have the same rotational correlation times.[28]

The Effect of Correlation Time and Internal Motions

The correlation time, associated with the molecular tumbling in solution, strongly influences the absolute magnitude of σ_{IS} and consequently of the NOE (Equation 5.3). In general, positive NOEs are observed for small, rapidly tumbling molecules, while negative NOEs result for large molecules. Under specific conditions (solvent, temperature, size of the molecular system) the NOEs can be very small or even vanish.[29]

Another layer of complexity comes from the presence of fast internal motions that can change the actual value of both τ_c and r_{IS}. For instance, several motions in non-covalently bonded compounds can be present: (1) overall rotation of the adducts and/or molecules constituting them, (2) adduct dissociation and formation, and (3) internal motions (Me group rotations around single bonds, chair-boat inversion of the six-member rings, etc.). From the formal point of view, the presence of motions different from the overall isotropic molecular tumbling can be described by substituting in the above equations σ_{IS}, r_{IS}^{-6}, and R with the corresponding mean values $\langle\sigma_{IS}\rangle$, $\langle r_{IS}^{-6}\rangle$, and $\langle R\rangle$. The type of expression that should be used for determining the mean values depends on the rate of the active motion and on the conformational distribution functions. The simplest situation appears in the presence of internal motions that only change the actual value of τ_c without altering the internuclear distances. In this case, only motions faster than overall motion contribute to the relaxation, but at the same time, they cause the correlation function to drop rapidly to lower values, making the relaxation process less efficient.[30] More complicated is the case in which the internal motions change the actual value of both τ_c and r_{IS}. When the motion is slower than the overall molecular tumbling, the corresponding effective distance sensed by the NOE is

$$r_{effective} = \left(\frac{1}{N}\sum_{\mu=1}^{N} r_{IS,\mu}^{-6}\right)^{-\frac{1}{6}} \tag{5.8}$$

the index μ indicating the different conformations assumed by the spin system. In this case, in which we assume that all the conformations are equally populated, the average distance is underestimated by the NOE measurements. On the other hand, when the motion is faster than the molecular tumbling the effective distance is

$$r_{effective} \geq \left(\frac{1}{N} \sum_{\mu=1}^{N} r_{IS,\mu}^{-3} \right)^{-\frac{1}{3}} \tag{5.9}$$

The symbol ≥ derives from the fact that the angular part of the correlation function can only reduce the size of the NOE.[31] In this case, assuming again an equal distribution of the different conformations, there are two different contributions to the actual value of $r_{effective}$ (i.e., of the NOE): The average on the radial part tends to underestimate the real average value of r_{IS}, while the angular part, as discussed above, tends to reduce the NOE and, consequently, overestimate the distance between I and S. Several models have been proposed in the literature:[32–36] While the conclusions are not at all consistent, all the authors agree that the number of cases where internal motions would lead to significant error is fairly small.[37]

Basic Pulse Sequences for Measuring NOEs

One of the basic pulse sequences used to measure the kinetics of NOE buildup is shown in Figure 5.9. Let us explain how it works. A shaped[38] selective π/2 radio frequency (rf) pulse followed by an unselective one inverts the population of the spin **S** levels and generates a transverse magnetization for **I**. During the mixing time (τ_m), the magnetization is transferred from **S** to **I** due to the competition between W_0 and W_2. Finally, NOE induced by magnetization transfer is read through the second unselective π/2 pulse. By alternating the phase of the initial selective and unselective pulses, the difference spectrum with NOE enhancement at **I** is directly obtained.[39] Both homonuclear (1D NOE)[40] and heteronuclear (1D HOE)[41] experiments can be carried out.

For complex systems with a large number of resonances and dipolar interactions it becomes preferable to use the 2D NOESY[42] experiments (or the heteronuclear analogue, called 2D HOESY[43]). The simplest and most commonly used pulse sequence for 2D NOESY experiments is shown in Figure 5.10.

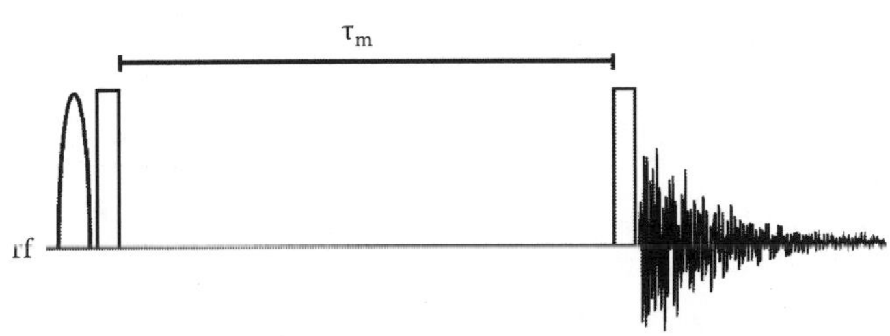

FIGURE 5.9 Basic pulse sequence for the 1D NOE experiment.

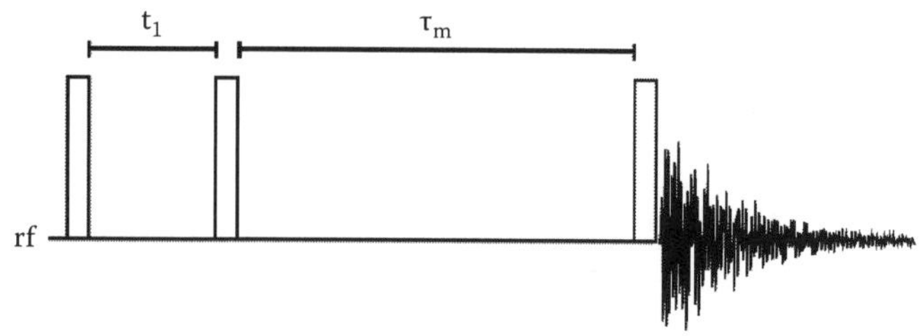

FIGURE 5.10 Basic pulse sequence for homonuclear 2D NOESY experiment.

The major advantage is that 2D NOESY experiments do not require selective irradiation, since frequency labeling (and consequently selective inversion) is obtained by appropriate variation of the t_1 period between the two unselective $\pi/2$ pulses. Thus a single experiment maps all the dipolar interactions in the system.

Examples on the Estimation of Internuclear Distances

Qualitative analysis of NOE data is usually sufficient to describe the intermolecular structure of supramolecular adducts. However, in particular cases, a more quantitative analysis could be needed to unambiguously describe the system. For instance, let us consider cationic ruthenium complexes **10** and **11** (Figure 5.11).[10] An NOE analysis based on triangulation reasoning indicates that the borate organic anions are localized on the side of the N-N ligand, as in the case of PF_6^- and BF_4^- described above; nevertheless, both *ortho*-aromatic and B-Me protons of the anion show NOE contacts with selected cationic resonances in the case of ion pair **11**. Thus it is not clear whether the unsymmetrical counterion orients the aryl or the methyl groups toward the cation. In order to solve this structural problem, the complete kinetics of NOE buildup have been recorded and average interionic distances determined by taking the *o*H–*m*H distance (2.47 Å) as a reference (Figure 5.11a). *o*H protons are 4.5–4.6 Å distant from the CH_2 moiety of the cation, while the methyl group of the unsymmetrical anion is, on average, 5.2 Å away from the same cationic group. The difference of 0.6–0.7 Å suggests that the unsymmetrical anion tends to direct the alkyl tail far away from the cation (Figure 5.11). In order to check the accuracy of the determined average interionic distances, rotational correlation times (τ_c) are estimated by using two independent NMR methodologies: ^{13}C relaxation measurements and the dependence of the ^1H-^1H cross-relaxation on the temperature.[10a] With both methodologies, it is found that intramolecular and interionic couples of nuclei have the same τ_c (ca. 100 ps) within the experimental error.

A similar structural problem has also been faced for metallocenium ion pairs, which are important catalysts for olefin polymerization.[44] Two kinds of ion pairs are involved in the catalytic process: an inner sphere ion pair (ISIP),[14] where the counterion occupies one of the coordination sites, corresponding to the resting state of the catalyst, and an outer sphere ion pair (OSIP), with the counterion in the second coordination sphere,[14] formed after the anion displacement by the monomer, corresponding to the propagating species. The structural interionic rearrangement occurring on passing from ISIP to OSIP is of fundamental importance to understanding

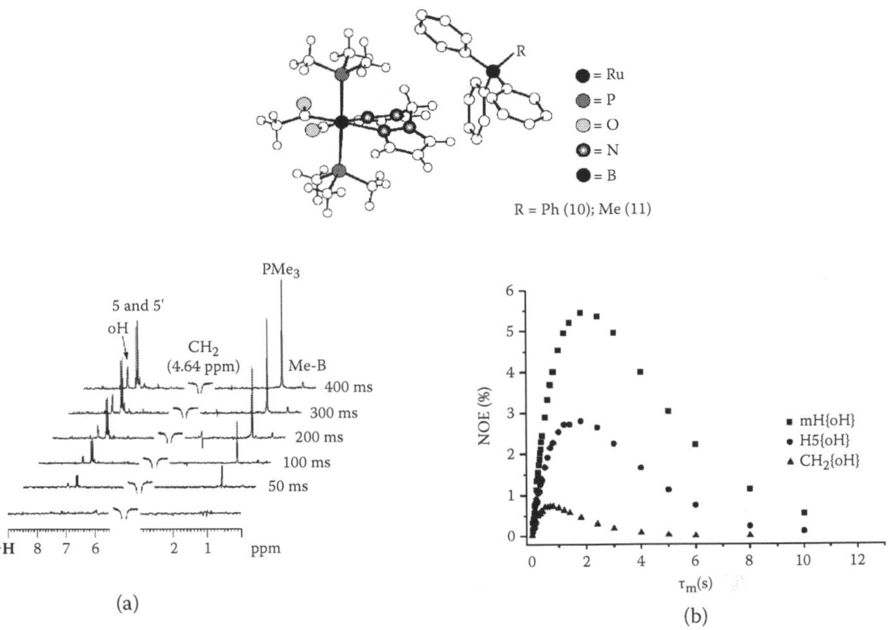

FIGURE 5.11 Top: Sketch of ion pairs **10** and **11**. (a) A series of 1D NOE spectra of complex **11** (400.13 MHz, 298 K, CD_2Cl_2) showing the inversion of CH_2 and the buildup of the enhancements on H5, H5′, and PMe_3 (intramolecular NOEs) and *o*H and B-Me (interionic NOEs). (b) Full NOE buildup curves on *m*H (intramolecular NOE) and H5 and CH_2 (interionic NOEs) after the irradiation of the *o*H resonance (complex **10**, 400.13 MHz, 298 K, CD_2Cl_2). (Adapted from Zuccaccia et al., *J. Am. Chem. Soc.*, 123, 11020–11028, 2001; Zuccaccia et al., *Organometallics*, 18, 1–3, 1999.)

the cation-anion structure-reactivity relationship.[45] Thus the interionic structures of the ISIP **12** and the corresponding OSIP **13**, in which THF has displaced the anion from the first coordination sphere, have been investigated by quantitative ^{1}H-NOESY and ^{19}F,^{1}H-HOESY experiments. Figure 5.12a shows the ^{1}H,^{19}F-HOESY spectrum of complex **12**.

o-F/H2 (4.1 Å) and B-Me/H2 (3.0 Å) are the shortest interionic distances, while interionic contacts were not observed for H4 and for the *m*-F and *p*-F nuclei of the anion. These data indicate that the anion steadily points the B-Me moiety toward the Zr center, as observed in the solid state for a number of zirconocenium ion pairs bearing the $MeB(C_6F_5)_3^-$ counterion. The 1D ^{1}H,^{19}F-HOESY spectrum of **13** is shown in Figure 5.12b. It is important to notice that (1) not only the *o*-F, but also the *m*-F fluorine atoms of the anion show interionic interactions with the cation in **13**, and (2) the shortest H-H interionic distance (4.1 Å) is between the B-Me and H1, approximately 1.1 Å longer than the shortest distance observed in **12** (B-Me/H2 = 3.0 Å). Despite the anion being much freer to assume various orientations with respect to the cation in **13**, the anion orientation in which the B-Me moiety points away from the metal center appears to be favored. Moreover, based on simple geometrical considerations, an average Zr-B distance of 7.2-7.3 Å is estimated for ion pair **13**. Dynamic

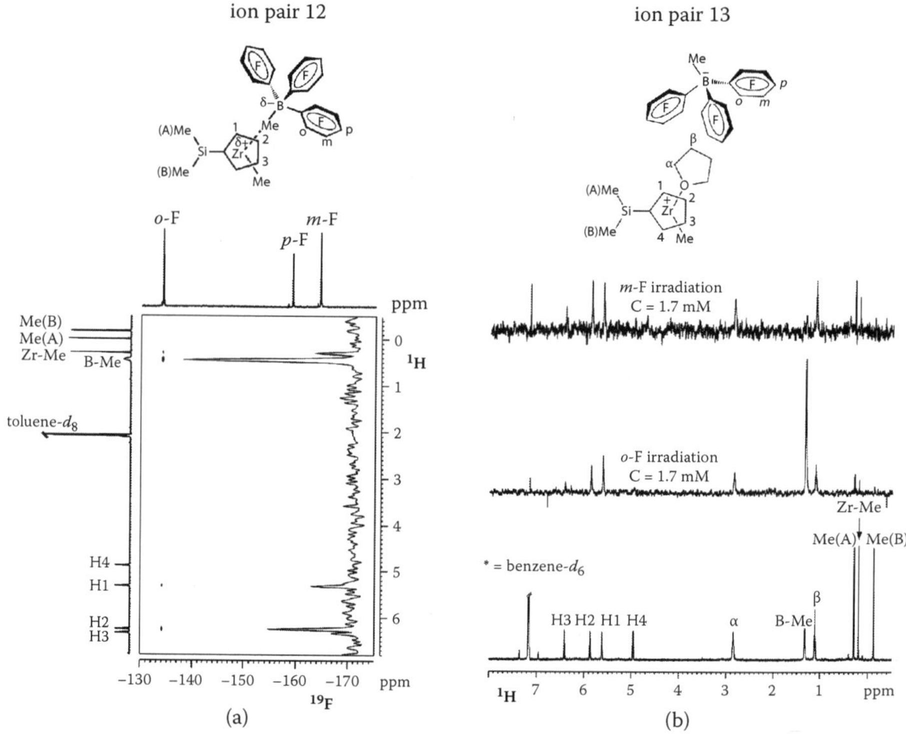

FIGURE 5.12 (a) A section of the ^{19}F,^{1}H-HOESY spectrum (376.4 MHz, 298 K, toluene-*d*8) of ion pair **12**. The F1 trace (indirect dimension) relative to the *o*-F resonance is reported on the right. (b) Proton (bottom) and 1D ^{1}H,^{19}F-HOESY spectra (middle: *o*-F irradiation; top: *m*-F irradiation) of ion pair **13** (399.94 MHz, 298 K, benzene-*d*6). (Adapted from Zuccaccia et al., *J. Am. Chem. Soc.*, 126, 1448–1464, 2004.)

atomistic simulations validate these conclusions.[46] Distribution functions of several metrical parameters are in excellent agreement with NOE data. Of particular interest is the value of the Zr-Me-B angle, which describes the relative orientation of the Me-B moiety with respect to the cationic fragment. At Zr-B distance of 4.2 Å, corresponding to ion pair **12**, the distribution function shows a sharp peak around 175°. A slight increase of the Zr-B distance to 5.25 Å causes only a modest broadening of the peak, which is now centered around 165° (Figure 5.13). At Zr-B = 6.25 Å, the distribution is considerably more flat, and angles around 90° become relatively populated. When the Zr-B distance is incremented to 7.25 Å, corresponding to ion pair **13**, a renarrowing of the angle distribution function is observed with a maximum around 10° (Figure 5.13). Consequently, the B-Me vector points away from the metal center, in remarkable agreement with the experimental conclusions based on NOE investigations.

The final example concerns the determination of H···F heteronuclear interionic distances and equilibrium constants between ion pairs, having two different anion-cation orientations, in square planar Pt(II) complexes bearing different

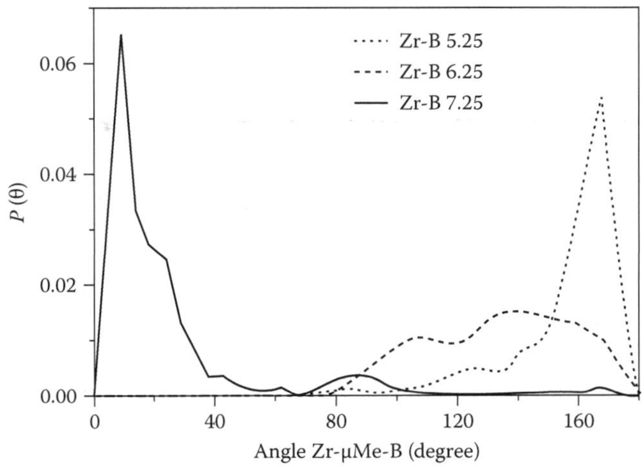

FIGURE 5.13 Distribution functions of the Zr-Me-B angle at various Zr-B distances, for an ion pair analogous to **12**, where hydrogen atoms have replaced the methyl groups on the silicon atom. (Reproduced from Correa and Cavallo, *J. Am. Chem. Soc.*, 128, 10952–10959, 2006. With permission.)

olefins (**15–18**, Figure 5.15) by means of quantitative $^{19}F,^{1}H$-HOESY NMR spectroscopy.[47,48] In these complexes, the anion is mainly located near the N=C-C=N moiety due to a partial accumulation of the positive charge in the backbone of the ligand and the steric protection of the metal center provided by the bulky isopropyl substituents. By using the CF_3···*m*-H distance as a reference, BF_4^-···H interionic distances are determined for the pseudo-cis cation-anion orientation (see below) of complex **16** recording a series of $^{19}F,^{1}H$-HOESY spectra as a function of the mixing time. It is concluded that an almost perpendicular disposition of the phenyl ring of the styrene allows the *ortho* proton to interact with the anion (Figure 5.14); an average BF_4^-···*o*-H distance of 4.4(8) Å is obtained in remarkable agreement with the value obtained by DFT calculations.[48]

In complex **15**, bearing propylene, the interionic interactions between BF_4^- and the four methyl groups that point toward the backbone of the ligand (indicated with b = backward in Figure 5.15) appear with the same intensity (Figure 5.15a). Consequently, the two sides, above and below the square planar coordination plane, are equally populated. On the contrary, the two methyl groups that lie on the same side with respect to the olefin substituent interact more strongly with the anion in complexes **16–18** (Figure 5.15b). This suggests that pseudo-trans (PT) and pseudo-cis (PC) cation-anion orientations are not equally populated in solution. By reasonably assuming that the interionic distances between the anion and all the methyl groups pointing backward are the same in PT and PC ion pairs, their relative abundance has been calculated by integrating the heteronuclear NOE cross-peaks. From the concentration, the equilibrium constant and the free energy gap between the two ion pairs can be evaluated. PC always results in being more stable than PT, with a difference in free energy of 1–3 kcal/mol in the temperature range 204–302 K.

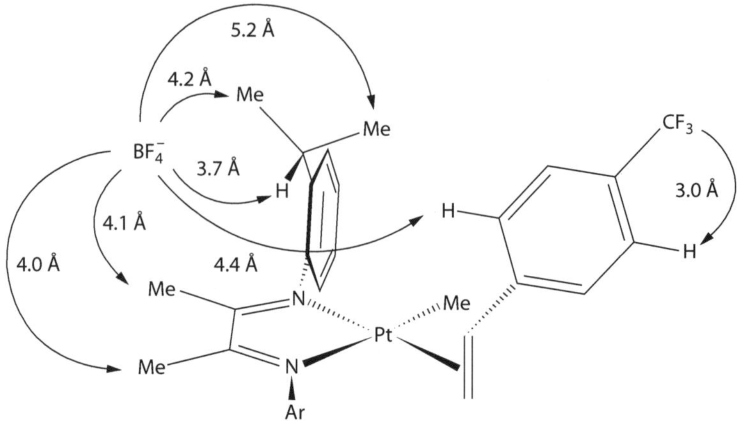

FIGURE 5.14 Sketch of representative average interionic distances in the ion pair of complex **16**. (From Macchioni et al., *New J. Chem.*, 27, 455–458. With permission of The Royal Society of Chemistry.)

EXCHANGE SPECTROSCOPY

As described above, NOE methods basically consist of the observation of an intensity variation caused by magnetization transfer under the presence of an external perturbation. Magnetization transfer can also occur by the physical transfer of a given nucleus between the two (or more) magnetically unequivalent sites. Under appropriate conditions it is possible to correlate the observed intensity variations with the rate of dynamic process that causes the chemical exchange.[5] This is the basis of the so-called exchange spectroscopy (or magnetization transfer spectroscopy). The methodology is similar to NOE: After the selective perturbation of the target resonance, intensity changes on both the inverted signal and the exchange-related resonance(s) are monitored as a function of the magnetization transfer delay. Thus the same (or similar) pulse sequences used for NOE investigations can be used for dynamic investigations.[49,50] The 2D EXSY (exchange spectroscopy) NMR technique is especially powerful for the investigation of chemical exchange in multisite systems, affording site-to-site rate constants, and thus providing indications on the exchange mechanism. For the simplest system consisting of two nonscalarly coupled spins, **A** and **B**, with equal longitudinal relaxation times, diagonal and off-diagonal cross-peak intensities are related to the rate constant k ($k = k_{AB} + k_{BA}$) by the following equation:

$$k = \frac{1}{\tau_m} \ln \frac{r+1}{r-1} \tag{5.10}$$

where τ_m is the evolution period and $r = (I_{AA} + I_{BB})/(I_{AB} + I_{BA})$ or $r = 4X_AX_B(I_{AA} + I_{BB})/[(I_{AB} + I_{BA}) - (X_A - X_B)^2]$ if the populations X_A and X_B of the exchanging sites are equal or unequal, respectively.[51] Magnetization transfer techniques are used to investigate dynamic processes whose rate constants are typically included between 10^{-2} and 10^2 s^{-1}. The upper limit is defined by the separation of the resonances in the

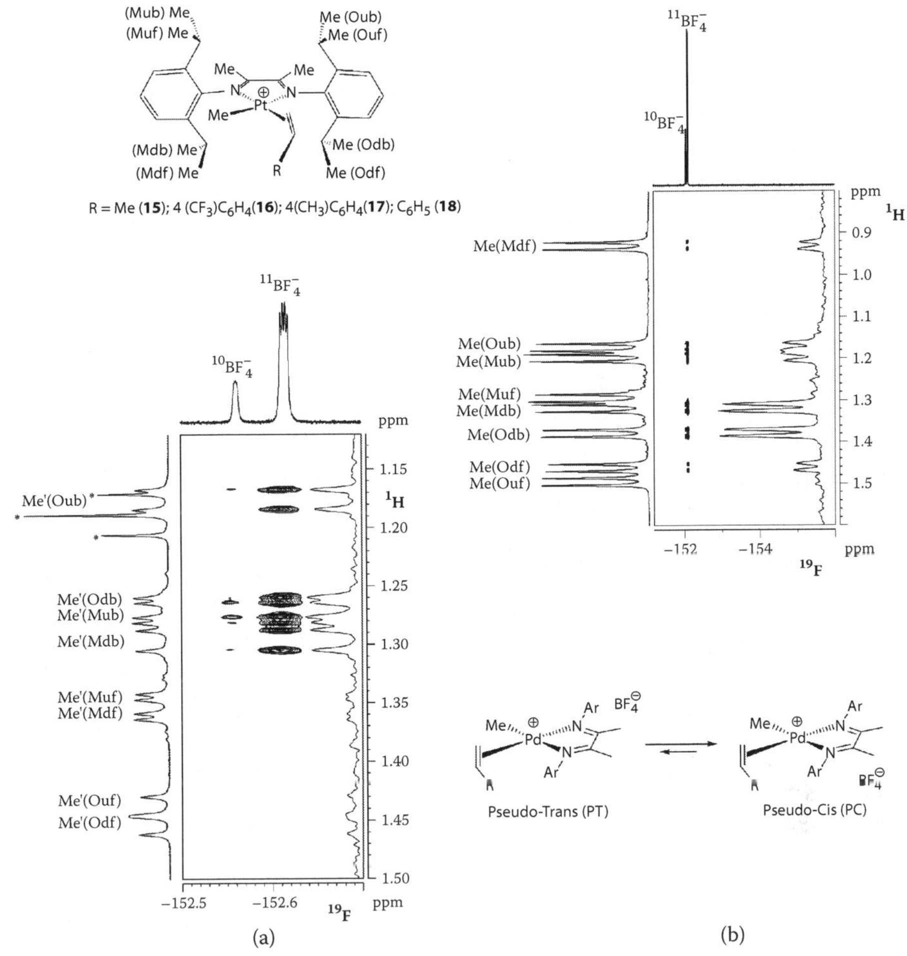

FIGURE 5.15 Top left: Sketch of ion pairs **15**, **16**, **17**, and **18**, including labeling scheme. M and O indicate groups that stay in the cis position with respect to methyl and olefin groups, respectively. u and d discriminate the up and down methyl orientations with respect to the olefin R group. Finally, b and f stand for backward and forward with respect to the plane containing the two phenyl groups (assumed to be coplanar). (a) A section of the $^{19}F,^{1}H$-HOESY spectrum (376.65 MHz, 302 K, $CDCl_3$) of ion pair **15**. The F1 trace (indirect dimension) relative to the $^{11}BF_4^-$ resonance is reported on the right. (b) A section of the $^{19}F,^{1}H$-HOESY spectrum (376.65 MHz, 302 K, $CDCl_3$) of ion pair **16**. The F1 trace (indirect dimension) relative to the $^{11}BF_4^-$ resonance is reported on the right. Bottom right: Proposed equilibrium between pseudo-cis (PC) and pseudo-trans (PT) ion pairs. (Adapted from Zuccaccia et al., *Organometallics*, 18, 4367–4372, 1999; Macchioni et al., *New J. Chem.*, 27, 455–458, 2003.)

spectrum, while the lower limit is dictated by the relative efficiency of magnetization transfer with respect to the longitudinal relaxation time. If the exchange rate is too slow, spins fully relax during the mixing period before any significant magnetization transfer can take place.

Examples of Supramolecular Host-Guest Exchange Processes

Dynamic properties play a fundamental role in determining the reactivity of supramolecular assemblies in solution. For instance, tetrahedral hosts of general formula M_4L_6 seen before[20] have been used as nanoscale molecular reactors to perform both stoichiometric[23,52] and catalytic[53,54] reactions with size, shape, and stereo-selectivity. In order for a reaction to occur inside the host structure, the catalyst (or the substrate) must remain encapsulated throughout the course of the transformation, and thus cannot exchange from the host cavity into the bulk solution faster than the rate of the reaction under consideration. Guest exchanges involving the substrate, the product, the catalyst, and any potential intermediates then become crucial in determining the overall performance of the systems. From the mechanistic point of view, the basic question pertinent to all self-assembled systems is whether the host undergoes partial (or total) breaking during the guest exchange reaction. Since thermodynamic parameters are kept constant, the latter information can be obtained by means of dynamic NMR investigations on guest exchange reactions involving the interchange of the encapsulated and nonencapsulated populations of the same chemical species (guest self-exchange). Selective inversion recovery methods have been utilized to investigate guest self-exchange reactions in a series of ammonium and phosphonium cations encapsulated within the Ga_4L_6 [L = *N,N′*-bis(2,3-dihydroxybenzoyl)-1,5-diaminonaphthalene] molecular host.[55] An example is given in Figure 5.16a: The resonance due to the terminal methyl groups belonging to the external NPr_4^+ cation (around 0.7 ppm) is selectively inverted and spectra recorded with increasing magnetization transfer delays (from bottom to top in Figure 5.16a). Magnetization is transferred from the exterior NPr_4^+ cation to the encapsulated one (triplet around −0.65 ppm) by chemical exchange, resulting in a reduced intensity of the latter resonance at intermediate delays. The corresponding rate constants are obtained by fitting the intensity profiles with the CIFIT computer program.[51] Moreover, experiments at different temperatures and pressures allow estimating the activation parameters ($\Delta H^{\ddagger}$, $\Delta S^{\ddagger}$, and $\Delta V^{\ddagger}$). The relatively low value of the activation enthalpies supports the idea that the M-L bond is not broken during the guest self-exchange reaction. In agreement, the activation entropies are all negative, indicating reduced degrees of freedom in the transition state. Finally, activation volumes are positive (higher values are observed for more sterically demanding guests; Figure 5.16b). From these findings, authors propose a convincing self-exchange mechanism in which the interior guest is squeezed out from one of the host apertures without rupture of the metal-ligand bond.

The last example of this section concerns the use of 2D EXSY NMR techniques for the characterization in solution of the self-assembling cage **19** (Figure 5.17).[56] Comparative ^1H, ^{19}F, and ^{31}P NMR indicate that the cage structure selectively and quantitatively encapsulates one of the eight $CF_3SO_3^-$ anions. Encapsulation is further confirmed by the appearance of selective contacts between the inner anion and the

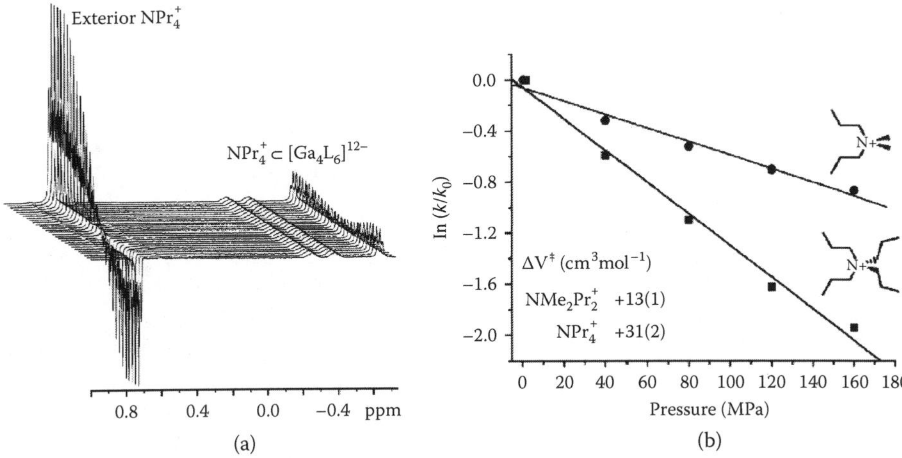

FIGURE 5.16 (a) ^{1}H-Selective inversion recovery (SIR) experiment showing the exchange of free (exterior) and $[Ga_4L_6]^{12-}$-encapsulated NPr_4^+ cations [L = *N,N′*-bis(2,3-dihydroxybenzoyl)-1,5-diaminonaphthalene]. The delay between the selective inversion pulse and spectrum acquisition increases from bottom to top. (b) Logarithmic plot of normalized exchange rate constants versus pressure for two different guests (NPr_4^+ and $NMe_2Pr_2^+$). Measurements were conducted in D_2O at basic pD and 303 K, 5 mM host ($[Ga_4L_6]^{12-}$). (Adapted from Davis et al., *J. Am. Chem. Soc.*, 128, 1324–1333, 2006.)

methylene hydrogen atoms oriented toward the interior of the cavity in the ^{19}F,^{1}H-HOESY NMR spectrum. The rate of anion self-exchange is measured for **19** by means of ^{19}F-EXSY spectroscopy, affording a rate constant $k_{in/out} = 0.32 \pm 0.08$ s^{-1} in $CDCl_3$ at 296 K. To extract information on the inclusion mechanism, the anion self-exchange rate is compared with the rate of the formation/dissociation process of the supramolecular assembly (kinetic stability). The latter can be obtained by investigating the chemical exchange between the free and assembled organic cavitand **20**. Thus ^{1}H-EXSY and ^{19}F-EXSY experiments are performed on a mixture of **19** and **20** (Figure 5.17). The results indicate that the rate constants for anion exchange are not affected by the addition of free **20** and, significantly, are the same as the exchange rate observed for free and coordinated cavitand **20**. Thus anion exchange and formation/dissociation processes are intrinsically related, suggesting that the anion can escape the interior of the cavity only when the cage dissociates. Further support comes from the observation that both processes are significantly and similarly accelerated ($k_{in/out} \approx k_{form/diss} \approx 5.2 \pm 0.8$ s^{-1}) on passing from a low polar solvent such as $CDCl_3$ to a high polar solvent such as CD_3CN. On the basis of the latter observation, the authors propose a dissociative polar transition state for the free/coordinated exchange of tetracoordinated cavitand ligand.[56]

DIFFUSION NMR

Translational self-diffusion[57] is the net result of the thermal motion induced by random-walk processes experienced by particles or molecules in solution. Diffusion

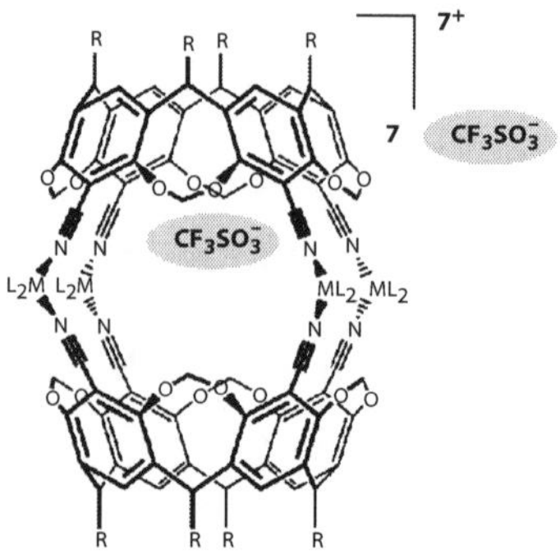

19 (R = $C_{11}H_{23}$; M = Pd; L = dppp)

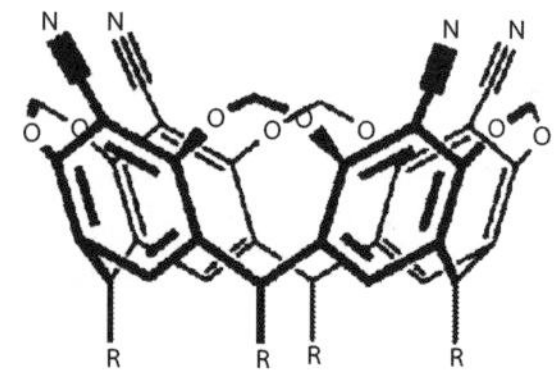

20 (R = $C_{11}H_{23}$)

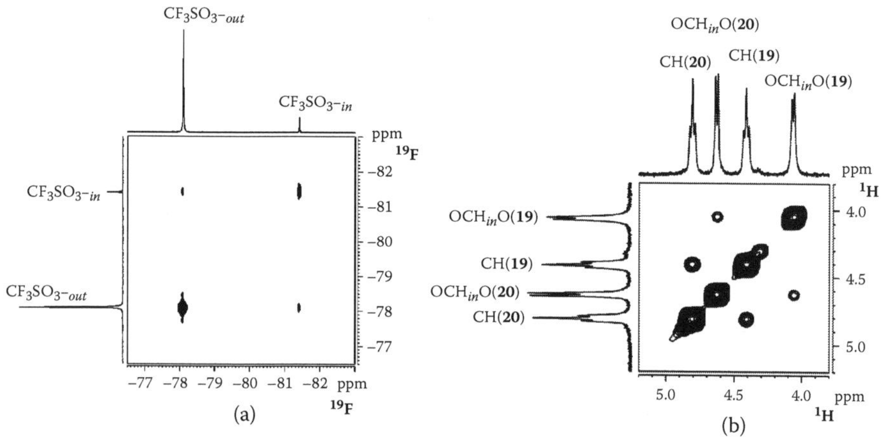

FIGURE 5.17 (a) ^{19}F-EXSY spectrum of the self-assembling cage **19** showing chemical exchange between free and encapsulated $CF_3SO_3^-$ anion (376.65 MHz, 296 K, $CDCl_3$). (b) Section of the ^{1}H-EXSY spectrum of **19** in the presence of **20**, showing the exchange between the assembled molecular cage and the free cavitand (400.13 MHz, 296 K, $CDCl_3$). (Adapted from Zuccaccia et al., *J. Am. Chem. Soc.*, 127, 7025–7032, 2005.)

NMR spectroscopy allows the self-diffusion translational coefficient (D_t), i.e., the translational diffusion coefficient of a species in the absence of a chemical potential gradient, to be determined. The advantages of using NMR methods to measure D_t rely on its noninvasive character, a relatively fast response, the opportunity to simultaneously investigate complex mixtures without the need of separation, and in general, the use of limited quantities of samples. The basic methodology is longstanding, and the subject has been periodically reviewed from both the theoretical and experimental points of view.[6]

As in the case of NOE and exchange spectroscopies, the experimental observable in NMR diffusion spectroscopy is an intensity variation. Specifically, the latter is caused by the controlled perturbation of the strength of the local magnetic field by means of pulsed field gradients. Let us consider the Larmor equation (Equation 5.11), which relates the precession or Larmor frequency (ω, radians s^{-1}) of the nuclear spin to the strength of the static magnetic field (B_0, T) along the z direction.

$$\omega = \gamma B_0 \tag{5.11}$$

When a spatially dependent magnetic field [G(z), (T m^{-1})] is present in addition to B_0, the actual precession frequency also becomes spatially dependent, according to

$$\omega(z) = \gamma(B_0 + G(z) \cdot z) \tag{5.12}$$

Consequently, a well-defined magnetic field gradient applied for a given period of time may be used to label the position of a given spin, providing the basis for measuring diffusion. To qualitatively illustrate the phenomenon, let us consider the simplest pulse sequence introduced by Stejskal and Tanner in the sixties[58] (Figure 5.18a) by modifying the Hahn spin-echo pulse sequence.[59]

The initial π/2 rf pulse rotates the magnetization from the z axis into the xy plane (Figure 5.18b); in the absence of pulsed field gradients, the transverse magnetization experiences a time-dependent phase shift due to chemical shift, hetero- and homonuclear J-coupling evolution, as well as spin-spin transverse relaxation (T_2). The application of a π rf pulse at time t = τ reverts the direction of the precession and an echo is formed at time t = 2τ. Acquisition of the echo decay and Fourier transformation afford the conventional NMR spectrum, where the intensities of each signal are weighted by their individual T_2 values and multiplets appear distorted due to homonuclear J-coupling. Let us now introduce the pulsed field gradients along z, assuming that molecules do not undergo self-diffusion (Figure 5.18c). A pulse gradient of duration δ and magnitude G is applied between the first π/2 and π pulses; it causes an additional dephasing of the magnetization that is now proportional to δ, G, and the z spatial position of each spin. Then, the precession of the magnetization is inverted by the π rf pulse at time t = τ, and a second pulse gradient, identical to the first, is applied. Since we have assumed no molecular diffusion, the second gradient pulse refocuses the magnetization, and the same echo as in the absence of gradient is obtained. In real systems, spins move during the period Δ (the delay between the two gradient pulses) as a consequence of self-diffusion; consequently, the refocusing is incomplete and a reduction of the echo amplitude is observed depending on the rate of diffusion (Figure 5.18d). This leads to an attenuation of the signals at time 2τ that depends on T_2, δ, G, Δ, γ, and the self-diffusion translational coefficient (D_t). In the case of rectangular-shaped gradient pulses, the signal intensity is described by[60]

$$I(2\tau) = I(0) \cdot e^{\left(-\frac{2\tau}{T_2}\right)} \cdot e^{\left(-\gamma^2 G^2 \delta^2 D_t \left(\Delta - \frac{\delta}{3}\right)\right)} \tag{5.13}$$

In Equation 5.13, I(2τ) is the signal amplitude at time 2τ and I(0) is the signal amplitude that would be observed immediately after the first π/2 rf pulse. In principle, D_t

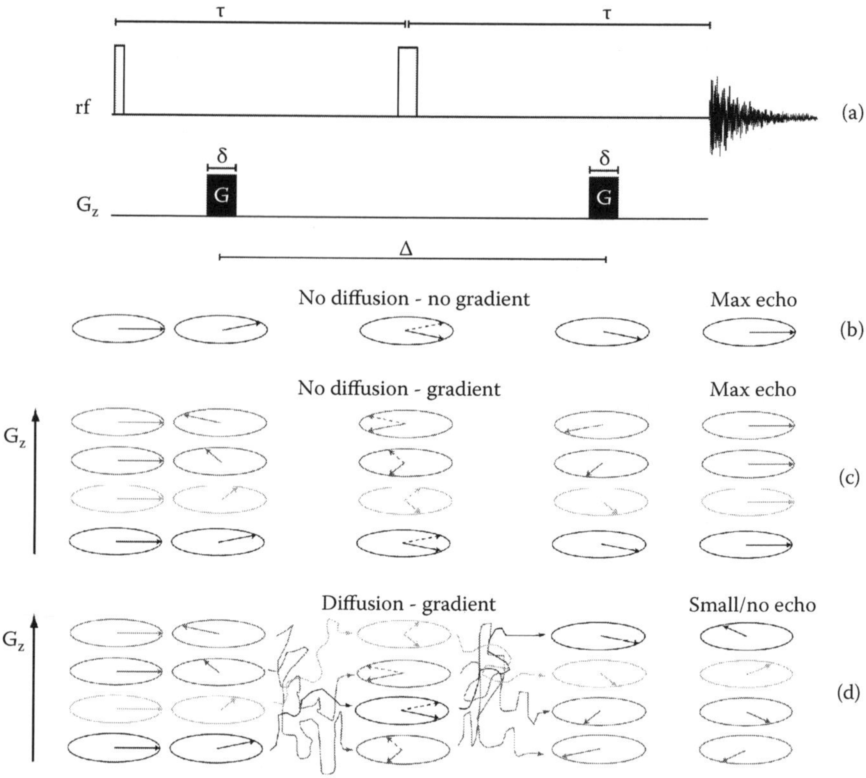

FIGURE 5.18 (See color insert) (a) Pulse sequence for the PGSE experiment. (b–d) Vectorial representation of the time evolution of the magnetization in the *xy* plane in the reference rotating frame in the absence of diffusion and gradients (b), absence of diffusion and presence of gradients (c), and presence of diffusion and gradients (d). Colors (black, green, red, and blue) represent hypothetical sample slices subjected to different values of the gradient strength.

can be obtained by varying one of the experimental parameters, namely, δ, Δ, or G. In practice, as evident from Equation 5.13, the attenuation due to relaxation can be more easily separated from that due to diffusion by performing a series of experiments keeping τ constant. Under these experimental conditions, the echo attenuation is expressed by Equation 5.14, which can be used to derive D_t:

$$I(2\tau) = I(2\tau)_{G=0} \cdot e^{\left(-\gamma^2 G^2 \delta^2 D_t \left(\Delta - \frac{\delta}{3}\right)\right)} \quad (5.14)$$

A number of pulse sequences more sophisticated than that originally introduced by Stejskal and Tanner have been developed. Excellent reviews have been reported describing them in detail.[6] Herein we just briefly mention the pitfalls the most commonly used sequences are able to bypass. Particularly, the most frequently used stimulated echo experiment[61–63] allows the distortion due to J-coupling and problems connected with the detection of nuclei with short T_2 to be partially overcome. The problem of eddy currents,[6d] generated by rapid rise and fall of gradient pulses, can be minimized by using (1) shielded gradient coils[64] and preemphasis,[65] or (2) longitudinal eddy current delay (LED) special sequences,[66] possibly incorporating self-compensating bipolar gradients.[67] Finally, special pulse sequences[68,69] have been developed in order to minimize convective motions (easily detectable by performing experiments with different Δ values[70,71]) due to temperature gradients derived from the temperature control of NMR machines. Convection motions can be also reduced by using NMR tubes with smaller diameters[72] or specially designed.[73] In optimal cases, diffusion coefficients can be measured with an overall uncertainty lower than 0.4%.[74]

Data Processing

A series of NMR free induction decays (FIDs) on varying G and maintaining δ and Δ constant are usually acquired to estimate D_t according to Equation 5.14 (or analogous ones derived for more sophisticated pulse sequences).[75] Their Fourier transformations lead to a series of spectra in the frequency domain (for instance, see Figure 5.19) where the resonance intensities are modulated by G and D_t.

D_t can be extracted by fitting I/I_0 or $\ln I/I_0$ versus G^2 trends by means of Equation 5.14. Linear regression ($\ln I/I_0$ versus G^2) is most frequently used since it immediately allows visualizing the relative D_t values (and thus size) of different species present in solution (Figure 5.20a). This is just looking at the slope of the obtained straight lines. On the other hand, nonlinear I/I_0 versus G fits are preferable when the signal/

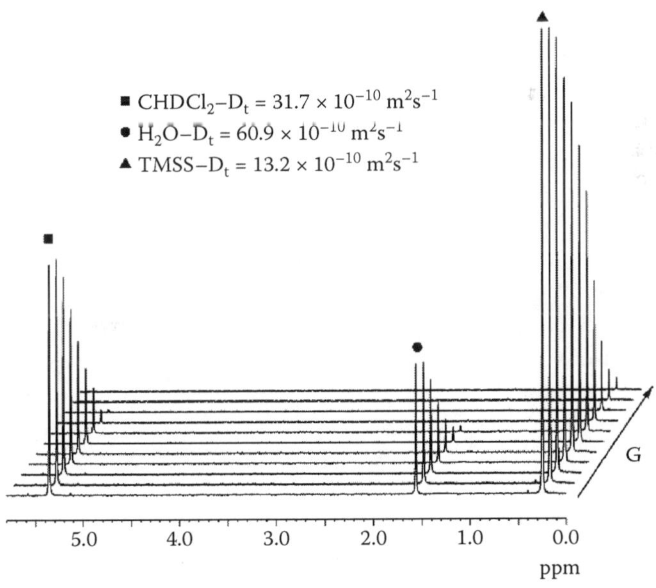

FIGURE 5.19 Dependence of the resonance intensities on G for a solution of TMSS and H_2O in CD_2Cl_2. (Reproduced from Macchioni et al., *Chem. Soc. Rev.*, 37, 479–489, 2008. With permission of the Royal Society of Chemistry.)

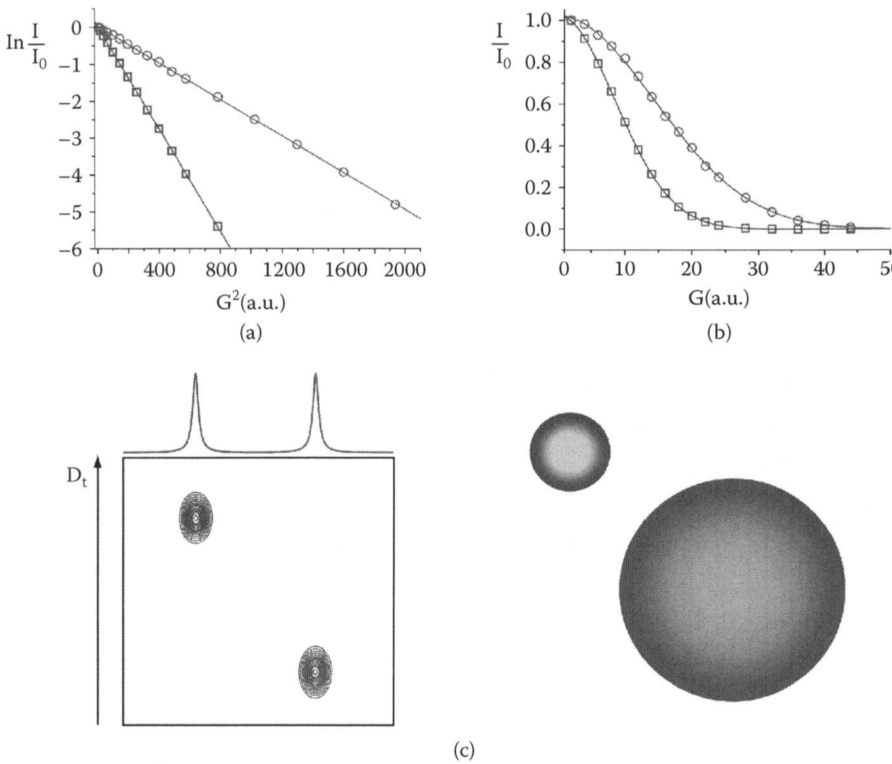

FIGURE 5.20 (See color insert) Schematization of PGSE data editing for a hypothetical mixture composed of a large (red), slow-diffusing species and a small (blue), fast-diffusing species.

noise ratios are low, in order to avoid incorrect weighting of standard deviations (Figure 5.20B).[76]

Diffusion NMR data can be processed in order to obtain a 2D map (Figure 5.20C) in which the horizontal axis encodes the chemical shift of the nucleus observed; the vertical dimension, however, encodes the diffusion constant D_t. This procedure of editing diffusion data is termed diffusion-ordered spectroscopy (DOSY) and is gaining popularity due to the default implementation of suitable processing software in modern NMR spectrometers. The analysis of the DOSY spectra for a mixture of samples of different sizes becomes rather simple. A sort of chromatographic map is obtained in which horizontal rows refer to species having the same D_t and, usually, mass (Figure 5.20c).

All of the above considerations assume that single molecules are present in solution. What happens if supramolecular adducts are present? In such a case, the sample is constituted by a distribution of non–covalently bounded species (monomers, dimers, trimers, etc.) that are in equilibrium with each other, and a single set of resonances is usually observed. If the rate of equilibration is higher than the timescale of diffusion experiments, little varies in terms of data processing: A straight line, a single exponential, or a single horizontal row is observed, processing the data as reported in Figure 5.20a–c, respectively. The derived D_t is an average numerical value of all D_t values of single species. On the contrary, if the rate of equilibration is slower than the timescale of diffusion experiments, nonlinear $\ln(I/I_0)$ versus G^2 and multiexponential I/I_0 versus G trends are observed. In DOSY spectra several horizontal rows are visible. Equation 5.14 (or analogous ones derived for more sophisticated pulse sequences) becomes inadequate to treat the data and has to be replaced by

$$I(2\tau)_{G,\upsilon} = \sum_{i=1}^{N} I(2\tau)_{i,G=0,\upsilon} \cdot e^{\left(-\gamma^2 G^2 \delta^2 D_{t(i)}\left(\Delta - \frac{\delta}{3}\right)\right)} \tag{5.15}$$

where $D_{t(i)}$ are the individual self-diffusion translational coefficients of the N supramolecular aggregates.[76]

From D_t to the Hydrodynamic Dimensions

In previous paragraphs, we illustrated how NMR methods can be used to measure the molecular self-diffusion translational coefficient D_t in solution. It is rather intuitive that smaller particles diffuse faster than larger ones. For rigid spherical particles of colloidal dimensions (much higher than those of the solvent), moving with uniform velocity in a fluid continuum of viscosity η, D_t is quantitatively related to the hydrodynamic dimensions through the Stokes-Einstein equation:

$$D_t = \frac{kT}{6\pi\eta r_H} \tag{5.16}$$

where T is the temperature, k is the Boltzmann constant, and r_H is the hydrodynamic radius of the diffusing particles. The precautions that must be taken to obtain accurate hydrodynamic dimensions have been described in detail in a recent review[77] and are herein just recalled. There are two general sources that contribute to increase the inaccuracy of diffusion NMR measurements:[78] (1) wrong evaluation of the experimental parameters (temperature, fluid viscosity, and strength of the pulsed gradient field) and (2) apparent inapplicability of the Stokes-Einstein equation connected to the investigation of smaller diffusing particles of nonspherical shape. Problem 1 can be solved by using an internal standard that can be the solvent itself or, more frequently, tetramethylsilane (TMS),[79] TMSS,[78] or tetrakis(para-tolyl)silane.[80] An internal standard must have at least one resonance falling in a free region of the NMR spectrum. It must be soluble in the most commonly used solvents, be thermally stable, be as chemically inert as possible, and it must not self-aggregate.[77] In order to bypass problem 2, derived from investigating systems that do not respect the requirements of the Stokes-Einstein equation,[81] correcting factors are used that account for nonspherical particles (shape factor, f_s) of comparable or slightly larger dimensions than the solvent (size factor, c). A modified Stokes-Einstein equation is obtained:

$$\left(\frac{D_t^{30}}{D_t^{29^{+/-}}}\right)^3 = \frac{V_H^{29^{+/-}}}{V_H^{30}} \tag{5.17}$$

The shape factor, f_s, is always larger than 1 and can be theoretically computed for prolate and oblate ellipsoids,[82] and even for ellipsoids with three different semi-axes.[83] The size factor, c, can be derived from the microfriction theory of Wirtz that takes into account the relative hydrodynamic dimensions of the diffusing species and solvent (r_{solv}).[84]

Examples of Applications of Diffusion NMR Techniques

Semiquantitative Investigation of the Dendrimer Size

Let us start with a simple PGSE NMR investigation of three different ferrocene phosphine dendrimers (Figure 5.21) carried out neglecting shape and size factors.[85] Increasing the size of the dendrimer, a decreasing of the absolute value of the slope of the $\ln(I/I_0)$ versus G^2 trends (and thus of D_t) is observed (Figure 5.21). If the

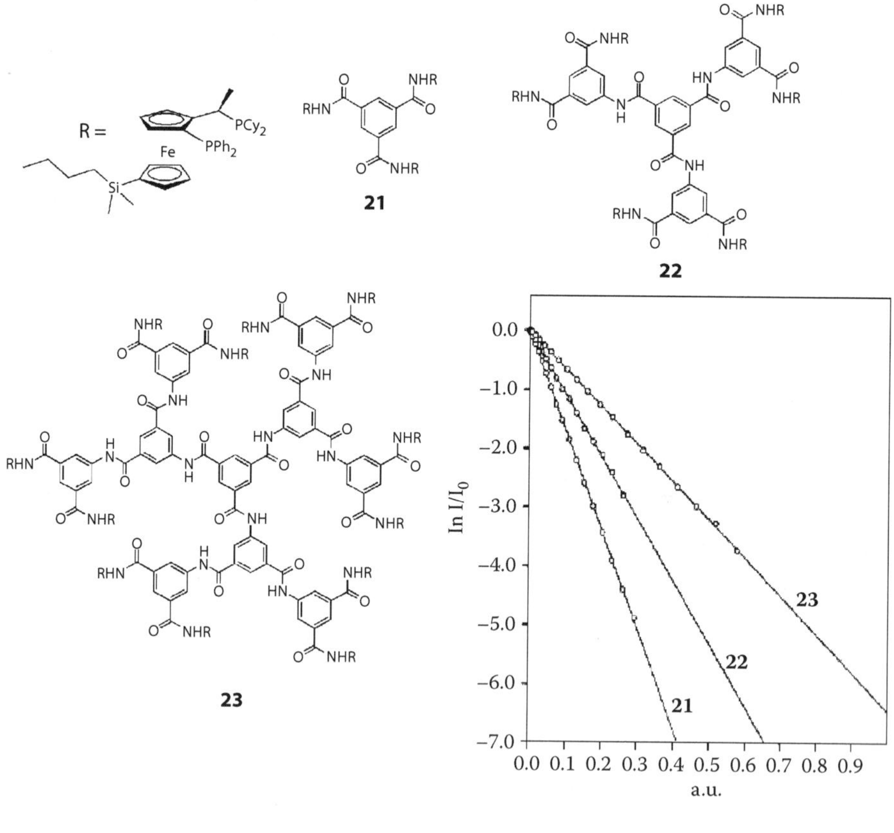

FIGURE 5.21 Plot of $\ln(I/I_0)$ versus G^2 in arbitrary units for **21**, **22**, and **23**. (Adapted from Valentini et al., *Organometallics*, 19, 2551–2555, 2000.)

Stokes-Einstein equation (Equation 5.16) is taken into account, a quick and relative semiquantification is possible. It results that the radius of **21** is about 1.6 times smaller than that of **22** and 2.2 times smaller than that of **23**.

Monomer-Dimer Discrimination

Another interesting application of diffusion NMR techniques concerns the determination of the nuclearity of complex **24**. While authors knew that **25** was a mononuclear complex,[86] they had no idea about that of **24**: In fact, two possible structures could be hypothesized (**24** and **24$_d$**), practically undistinguishable by ^{1}H NMR spectra. The measured D_t value is 1.22 times smaller than that for **25**, supporting the **24$_d$** structure (Figure 5.22).

Noncovalent Dimerization

Complexes **26** and **28** (Figure 5.23) are effective catalysts for the reduction of ketones via transfer hydrogenation.[87] The key of their catalytic activity lies in the presence of both alcoholic and hydridic hydrogen atoms. It is believed that hydrogenation occurs in the second coordination sphere of the catalyst due to the simultaneous interaction of the oxygen and carbon of the ketone with the alcoholic and hydridic functionalities, respectively. In-depth kinetic studies on the loss of H_2 by **28** in toluene at high temperature, leading to the formation of **26**, indicate that the reaction mechanism is different in the presence and absence of ethanol.[88] In particular, they suggest the formation of a dimeric intermediate in the latter case. This has been demonstrated by performing diffusion NMR measurements. In fact, **28** in toluene shows a D_t comparable with that of **26** and smaller than the one of **27** that can be reasonably taken as a monomeric reference. A noncovalent dimer with a structure similar to that indicated as **28$_d$** forms in toluene solution. D_t of **28** becomes similar to that of **27** when ethanol is added to toluene, probably due to breaking of intermolecular hydrogen bonds.

Ion Pairing and Further Aggregation

Diffusion NMR techniques have been extensively used to investigate the level of aggregation in solution of organic[89] and organometallic[90–93] salts in recent years. For instance, let us consider the ionic compound **29** having the cation (**29$^+$**) and anion (**29$^-$**) of similar sizes, and its neutral analogue **30**, which is isosteric with **29$^+$** (Figure 5.24).[94]

In nitromethane (an aprotic, little coordinating, and polar solvent) all the species have more or less the same D_t, while in chloroform the D_t of **29$^+$** and **29$^-$** are the same but markedly smaller than that of **30** (Figure 5.24). Reasonably assuming that **30** does not undergo associative processes, the level of aggregation can be evaluated by the following equation:

$$\left(\frac{D_t^{30}}{D_t^{29^{+/-}}}\right)^3 = \frac{V_H^{29^{+/-}}}{V_H^{30}} \tag{5.18}$$

where V_H is the hydrodynamic volume (calculated assuming spherical objects). The ratio in Equation 5.18 gives an approximated evaluation of the average size of the

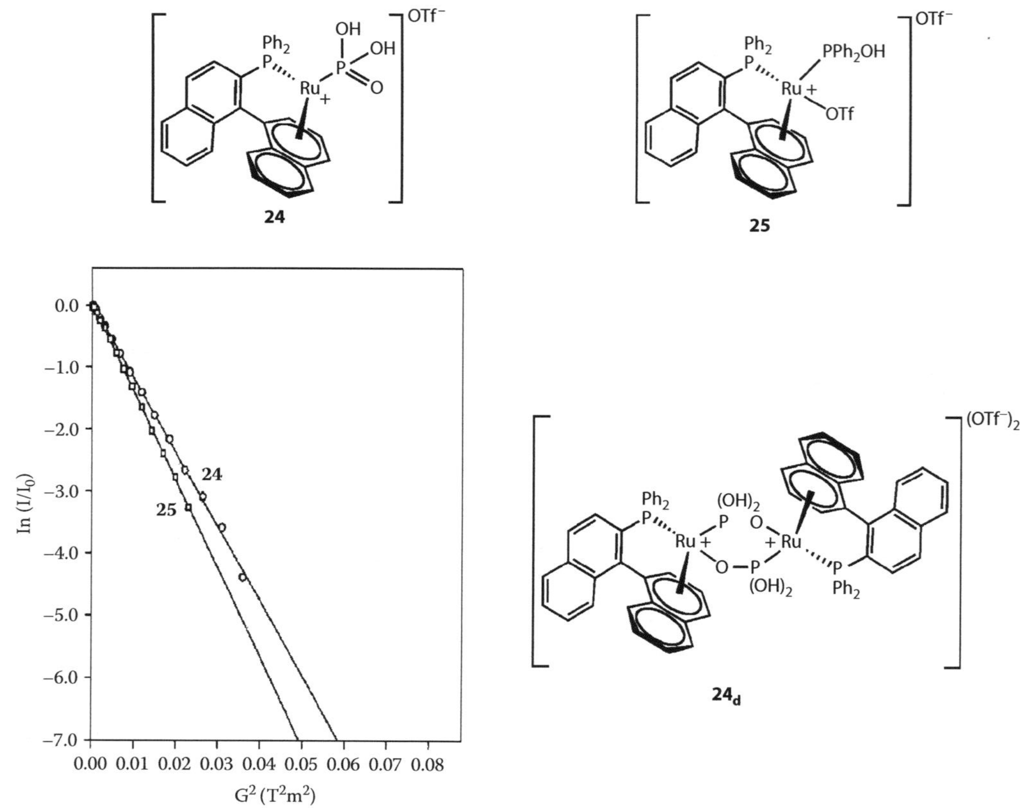

FIGURE 5.22 Plot of $\ln(I/I_0)$ versus G^2 for **24** and **25**. **24_d** is a dimeric structure of **24**. (Adapted from Valentini et al., *J. Chem. Soc. Dalton Trans.*, 4507–4510, 2000.)

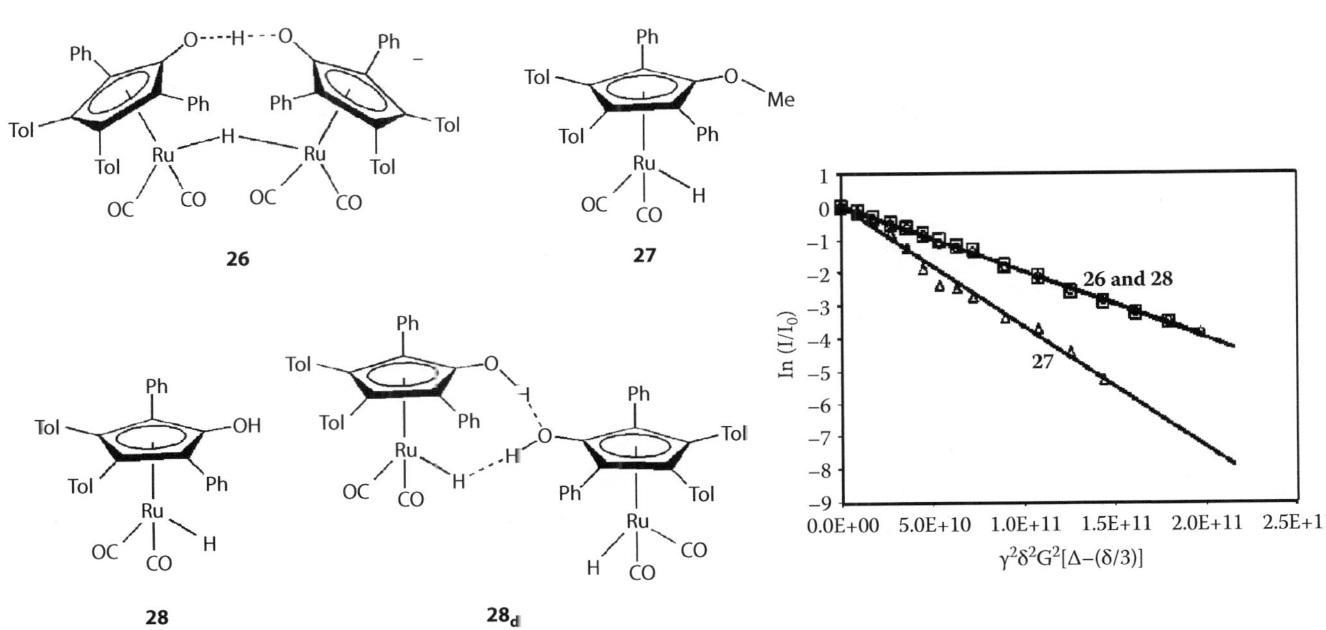

FIGURE 5.23 Plot of $\ln(I/I_0)$ versus G^2 for **26**, **27**, and **28** (Adapted from Casey et al., *J. Am. Chem. Soc.*, 127, 3100–3109, 2005.)

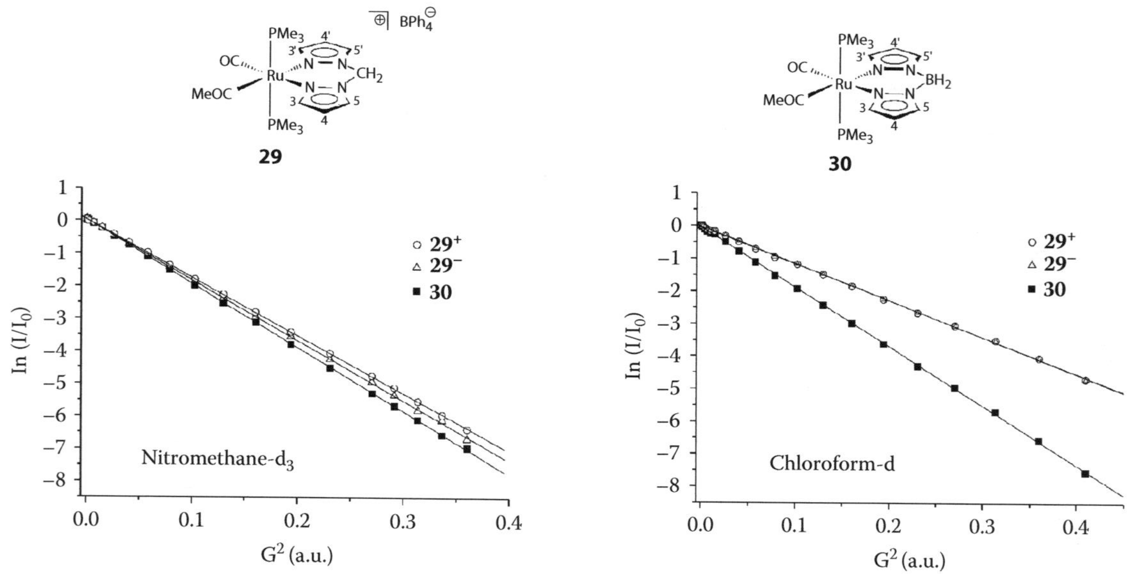

FIGURE 5.24 Plot of $\ln(I/I_0)$ versus G^2 (a.u. = arbitrary unit) for the PMe_3 resonances of **29**$^+$ and **30** and for the *m*-H resonance of **29**$^-$. (Adapted from Zuccaccia et al., *Organometallics*, 19, 4663–4665, 2000.)

ionic species: Since each ion has a size similar to that of the standard, if the ratio is equal to 2, ion pairs are mainly present. Applying this equation, it is deduced that **29** is mainly present as an ion pair in chloroform down to a concentration of 0.47 mM; ion quadruples are relevant at 84 mM, while free ions are predominant in nitromethane.

Quantum Dots

Diffusion NMR has been used to evaluate the size of functionalized nanoparticles and quantum dots (QDs), obtaining results that are consistent with those found with other techniques, such as small-angle x-ray scattering (SAXS), and scanning tunneling (STM) and atomic force microscopy (AFM).[95] For instance, let us consider the displacement of the trioctylphospine oxide (TOPO) bound to the surface of CdSe/TOPO QDs by poly(2-(*N,N*-dimethylamino)ethyl-methacrylate (PDMA).[96]

While a solution of CdSe/TOPO QDs exhibits a single D_t, a biexponential trend is observed when free TOPO is present in addition to that bound to the QDs' surface (Figure 5.25). This means that the exchange between free and bounded TOPO is slow compared to the diffusion experiment timescale, making it possible to estimate the fraction of free and bounded TOPO from diffusion experiments.

The latter possibility has been used to quantitatively evaluate the displacement of TOPO by PDMA (Figure 5.26). If the concentration of PDMA is twice that of QD, the displacement is nearly complete. Interestingly, at low polymer binding, PDMA enhanced the colloidal stability of the QD, and the hydrodynamic radius is only slightly increased. On the contrary, when the polymer loading is high, the hydrodynamic radius is substantially increased. Authors propose a different binding mode of the polymer in the two cases:[96] When the polymer loading is low, each polymer binds the QD with different anchor points, and thus it is smeared on the surface of the QD; when the polymer loading is high, the number of anchor points decreases and the polymer forms small loops and tails, which contribute to the hydrodynamic radius (Figure 5.27).

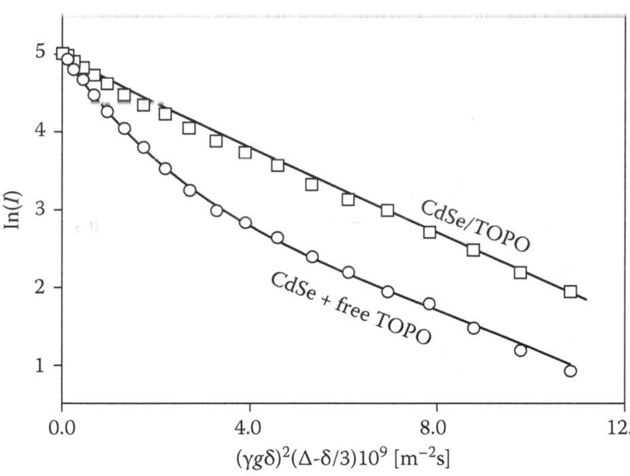

FIGURE 5.25 Plot of ln(I) versus G^2 of the TOPO NMR signal for the functionalized CdSe/TOPO QD alone and the mixture CdSe/TOPO QD + free TOPO. (Reproduced from Shen et al., *J. Phys. Chem. B*, 112, 1626–1633, 2008. With permission of the American Chemical Society.)

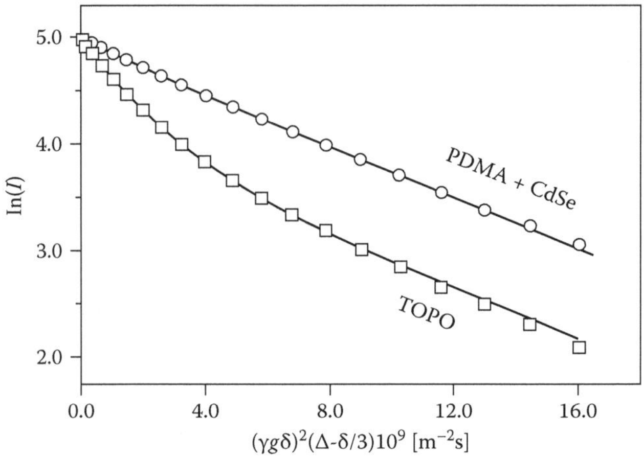

FIGURE 5.26 Plot of ln(I) versus G^2 of the PDMA and TOPO NMR signals for the QD/PDMA mixture. (Reproduced from Shen et al., *J. Phys. Chem. B*, 112, 1626–1633, 2008. With permission of the American Chemistry Society.)

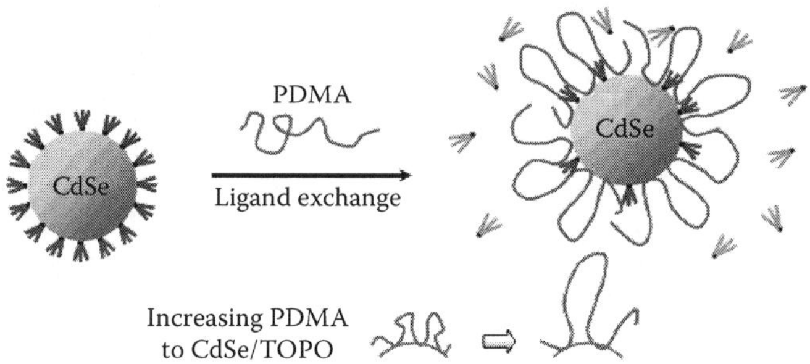

FIGURE 5.27 (See color insert) Sketch of the replacement of TOPO by PDMA on the surface of the QD and different binding conformation of PDMA. (Reproduced from Shen et al., *J. Phys. Chem. B*, 112, 1626–1633, 2008. With permission of the American Chemistry Society.)

Supramolecular Self-Assembly

Diffusional NMR techniques are an essential tool for characterizing supramolecular self-assemblies,[75,97,98] i.e., adducts held together by weak interactions (e.g., hydrogen bonds) or labile coordinative bonds (e.g., Pd-N bonds).

For example,[99] the cuboctahedron is made up of six square faces and eight equilateral triangular faces. Overall it has 12 vertices, 24 edges, and a dihedral angle between the triangular and square surfaces of 125°. A good approximation of a cuboctahedron can be achieved chemically by making use of the face-directed design strategy. When a 108° ditopic tecton (compound **31**, Scheme 5.4) is reacted with a complementary planar, tritopic, 120° building block (compound **32**, Scheme 5.4) in a 12:8 ratio, respectively, the cuboctahedral cage 33 results. When these two building blocks are mixed in a 3:2 ratio, respectively, cuboctahedron **33** results. Analogously, cuboctahedron **36** can be formed from compounds **34** and **35** (Scheme 5.4). Diffusion

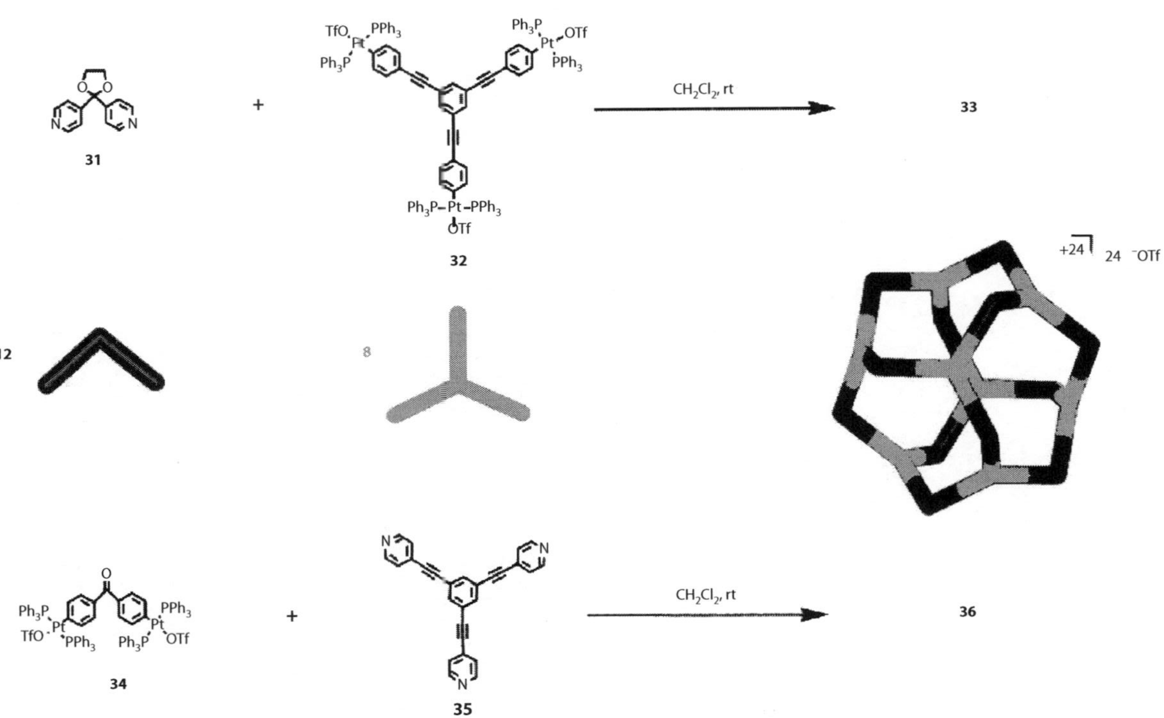

SCHEME 5.4 (See color insert)

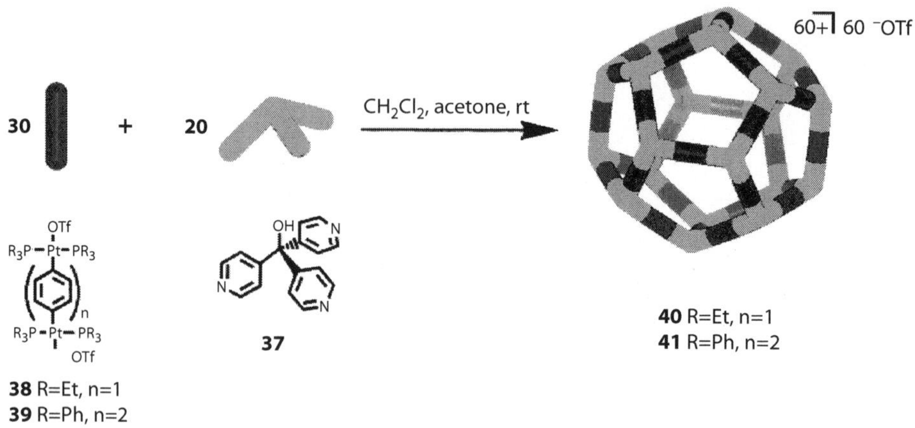

SCHEME 5.5 (See color insert)

NMR measurements show a self-diffusion coefficient of (1.92 ± 0.072) × 10^{-6} cm^2 s^{-1} at 25°C, providing a hydrodynamic diameter of 5.0 nm for cuboctahedron **36**. This value is consistent with the molecular model of **36**, making it one of the largest self-assembled species. The most complex of the five Platonic polyhedra, the dodecahedron, contains 12 fused 5-membered rings that comprise the highest symmetry group I_h. Its 12 pentagonal faces are formed from 20 vertices and 30 edges, hence it can be prepared via edge-directed assembly from 20 tridentate angular subunits with approximately 108° directing angles (compound **37**, Scheme 5.5) combined with 30 bidentate linear subunits (compounds **38** and **39**, Scheme 5.5). The experimental size of **40** and **41** is assessed by measuring their self-diffusion coefficients through the PGSE NMR technique. The experimental self-diffusion coefficients at 25°C are (1.80 ± 0.05) × 10^{-6} cm^2/s for **40** and (1.32 ± 0.06) × 10^{-6} cm^2/s for **41**, which give experimental hydrodynamic diameters of 5.2 nm for **40** and 7.5 nm for **41** that perfectly match the simulation results. As a final example, let us consider a multicromophore supramolecular system synthesized through self-assembly directed by metal-ion coordination (Figure 5.28).[100] In the ^{1}H-DOSY spectrum shown in Figure 5.28, it can be clearly seen that the product of the self-assembly is much larger than tetramethylsilane (TMS). The use of an internal standard allows the authors to determine the size of the complex with respect to that of the free ligand; the results indicate that the nuclearity of the main product is four.

Self-Aggregation of Spherical Organometallic Particles

The importance of the size factor c of Equation 5.17 can be appreciated considering the self-aggregation of (η^6-arene)RuCl(AA) complexes,[101] where AA = α-aminoacidate ligand. Diffusion NMR measurements ($CDCl_3$, using TMSS as an internal standard) indicate that the latter complexes have a remarkable tendency to self-aggregate, forming spherical supramolecular adducts held together by the cooperation of several noncovalent interactions, H-bonding being the most important one. A schematization of the supramolecular adduct, as observed in the solid state, is reported in Figure 5.29.[102,103] Since both the monomer and supramolecular adducts

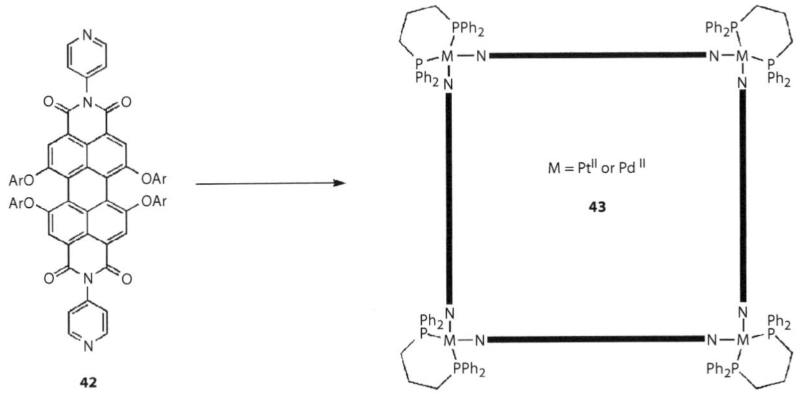

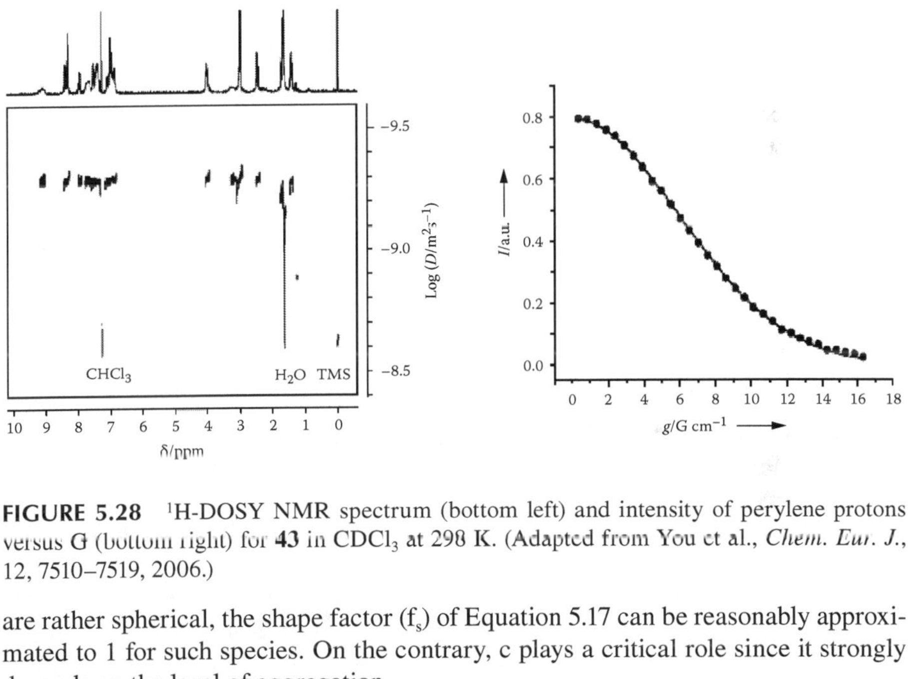

FIGURE 5.28 ^{1}H-DOSY NMR spectrum (bottom left) and intensity of perylene protons versus G (bottom right) for **43** in $CDCl_3$ at 298 K. (Adapted from You et al., *Chem. Eur. J.*, 12, 7510–7519, 2006.)

are rather spherical, the shape factor (f_s) of Equation 5.17 can be reasonably approximated to 1 for such species. On the contrary, c plays a critical role since it strongly depends on the level of aggregation.

In particular, when arene = *p*-cymene and AA = *S*-alaninate (**44**), the c factor spans from 5.3 (r_H = 5.5 Å) to 6.0 (r_H = 20.8 Å). Consequently, the utilization of the standard Stokes-Einstein equation (c = 6) would be appropriated for the largest adduct, but it would lead to an error higher than 10% for the smallest ones.

Self-Aggregation of Spherical Building Blocks into Nonspherical Supramolecules

In this final example a case is shown where both the size and shape factors have been used in order to determine accurate hydrodynamic dimensions according to Equation 5.17. Once activated with KOH, complex **45** (Figure 5.30) is a good catalyst for the asymmetric transfer hydrogenation of ketones occurring through a bifunctional

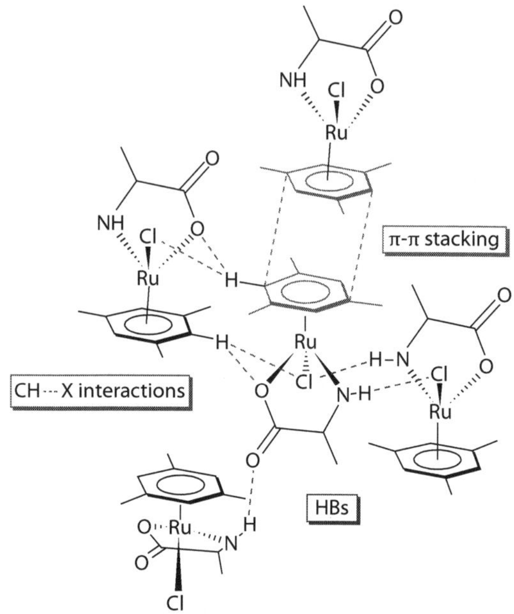

FIGURE 5.29 Noncovalent interactions observed in the x-ray single-crystal structure of RuCl(S-Ala)(Mes). (Reproduced from Ciancaleoni et al., *Organometallics*, 26, 489–496, 2007. With permission of the American Chemistry Society.)

mechanism.[104] All the species involved in the catalytic cycle (**46** and **47**) have been recently isolated and studied by means of PGSE NMR since,[105] due to the presence of an electronegative moiety (the sulfonyl group) and an electropositive moiety (the protic N–H moiety), they are suitable for establishing intermolecular hydrogen bonds that can lead to their self-aggregation in solution. Both the crystal structure[104b] of **45** and **46** and ONIOM(B3PW91/HF) calculations[105] indicate the presence of a head-to-tail self-aggregation pattern (Figure 5.30), leading to nonspherical adducts, so that f_s has to be considered. At the same time, since the monomer is rather small, the c factor also has to be taken into account. Another layer of complexity is due to the presence of a deep inlet on the surface of the monomer generated by the two phenyls in the back of the N,N ligand. This asks for a careful evaluation of the hydrodynamic volume of the monomer (an in-depth investigation carried out by combining conductometric and diffusion NMR[106] studies allows us to conclude that the ratio V_H^0/V_{vdW} is 1.8 for **45** and **46** and 1.4 for **47**).

Using the appropriate size and shape factors (spanning from f_s = 1.00 and c = 5.1 for the monomer to f_s = 1.13 and c = 5.8 for the tetramer) and the volumes of the monomers, accurate aggregation numbers[107] are derived according to

$$N = \frac{V_H}{V_H^0} \tag{5.19}$$

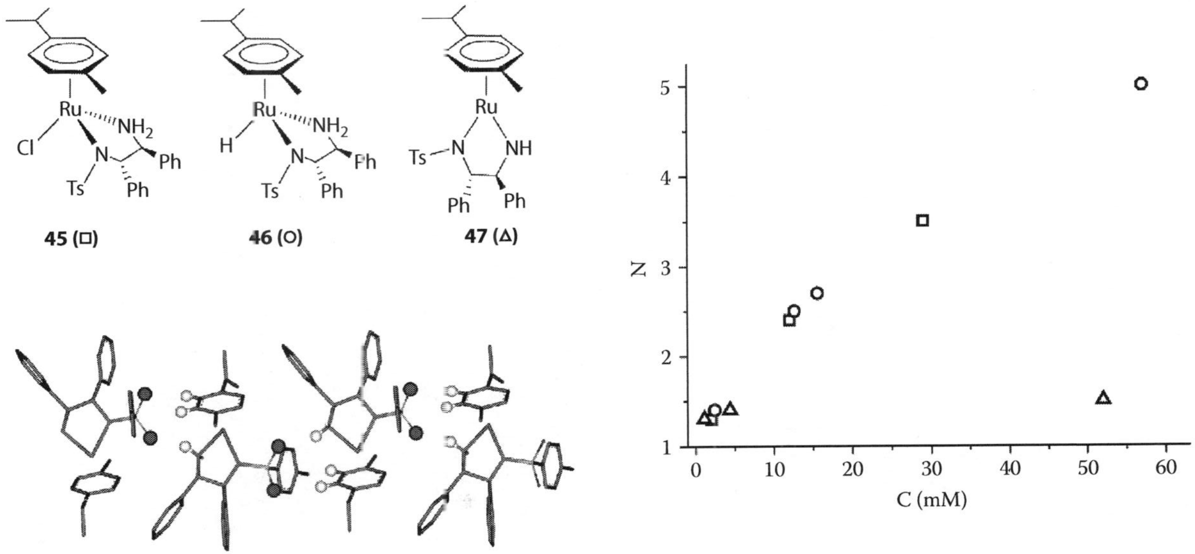

FIGURE 5.30 A linear noncovalent tetramer of **46**, as found in the solid-state structure, is shown (bottom left). Trends of N versus C are shown on the right. (Adapted from Ciancaleoni et al., *Organometallics*, 28, 960–967, 2009.)

Ns are then used to estimate the equilibrium constant of the aggregative process (K_{agg}), taking advantage of the linear dependence of N(N – 1) on C (concentration), according to an indefinite isodesmic self-aggregation model.[108]

$$N(N-1) = K_{agg}C \tag{5.20}$$

The overall study reveals that dimers and higher aggregates are absent in isopropanol (C = 0.1 mM), where the reduction of ketones is carried out, but they are predominant in toluene, where **47** is used for the asymmetric Micheal reaction.[109]

COMBINING NOE AND DIFFUSION NMR MEASUREMENTS

The potentialities of NOE and diffusion NMR techniques have been separately illustrated in previous paragraphs, but they offer their best when applied contemporary to study the same system. They are, in fact, complementary because while diffusion NMR allows the average dimension of the aggregates to be determined, NOE experiments clarify the intermolecular structure within the aggregates.[75,110,111] Let us conclude this chapter by showing one example where the change of aggregation leads to a different pattern of NOEs. It concerns complexes [(*p*-cymene)Ru(κ^3-dpk-OR)]X (dpk = di-pyridyl-ketone, R = H (**48**) or Me (**49**), X = BF_4^- or PF_6^-).[92a] Diffusion NMR measurements indicate that in CD_2Cl_2 **49** undergoes mainly ion pairing phenomenon (N^+ and N^- are close to 1 at concentration of 3 mM), while **48** is present as a mixture of ion triples and ion quadruples (N^+ = 2.2 and N^- = 1.7 at 1.7 mM). $^{19}F,^1H$-HOESY experiments clarify how the aggregation higher than ion pairing occurs.

In fact, the patterns of NOEs are slightly different for the two complexes (Figure 5.31). In **49**PF_6 the counterion strongly interacts with 8, 9, 4, and 5 protons (Figure 5.31). Weak NOEs are also detected with 1, 2, and 7 cymene resonances. No interaction is observed between the counterion and 10 and 11 pyridyl protons and the protons of the OMe-tail. All of these observations indicate that the anion is located close to the nitrogen of the N,N,O ligand. In **48**BF_4, two strong additional NOEs are detected between the counterion and 10 and 11 pyridyl resonances. These new contacts combined with PGSE evidence suggest that a second cationic unit is close to the ion pair forming a cationic ion triple (Figure 5.31). The formation of ion triples is attributed to the establishment of two intercationic hydrogen bonds, enforced by a π-π stacking between the pyridyl rings. The solid-state structure of **48**BF_4 is perfectly consistent with these results.[92a]

CONCLUSIONS

NOE and diffusion NMR techniques allow the supramolecular structure of transition metal complexes in solution to be carefully determined. The relative orientation(s) of interacting units constituting the supramolecule can be easily derived by the detection of intermolecular NOEs based just on triangulation reasoning. A more in-depth description of the system can be obtained by the quantification of the NOEs leading to the intermolecular distances. This procedure asks for a suitable reference distance

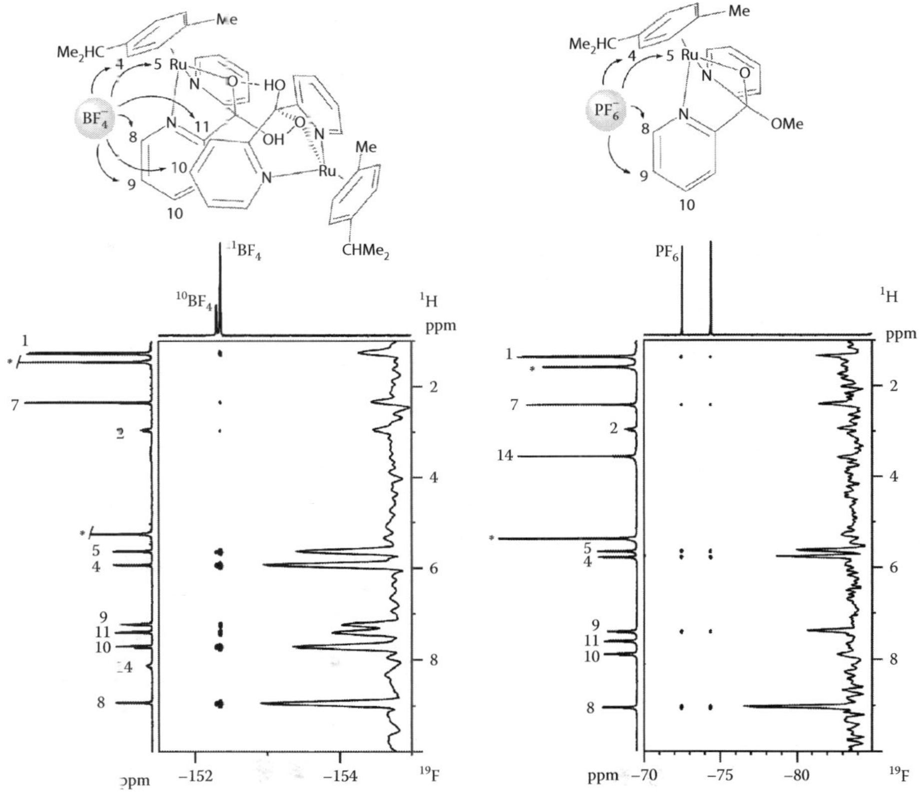

FIGURE 5.31 (See color insert) $^{19}F,^{1}H$-HOESY NMR spectra (376.65 MHz, 296 K) of complexes **48** (left) and **49** (right) in CD_2Cl_2. The asterisk denotes the residues of nondeuterated solvent and water. (Reproduced from Zuccaccia et al., *Organometallics*, 26, 6099–6105, 2007. With permission of the American Chemical Society.)

and a proper checking of the rotational correlation times of the two vectors defined by the two couples of nuclei (of interest and reference). Diffusion NMR experiments afford the other crucial piece of information, i.e., the average size of the supramolecular adducts. Consequently, it is possible to derive the nuclearity of the supramolecule (monomer, dimer, trimer, etc., for neutral molecules, or ions, ion pairs, ion triples, ion quadruples, etc., for ionic molecules) and obtain the thermodynamic parameters of the self-aggregation processes. Preliminary to this is the correct evaluation of the average hydrodynamic size of the supramolecular adducts, which requires some knowledge (or hypotheses) about the size and shape of the adducts.

ACKNOWLEDGMENTS

This work was supported by grants from the Ministero dell'Istruzione, dell'Università e della Ricerca (MIUR, Rome, Italy), Programma di Rilevante Interesse Nazionale, Cofinanziamento 2007.

REFERENCES

1. Müller-Dethlefs, K., and Hobza, P. 2000. Noncovalent interactions: A Challenge for experiment and theory. *Chem. Rev.* 100:143–167. Williams, D. H., and Westwell, M. S. 1998. Aspects of weak interactions. *Chem. Soc. Rev.* 27:57–63.
2. Crabtree, R. H., Siegbahn, P. E. M., Eisenstein, O., Rheingold, A. L., and Koetzle, T. F. 1996. A new intermolecular interaction: Unconventional hydrogen bonds with element-hydride bonds as proton acceptor. *Acc. Chem. Res.* 29:348–354. Custelcean, R., and Jackson, J. E. 2001. Dihydrogen bonding: Structures, energetics, and dynamics. *Chem. Rev.* 101:1963–1980.
3. Braga, D., Grepioni, F., and Desiraju, G. R. 1998. Crystal engineering and organometallic architecture. *Chem. Rev.* 98:1375–1406. Borovik, A. S. 2002. The use of non-covalent interactions in the assembly of metal/organic supramolecular arrays. *Comments Inorg. Chem.* 23:45–78.
4. Pastor, A., and Martınez-Viviente, E. 2008. NMR spectroscopy in coordination supramolecular chemistry: A unique and powerful methodology. *Coord. Chem. Rev.* 252:2314–2345.
5. Neuhaus, D., and Williamson, M. 2000. *The Nuclear Overhauser Effect in Structural and Conformational Analysis*. New York: WILEY-VCH.
6. (a) Kärger, J., Pfeifer, H., and Heink, W. 1988. In *Advances in Magnetic Resonance*, ed. W. S. Warren, 12. New York: Academic Press. (b) Stilbs, P. 1987. Fourier transform pulsed-gradient spin-echo studies of molecular diffusion. *Prog. Nucl. Magn. Reson. Spectros.* 19:1–45. (c) Price, W. S. 1997. Pulsed-field gradient nuclear magnetic resonance as a tool for studying translational diffusion. Part I. Basic theory. *Concepts Magn. Res.* 9:299–336. (d) Price, W. S. 1998. Pulsed-field gradient nuclear magnetic resonance as a tool for studying translational diffusion. Part II. Experimental aspects. *Concepts Magn. Res.* 10:197–237.
7. Noggle, J. H., and Schirmer, R. E. 1971. *The Nuclear Overhauser Effect*. New York: Academic Press.
8. Pochapsky, T. C., Wang, A., and Stone, P. M. 1993. Closed-shell ion pair structure and dynamics: Steady-state $^{1}H\{^{1}H\}$, $^{10}B\{^{1}H\}$, and $^{11}B\{^{1}H\}$ nuclear Overhauser effects and ^{10}B, ^{11}B nuclear relaxation of tetraalkylammonium tetrahydridoborates in ion-pairing and dissociative solvents. *J. Am. Chem. Soc.* 115:11084–11091.

9. (a) Bellachioma, G., Cardaci, G., Macchioni, A., Reichenbach, G., and Terenzi, S. 1996. Application of ^{1}H-NOESY NMR spectroscopy to the investigation of ion pair solution structures of organometallic complexes by the detection of interionic contacts. *Organometallics* 15:4349–4351. (b) Macchioni, A., Bellachioma, G., Cardaci, G., Gramlich, V., Rüegger, H., Terenzi, S., and Venanzi, L. M. 1997. Cationic bis- and tris(η^2-(pyrazol-1-yl)methane) acetyl complexes of iron(II) and ruthenium(II): Synthesis, characterization, reactivity, and interionic solution structure by NOESY NMR spectroscopy. *Organometallics* 16:2139–2145.
10. (a) Zuccaccia, C., Bellachioma, G., Cardaci, G., and Macchioni, A. 2001. Solution structure investigation of Ru(II) complex ion pairs: Quantitative NOE measurements and determination of average interionic distances. *J. Am. Chem. Soc.* 123:11020–11028. (b) Zuccaccia, C., Bellachioma, G., Cardaci, G., and Macchioni, A. 1999. Specificity of interionic contacts and estimation of average interionic distances by NOE NMR measurements in solution of cationic Ru(II) organometallic complexes bearing unsymmetrical counterions. *Organometallics* 18:1–3.
11. Macchioni, A., Bellachioma, G., Cardaci, G., Cruciani, G., Foresti, E., Sabatino, P., and Zuccaccia, C. 1998. Synthesis and structural studies of cationic bis- and tris(pyrazol-1-yl)methane acyl and methyl complexes of ruthenium(II): Localization of the counterion in solution by NOESY NMR spectroscopy. *Organometallics* 17:5549–5556.
12. (a) Schneider, H.-J., Schiestel, T., and Zimmermann, P. 1992. Host-guest supramolecular chemistry. 34. The incremental approach to noncovalent interactions: Coulomb and van der Waals effects in organic ion pairs. *J. Am. Chem. Soc.* 114:7698–7703. (b) Schneider, H.-J. 1994. Linear free energy relationships and pairwise interactions in supramolecular chemistry. *Chem. Soc. Rev.* 23:227–234.
13. Binotti, B., Macchioni, A., Zuccaccia, C., and Zuccaccia, D. 2002. Application of NOE and PGSE NMR methodologies to investigate non-covalent intimate inorganic adducts in solution. *Comments Inorg. Chem.* 23:417–450.
14. Macchioni, C. 2005. Ion pairing in transition-metal organometallic chemistry. *Chem. Rev.* 105:2039–2073.
15. Binotti, B., Bellachioma, G., Cardaci, G., Carfagna, C., Zuccaccia, C., and Macchioni, A. 2007. The effect of counterion/ligand interplay on the activity and stereoselectivity of palladium(II)-diimine catalysts for CO/*p*-methylstyrene copolymerization. *Chem. Eur. J.* 13:1570–1582.
16. Romeo, R., Nastasi, N., Monsù Scolaro, L., Plutino, M. R., Albinati, A., and Macchioni, A. 1998. Molecular structure, acidic properties, and kinetic behavior of the cationic complex (methyl)(dimethyl sulfoxide)(bis-2-pyridylamine) platinum(II) Ion. *Inorg. Chem.* 37:5460–5466.
17. Fielding, L. 2000. Determination of association constants (Ka) from solution NMR Data. *Tetrahedron* 56:6151–6170.
18. Goshe, A. J., Steele, I. M., and Bosnich, B. 2003. Supramolecular recognition. Terpyridyl palladium and platinum molecular clefts and their association with planar platinum complexes. *J. Am. Chem. Soc.* 125:444–451.
19. Meyer, A. S., Jr., and Ayres, G. H. 1957. The mole ratio method for spectrophotometric determination of complexes in solution. *J. Am. Chem. Soc.* 79: 49–53.
20. Fiedler, D., Leung, D. H., Bergman, R. G., and Raymond, K. N. 2005. Selective molecular recognition, C-H bond activation, and catalysis in nanoscale reaction vessels. *Acc. Chem. Res.* 38:351–360.
21. Pluth, M. D., Bergman, R. G., and Raymond, K. N. 2007. Making amines strong bases: Thermodynamic stabilization of protonated guests in a highly-charged supramolecular host. *J. Am. Chem. Soc.* 129:11459–11467.

22. Fiedler, D., Pagliero, D., Brumaghim, J. L., Bergman, R. G., and Raymond, K. N. 2004. Encapsulation of cationic ruthenium complexes into a chiral self-assembled cage. *Inorg. Chem.* 43:846–848.
23. Leung, D. H., Bergman, R. G., and Raymond, K. N. 2006. Scope and mechanism of the C-H bond activation reactivity within a supramolecular host by an iridium guest: A stepwise ion pair guest dissociation mechanism. *J. Am. Chem. Soc.* 128:9781–9797.
24. Bourgeois, J.-P., Fujita, M., Kawano, M., Sakamoto, S., and Yamaguchi, K. 2003. A cationic guest in a 24^+ cationic host *J. Am. Chem. Soc.* 125:9260–9261.
25. The efficiency (or transition probability W) of a particular relaxation mechanism depends on the square of an interaction energy constant multiplied by a function of the correlation time (spectral density function):

$$W_{ij} = J(\omega)\left|\left(\langle\iota|\mathcal{H}_{relaxation}|j\rangle\right)^2\right|_{average} \qquad J(\omega) = \frac{2\tau_C}{1+\omega^2\tau_C^2}$$

26. In real systems other relaxation mechanisms, different from dipole-dipole interaction, are active and, with the exception of relaxation coming from modulated scalar coupling, they only contribute to the single-quantum transition probabilities. As a consequence, the steady-state NOE can now be defined as

$$NOE_I\{S\} = \frac{\gamma_S}{\gamma_I}\frac{\sigma_{IS}}{\rho_{IS}+\rho_{IS}^*}$$

where ρ_{IS}^* indicates the contribution to the total longitudinal relaxation rate coming from sources different from dipole-dipole interaction. In this case, the steady-state NOE for a two-spin system depends on the internuclear distance.
27. The choice of the reference distance r_{AB} is critical. It is preferable that (a) **A** and **B** are not scalarly coupled, (b) r_{AB} is comparable with the distance of interest, and (c) **A** and **B** do not undergo dynamic processes.
28. The rotational correlation time can be estimated using both homonuclear and heteronuclear dipolar interactions (see Neuhaus and Williamson[5]).
29. Under these circumstances, ROE experiments should be used: Bax, A., and Davis, D. G. 1985. Practical aspects of two-dimensional transverse NOE spectroscopy. *J. Magn. Reson.* 63:207–213.
30. (a) Lipari, G., and Szabo, A. 1982. Model-free approach to the interpretation of nuclear magnetic resonance relaxation in macromolecules. 1. Theory and range of validity. *J. Am. Chem. Soc.* 104:4546–4559. (b) Lipari, G., and Szabo, A. 1982. Model-free approach to the interpretation of nuclear magnetic resonance relaxation in macromolecules. 2. Analysis of experimental results. *J. Am. Chem. Soc.* 104:4559–4570.
31. (a) Tropp, J. 1980. Dipolar relaxation and nuclear Overhauser effects in nonrigid molecules: The effect of fluctuating internuclear distances. *J. Chem. Phys.* 72:6035–6043. (b) Yip, P. F., Case, D. A., Hoch, J. C., Poulsen, F. M., and C. Redfield, eds. 1991. *Computational Aspect of the Study of Biological Macromolecules by Nuclear Magnetic Resonance Spectroscopy*, 317–330. New York: Plenum Press.
32. Olejniczak, E. T., Dobson, C. M., Karplus, M., and Levy, R. M. 1984. Motional averaging of proton nuclear Overhauser effects in proteins. Predictions from a molecular dynamics simulation of lysozyme. *J. Am. Chem. Soc.* 106:1923–1930.
33. Post, C. B. J. 1992. Internal motional averaging and three-dimensional structure determination by nuclear magnetic resonance. *Mol. Biol.* 224:1087–1101.

34. LeMaster, D. M., Kay, L. E., Brünger, A. T., and Prestegard, J. H. 1988. Protein dynamics and distance determination by NOE measurements. *FEBS Lett.* 236:71–76.
35. Abseher, R., Lüdemann, S., Schreiber, H., and Steinhauser, O. 1994. Influence of molecular motion on the accuracy of NMR-derived distances. A molecular dynamics study of two solvated model peptides. *J. Am. Chem. Soc.* 116:4006–4018.
36. Edmondson, S. P. 1994. Molecular dynamics simulation of the effects of methyl rotation and other protein motions on the NOE. *J. Magn. Reson. B* 103:222–233.
37. A simple way to take into account the dynamical motions of **I** and **S** is to increase by 10% the r_{IS} values obtained from Equation 5.7. This should correct, at least partially, for the overestimation of short distance contribution to the observed NOE.
38. Bauer, C. J., Freeman, R., Frenkiel, T., Keeler, J., and Shaka, J. 1984. Gaussian pulses. *J. Magn. Res.* 58:442–457.
39. Kessler, H., Oschkinat, H., Griesinger, G., and Bermel, W. 1986. Transformation of homonuclear two-dimensional NMR techniques into one-dimensional techniques using Gaussian pulses. *J. Magn. Res.* 70:106–133.
40. (a) Stonehouse, J., Adel, P., Keeler, J., and Shaka, J. 1994. Ultrahigh-quality NOE spectra. *J. Am. Chem. Soc.* 116:6037–6038. (b) Stott, K., Stonehouse, J., Keeler, J., Hwang, T.-L., and Shaka, J. 1995. Excitation sculpting in high-resolution nuclear magnetic resonance spectroscopy: Application to selective NOE experiments. *J. Am. Chem. Soc.* 117:4199–4200.
41. See, for example: Gerig, G. T. 1999. Gradient-enhanced proton–fluorine NOE experiments. *Magn. Reson. Chem.* 37:647–652.
42. Jeener, J., Meier, B. H., Bachmann, P., and Ernst, R.R. 1979. Investigation of exchange processes by two-dimensional NMR spectroscopy. *J. Chem. Phys.* 71:4546–4553.
43. (a) Rinaldi, P. L. 1993. Heteronuclear 2D-NOE spectroscopy. *J. Am. Chem. Soc.* 105:5167–5168. (b) Lix, B., Sönnichsen, F. D. and Sykes, B. D. 1996. The role of transient changes in sample susceptibility in causing apparent multiple-quantum peaks in HOESY spectra. *J. Magn. Reson. Series A* 121:83–87. (c) Alam, T. M., Pedrotty, D. M., and Boyle, T. J. 2002. Modified, pulse field gradient-enhanced inverse-detected HOESY pulse sequence for reduction of t1 spectral artefacts. *Magn. Reson. Chem.* 40:361–365.
44. Zuccaccia, C., Stahl, N. G., Macchioni, A., Chen, M. C., Roberts, J. A., and Marks, T. J. 2004. NOE and PGSE NMR spectroscopic studies of solution structure and aggregation in metallocenium ion-pairs. *J. Am. Chem. Soc.* 126:1448–1464.
45. Chen, E. Y.-X., and Marks, T. J. 2000. Cocatalysts for metal-catalyzed olefin polymerization: Activators, activation processes, and structure-activity relationships. *Chem. Rev.* 100:1391–1434.
46. Correa, A., and Cavallo, L. 2006. Dynamic properties of metallocenium ion pairs in solution by atomistic simulations. *J. Am. Chem. Soc.* 128:10952–10959.
47. Zuccaccia, C., Macchioni, A., Orabona, I., and Ruffo, F. 1999. Interionic solution structure of $[PtMe(\eta^2\text{-olefin})(N,N\text{-diimiane})]BF_4$ complexes by $^{19}F\{^1H\}$-HOESY NMR spectroscopy: Effect of the substituent on the accessibility of the counterion to the metal. *Organometallics* 18:4367–4372.
48. Macchioni, A., Magistrato, A., Orabona, I., Ruffo, F., Rothlisberger, U., and Zuccaccia, C. 2003. Direct observation of an equilibrium between two anion-cation orientations in olefin Pt(II) complex ion pairs by HOESY NMR spectroscopy. *New J. Chem.* 27:455–458.
49. (a) Bain, A., and Cramer, J. A. 1993. A method for optimizing the study of slow chemical exchange by NMR spin-relaxation measurements. Application to tripodal carbonyl rotation in a metal complex. *J. Magn. Reson. A* 103:217–222. (b) Bain, A. D., and Cramer, J. A. 1996. Slow chemical exchange in an eight-coordinated bicentered ruthenium complex studied by one-dimensional methods. Data fitting and error analysis. *J. Magn. Reson. A* 118:21–27.

50. Perrin, T. C., and Dwyer, T. J. 1990. Application of two-dimensional NMR to kinetics of chemical exchange. *Chem. Rev.* 90:935–967.
51. Solutions of more complicated equations are indeed necessary in the case of multisite exchange and are usually carried out with the help of a computer. For free software to analyze EXSY data, see http://nmr-analysis.blogspot.com/2008/11/exsycalc-free-software-for-nmr-analysis.html. Software for the analysis of selective inversion recovery (SIR) data is also available from the Bain group: http://www.chemistry.mcmaster.ca/bain/.
52. Leung, D. H., Fiedler, D., Bergman, R. G., and Raymond, K. N. 2004. Selective C-H bond activation by a supramolecular host–guest assembly. *Angew. Chem. Int. Ed.* 43:963–963.
53. Leung, D. H., Bergman, R. G., and Raymond, K. N. 2007. Highly selective supramolecular catalyzed allylic alcohol isomerization. *J. Am. Chem. Soc.* 129:2746–2747.
54. Fiedler, D., van Halbeek, H., Bergman, R. G., and Raymond, K. N. 2006. Supramolecular catalysis of unimolecular rearrangements: Substrate scope and mechanistic insights. *J. Am. Chem. Soc.* 128:10240–10252.
55. Davis, A. V., Fiedler, D., Seeber, G., Zahl, A., van Eldik, R., and Raymond K. N. 2006. Guest exchange dynamics in an M4L6 tetrahedral host. *J. Am. Chem. Soc.* 128:1324–1333.
56. Zuccaccia, C., Pirondini, R., Pinalli, L., Dalcanale, E., and Macchioni, A. 2005. Dynamic and structural NMR studies of cavitand-based coordination cages. *J. Am. Chem. Soc.* 127:7025–7032.
57. (a) Crank, J. 1975. *The Mathematics of Diffusion*. 2nd ed. Oxford: Clarendon Press. (b) Einstein, A. 1956. *Investigations in the Theory of Brownian Movements*. New York: Dover.
58. Stejskal, E. O., and Tanner, J. E. 1965. Spin diffusion measurements: Spin echoes in the presence of a time-dependent field gradient. *J. Chem. Phys.* 42:288–292.
59. Hahn, E. L. 1950. Spin echoes. *Phys. Rev.* 80:580–594.
60. In the case of gradient pulses with different shapes, corresponding equations can be derived. See, for example: Price, W. S., and Kuchel P. W. 1991. Effect of nonrectangular field gradient pulses in the Stejskal and Tanner (diffusion) pulse sequence. *J. Magn. Reson.* 94:133–139.
61. Tanner, J. E. 1970. Use of the stimulated echo in NMR diffusion studies. *J. Chem. Phys.* 52:2523–2526.
62. Dehner, A., and Kessler, H. 2005. Diffusion NMR spectroscopy: Folding and aggregation of domains in p53. *ChemBioChem* 6:1550–1565.
63. More recently a method to efficiently remove J-coupling peak distortion has been developed by Torres and coworkers. Torres, A. M., Dela Cruz, R., and Price, W. S. 2008. Removal of J-coupling peak distorsion in PGSE experiments. *J. Magn. Reson.* 193:311–316.
64. Burl, M., and Young, I. R. 1996. *Encyclopedia of Nuclear Magnetic Resonance*, ed. D. M. Grant and R. K. Harris, 1841. Vol. 3. New York: Wiley.
65. (a) Jehenson, P., Westphal, M., and Schuff, N. 1990. Analytical method for the compensation of eddy-current effects induced by pulsed magnetic field gradients in NMR systems. *J. Magn. Reson.* 90:264–278. (b) Van Vaals, J. J., and Bergman, H. A. 1990 Optimization of eddy-current compensation. *J. Magn. Reson.* 90:52–70. (c) Majors, P. D., Blackley, J. L., Altobelli, S. A., Caprihan, A., and Fukushima, E. 1990. Eddy current compensation by direct field detection and digital gradient modification. *J. Magn. Reson.* 87:548–553.

66. Gibbs, S. J., and Johnson, C. S. 1991. A PFG NMR experiment for accurate diffusion and flow studies in the presence of eddy currents. *J. Magn. Reson.* 93:395–402.
67. (a) Wu, D. H., Chen, A. D., and Johnson, C. S. 1995. An improved diffusion-ordered spectroscopy experiment incorporating bipolar-gradient pulses. *J. Magn. Reson. A* 115:260–264. (b) Wider, G., Dötsch, V., and Wüthrich, K. 1994. Self-compensating pulsed magnetic-field gradients for short recovery times. *J. Magn. Reson. A* 108:255–258.
68. Hedin N., Yu T. Y., and Furó, I. 2000. Growth of C12E8 micelles with increasing temperature. A convection-compensated PGSE NMR study. *Langmuir* 16:7548–7550.
69. Jerschow, A., and Müller, N. 1997. Suppression of convection artifacts in stimulated-echo diffusion experiments. Double-stimulated-echo experiments. *J. Magn. Reson.* 125:372–375.
70. Δ-Dependent D values may also be observed in the presence of chemical exchange (Chen, A., Johnson, C. S., Jr., Lin, M., and Shapiro, M. J. 1998. Chemical exchange in diffusion NMR experiments. *J. Am. Chem. Soc.* 120:9094–9095) or in the presence of intermolecular NOE (Avram, L., and Cohen, Y. 2005. Diffusion measurements for molecular capsules: Pulse sequences effect on water signal decay. *J. Am. Chem. Soc.* 127:5714–5719).
71. Callaghan, P. T. 1991. *Principles of Nuclear Magnetic Resonance Microscopy*. Oxford: Clarendon.
72. Martìnez-Viviente, E., and Pregosin, P. S. 2003. Low temperature ^{1}H-, ^{19}F-, and ^{31}P-PGSE diffusion measurements. Applications to cationic alcohol complexes. *Helv. Chim. Acta* 86:2364–2378.
73. Hayamizu, K., and Price, W. S. 2004. A new type of sample tube for reducing convection effects in PGSE-NMR measurements of self-diffusion coefficients of liquid samples. *J. Magn. Reson.* 167:328–333.
74. Kato, H., Saito, T., Nabeshima, M., Shimada, K., and Kinugasa, S. 2006. Assessment of diffusion coefficients of general solvents by PFG-NMR: Investigation of the sources error. *J. Magn. Reson.* 180:266–273.
75. Brand, T., Cabrita, E. J., and Berger, S. 2005. Intermolecular interaction as investigated by NOE and diffusion studies. *Prog. Nucl. Magn. Reson. Spectrosc.* 46:159–196.
76. Johnson, C. S., Jr. 1999. Diffusion ordered magnetic resonance spectroscopy: Principles and applications. *Prog. Nucl. Magn. Reson. Spectrosc.* 34:203–256.
77. Macchioni, A., Ciancaleoni, G., Zuccaccia, C., and Zuccaccia, D. 2008. Determining accurate molecular sizes in solution through NMR diffusion spectroscopy. *Chem. Soc. Rev.* 37:479–489.
78. Zuccaccia, D., and Macchioni, A. 2005. An accurate methodology to identify the level of aggregation in solution by PGSE NMR measurements: The case of half-sandwich diamino ruthenium(II) salts. *Organometallics* 24:3476–3486.
79. Xie, X., Auel, C., Henze, W., and Gschwind, R. M. 2003. Dimethyl- and bis[(trimethylsilyl)methyl]cuprates show aggregates higher than dimers in diethyl ether: Molecular diffusion studies by PFG NMR and aggregation–reactivity correlations. *J. Am. Chem. Soc.* 125:1595–1601.
80. Stahl, N. G., Zuccaccia, C., Jensen, T. R., and Marks, T. J. 2003. Metallocene polymerization catalyst ion-pairs aggregation by cryoscopy and pulsed field gradient spin-echo NMR diffusion measurements. *J. Am. Chem. Soc.* 125:5256–5257.
81. Edward, J. T. 1970. Molecular volumes and the Stokes-Einstein equation. *J. Chem. Educ.* 47:262–270.
82. For prolate ($a > b = d$) and oblate ($a < b = d$) ellipsoids shape factors are:

$$f_S(\text{prolate}) = \frac{\sqrt{1-\left(\frac{a}{b}\right)^2}}{\left(\frac{b}{a}\right)^{\frac{2}{3}} \ln \frac{1+\sqrt{1-\left(\frac{b}{a}\right)^2}}{\left(\frac{b}{a}\right)}} \qquad f_S(\text{oblate}) = \frac{\sqrt{\left(\frac{b}{a}\right)^2 - 1}}{\left(\frac{b}{a}\right)^{\frac{2}{3}} \arctan\sqrt{\left(\frac{b}{a}\right)^2 - 1}}$$

Perrin, F. 1936. Mouvement Brownien d'un ellipsoide (II). Rotation libre et dépolarisation des fluorescences. Translation et diffusion de molécules ellipsoidales. *J. Phys. Radium.* 7:1–11. For cylindrical geometries see Li, G. L., and Tang, J. X. 2004. Diffusion of actin filaments within a thin layer between two walls. *Phys. Rev. E* 69:061921-1-061921-5, and references therein.

83. Elworthy, P. H. 1962. A test of Perrin's relationships for small molecules. *J. Chem. Soc.* 3718–3723.

$$c = \frac{6}{\frac{3r_{solv}}{2r_H} + \frac{1}{1+\frac{r_{solv}}{r_H}}}$$

84. Gierer, A., and Wirtz, K. 1953. Microfriction in liquids. *Z. Naturforsch. A* 8:522–532; Spernol, A., and Wirtz, K. 1953. Molekulare Theorie der Mikroreibung. *Z. Naturforsch. A* 8:532–538. A semiempirical improvement has been proposed:

$$c = \frac{6}{1+0.695\left(\frac{r_{solv}}{r_H}\right)^{2.234}}$$

Chen, H.-C., and Chen, S.-H. 1984. Diffusion of crown ethers in alcohols. *J. Phys. Chem.* 88:5118–5121.

85. Valentini, M., Pregosin, P. S., and Rüegger, H. 2000. Applications of pulsed field gradient spin-echo measurements to the determination of molecular diffusion (and thus size) in organometallic chemistry. *Organometallics* 19:2551–2555.
86. Valentini, M., Pregosin, P. S., and Rüegger, H. 2000. Applications of pulsed field gradient spin-echo measurements for the determination of molecular volumes of organometallic ionic complexes. *J. Chem. Soc. Dalton Trans.* 4507–4510.
87. (a) Shvo, Y., Czarkie, D., Rahamim, Y., and Chodash, D. F. 1986. A new group of ruthenium complexes: Structure and catalysis. *J. Am. Chem. Soc.* 108:7400–7402. (b) Menashe, N., and Shvo, Y. 1991. Catalytic disproportionation of aldehydes with ruthenium complexes. *Organometallics* 10:3885–3891. (c) Menashe, N., Salant, E., and Shvo, Y. 1996. Efficient catalytic reduction of ketones with formic acid and ruthenium complexes. *J. Organomet. Chem.* 514:97–102.
88. Casey, C. P., Johnson, J. B., Singer, S. W., and Cui, Q. 2005. Hydrogen elimination from a hydroxycyclopentadienyl ruthenium(II) hydride: Study of hydrogen activation in a ligand–metal bifunctional hydrogenation catalyst. *J. Am. Chem. Soc.* 127:3100–3109.

89. (a) Mo, H., and Pochapsky, T. C. 1997. Self-diffusion coefficients of paired ions. *J. Phys. Chem. B* 101:4485–4486. (b) Pochapsky, S. S., Mo, H., and Pochapsky, T. C. 1995. Closed-shell ion pair aggregation in non-polar solvents characterized by NMR diffusion measurements. *J. Chem. Soc. Chem. Commun.* 2513–2514.
90. (a) Pregosin, P. S. 2006. Ion pairing using PGSE diffusion methods. *Prog. Nucl. Magn. Reson. Spectrosc.* 49:261–288. (b) Moreno, A., Pregosin, P. S., Veiros, L. F., Albinati, A., and Rizzato, S. 2008. PGSE NMR diffusion Overhauser studies on Ru(Cp*)(η^6-arene)][PF_6], plus a variety of transition-metal, inorganic, and organic salts: An overview of ion pairing in dichloromethane. *Chem. Eur. J.* 14:5617–5629. (c) Fernandez, I., and Pregosin, P. S. 2006. H-1 and F-19 PGSE diffusion and HOESY NMR studies on cationic palladium(II) 1,3-diphenylallyl complexes in THF solution. *Magn. Reson. Chem.* 44:76–82.
91. Bellachioma, G., Ciancaleoni, G., Zuccaccia, C., Zuccaccia, D., and Macchioni, A. 2008. NMR investigation of non-covalent aggregation of coordination compounds ranging from dimers and ion pairs up to nano-aggregates. *Coord. Chem. Rev.* 252:2224–2238.
92. (a) Zuccaccia, D., Foresti, E., Pettirossi, S., Sabatino, P., Zuccaccia, C., and Macchioni, A. 2007. From ion pairs to ion triples through a hydrogen bonding-driven aggregative process. *Organometallics* 26:6099–6105. (b) Bolaño, S., Ciancaleoni, G., Bravo, J., Gonsalvi, L., and Macchioni, A. 2008. PGSE NMR studies on RAPTA derivatives: Evidence for the formation of H-bonded dicationic species. *Organometallics* 27:1649–1652.
93. Zuccaccia, D., Bellachioma, G., Cardaci, G., Ciancaleoni, G., Zuccaccia, C., Clot, E., and Macchioni, A. 2007. Interionic structure of ion pairs and ion quadruples of half-sandwich Ru(II) salts bearing α-diimine ligands. *Organometallics* 26:3930–3946.
94. Zuccaccia, C., Bellachioma, G., Cardaci, G., and Macchioni, A. 2000. Self-diffusion coefficients of transition-metal complex ions, ion pairs, and higher aggregates by pulsed field gradient spin-echo NMR measurements. *Organometallics* 19:4663–4665.
95. Terrill, R. H., Postlethwaite, T. A., Chen, C.-H., Poon, C.-D., Terzis, A., Chen, A., Hutchison, J. E., Clark, M. R., Wignall, G., Londono, J. D., Superfine, R., Falvo, M., Johnson, C. S., Jr., Samulski, E. T., and Murray, R. W. 1995. Monolayers in three dimensions: NMR, SAXS, thermal, and electron hopping studies of alkanethiol stabilized gold clusters. *J. Am. Chem. Soc.* 117:12537–12548.
96. Shen, L., Soong, R., Wang, M., Lee, A., Wu, C., Scholes, G. D., Macdonald, P. M., and Winnik, M. A. 2008. Pulsed field gradient NMR studies of polymer adsorption on colloidal CdSe quantum dots. *J. Phys. Chem. B* 112:1626–1633.
97. Cohen, Y., Avram, L., and Frish, L. 2005. Diffusion NMR spectroscopy in supramolecular and combinatorial chemistry: An old parameter-new insights. *Angew. Chem. Int. Ed.* 44:520–554.
98. Bagno, A., Rastrelli, F., and Saielli, G. 2005. NMR techniques for the investigation of solvation phenomena and non-covalent interactions. *Prog. Nucl. Magn. Reson. Spectros.* 47:41–93.
99. Seidel, S. R., and Stang, P. J. 2002. High-symmetry coordination cages via self-assembly. *Acc. Chem. Res.* 35:972–983.
100. You, C.-C., Hippius, C., Grüne, M., and Würthner, F. 2006. Light-harvesting metallo-supramolecular squares composed of perylene bisimide walls and fluorescent antenna dyes. *Chem. Eur. J.* 12:7510–7519.
101. Ciancaleoni, G., Di Maio, I., Zuccaccia, D., and Macchioni, A. 2007. Self-aggregation of amino-acidate half-sandwich ruthenium(II) complexes in solution: From monomers to nanoaggregates. *Organometallics* 26:489–496.
102. Carter, L. C., Davies, D. L., Duffy, K. T., Fawcett, J., and Russell, D. R. 1994. [(η^6-$C_6H_3Me_3$)Ru(L-Ala)Cl]. *Acta Crystallogr. C* 50:1559–1561.

103. Sheldrick, W. S., and Heeb, S. 1990. Synthesis and structural characterization of η6-arene-ruthenium(II) complexes of alanine and guanine derivatives. *Inorg. Chim. Acta* 168:93–100.
104. (a) Hashiguchi, S., Fujii, A., Takehara, J., Ikariya, T., and Noyori, R. 1995. Asymmetric transfer hydrogenation of aromatic ketones catalyzed by chiral ruthenium(II) complexes. *J. Am. Chem. Soc.* 117:7562–7563. (b) Haack, K. J., Hashiguchi, S., Fukii, A., Ikariya, T., and Noyori, R. 1997. The catalyst precursor, catalyst, and intermediate in the Ru^{II}-promoted asymmetric hydrogen transfer between alcohols and ketones. *Angew. Chem. Int. Ed. Engl.* 36:285–288.
105. Ciancaleoni, G., Zuccaccia, C., Zuccaccia, D., Clot, E., and Macchioni, A. 2009. Self-aggregation tendency of all species involved in the catalytic cycle of bifunctional transfer hydrogenation. *Organometallics* 28:960–967.
106. Ciancaleoni, G., Zuccaccia, C., Zuccaccia, D., and Macchioni, A. 2007. Combining diffusion NMR and conductometric measurements to evaluate the hydrodynamic volume of ions and ion pairs. *Organometallics* 26:3624–3626.
107. For ionic species, the equation becomes $N^{+/-} = V_H^{+/-} / V_H^{0,ip}$, where $V_H^{+/-}$ is the measured hydrodynamic volume of the cation or the anion and $V_H^{0,ip}$ is the hydrodynamic volume of the ion pair.
108. (a) Ts'O, P. O. P., Melvin, I. S., and Olson, A. C. 1963. Interaction and association of bases and nucleosides in aqueous solutions. *J. Am. Chem. Soc.* 85:1289–1296. (b) Martin, R. B. 1996. Comparisons of indefinite self-association models. *Chem. Rev.* 96:3043–3064.
109. (a) Watanabe, M., Murata, K., and Ikariya, T. 2003. Enantioselective Michael reaction catalyzed by well-defined chiral ru amido complexes: Isolation and characterization of the catalyst intermediate, Ru malonato complex having a metal–carbon bond. *J. Am. Chem. Soc.* 125:7508–7509. (b) Watanabe, M., Ikagawa, A., Wang, H., Murata, K., and Ikariya, T. 2004. Catalytic enantioselective Michael addition of 1,3-dicarbonyl compounds to nitroalkenes catalyzed by well-defined chiral Ru amido complexes. *J. Am. Chem. Soc.* 126:11148–11149.
110. Burini, A., Fackler, J. P., Jr., Galassi, R., Macchioni, A., Omary, M. A., Rawashdeh-Omary, M. A., Pietroni, B. R., Sabatini, S., and Zuccaccia, C. 2002. ^{49}F, ^{1}H-HOESY and PGSE NMR studies of neutral trinuclear complexes of AuI and HgII: Evidence for acid-base stacking in solution. *J. Am. Chem. Soc.* 124:4570–4571.
111. Pregosin, P. S., Kumar, G. A., and Fernandez, I. 2005. Pulsed gradient spin-echo (PGSE) diffusion and ^{1}H,^{19}F heteronuclear Overhauser spectroscopy (HOESY) NMR methods in inorganic and organometallic chemistry: Something old and something new. *Chem Rev.* 105:2977–2998.

6 Pressure-Induced Change of *d-d* Luminescence Energies, Vibronic Structure, and Band Intensities in Transition Metal Complexes

Christian Reber, John K. Grey, Etienne Lanthier, and Kari A. Frantzen

CONTENTS

Abstract: The effects of hydrostatic pressure on the luminescence spectra of tetragonal transition metal complexes with nondegenerate electronic ground states are analyzed quantitatively using models based on potential energy surfaces defined along normal coordinates. Pressure-induced changes of intensity distributions within vibronic progressions, band maxima, electronic origins, and relaxation rates are discussed for metal-oxo complexes of rhenium(V) and molybdenum(IV) (d^2 electron configuration) and for square-planar complexes of palladium(II) and platinum(II) (d^8 electron configuration).

Keywords: Luminescence spectroscopy, *d-d* transitions, pressure, metal-oxo complexes, square-planar complexes, molybdenum(IV), palladium(II), platinum(II), rhenium(V)

INTRODUCTION

External pressure provides an important pathway to explore the variation of many aspects of solid-state structures, from electrostatic effects to covalent bonds and relatively weak intermolecular interactions, providing an intriguing field for both experimental and theoretical research in a variety of disciplines.[1–3] A large number of pressure-dependent physical properties for many different materials have been reported, as described in a number of extensive reviews with detailed bibliographies covering applications in chemistry, materials science, and physics, as well as the experimental methodology, in particular for spectroscopic measurements.[2–8]

Transition metal complexes are particularly attractive for the study of pressure effects, due to their high-symmetry structures, with many possibilities for subtle variations induced by relatively modest pressures, and their electronic structure, with degenerate and nondegenerate electronic states. Luminescence and absorption spectra of many transition metal compounds, including organometallic molecules, have been measured, and pressure-induced variations have been reported for the energies of their band maxima.[2,4,5,7,8] Excited states with different multiplicities are often close in energy, as illustrated by numerous studies of octahedral chromium(III) complexes, where a pressure-induced emitting state crossover from a quartet to a doublet state has been observed.[7] The corresponding change from a triplet to a singlet emitting state for an octahedral vanadium(III) complex was recently reported.[9] Many literature studies focus on pressure-induced spectroscopic effects caused by ground-state metal-ligand bond length changes, illustrated, for example, by spin-crossover complexes, where pressure can lead to very large metal-ligand bond length changes and even to crystallographic phase transitions.[10,11] Large spectroscopic changes have been observed as a consequence of a few intermolecular effects, involving, for example, the stacking of square-planar d^8 complexes,[12] where pressure-dependent luminescence and triboluminescence phenomena have been compared.[13] Pressure effects on intermolecular distances in luminescent gold(I) cyanides have been reported to lead to significant red shifts of the luminescence maxima due to shorter metal-metal distances.[7,14] A middle ground between intra- and intermolecular effects of pressure is occupied by exchange-coupled polymetallic complexes[6,15,16] and materials of interest as molecular magnets.[17] Phenomena such as piezochromism, mechanochromism, and their characterization through luminescence spectroscopy and other properties, such as electrical conductivity measurements, have been reviewed recently.[18–20] There are many effects where small changes to the environment of transition metal compounds create large changes of their properties, reported as tribochromism[21] and vapochromism.[22,23] Several of the compounds showing these phenomena can be probed by luminescence spectroscopy, and adjustable external pressure provides an important tool to study and control such effects.

This chapter addresses an apparent gap in the literature: On the one hand, structural changes occupy a prominent place in high-pressure research; on the other hand, the vast majority of literature reports on high-pressure luminescence and absorption spectroscopy focus on band maxima and rationalize the observed variations entirely in terms of electronic energy levels, neglecting effects due to pressure-induced changes of the structural differences between the initial and final states of the transition. The traditional approach allows transitions to be classified based on the pressure-induced shifts of band maxima,[4,5,24,25] but more detailed comparisons have to take into account the vibronic nature of electronic transitions, leading to resolved structure and broad bands for many transition metal compounds. In ambient pressure spectroscopy, often carried out at low temperature, the vibronic spectra obtained with different spectroscopic techniques have been quantitatively analyzed with theoretical models,[26–30] and the chemical and photochemical consequences of excited-state distortions have been discussed.[27,31] Only a small number of recent studies link these two aspects and determine pressure-dependent structural changes between the ground and excited states of transition metal complexes, for example, in octahedral halide complexes of chromium(III)[32] and vanadium(III),[9] where luminescence spectra were used, and in permangante, where pressure-dependent absorption and resonance Raman spectra have been analyzed.[33]

Theoretical work on pressure-dependent electronic spectra is based on potential energy surfaces. Traditionally, assumptions are made that lead to calculated variations of band maxima and widths using quantities that are not obvious to determine experimentally, such as local compressibilities and assumptions on the variation of crystal field parameters with metal-ligand bond distance.[34–36] Most often, experimental results with sufficient detail to determine potential energy surfaces were not available for these studies. Recent work based on electronic structure calculations shows interesting trends for relatively simple transition metal complexes, such as the octahedral VCl_6^{3-} anion,[37] and new general approaches have been described,[38] but not yet applied to transition metal complexes.

In the following, we summarize recent work on the combination of experimental and calculated spectra based on potential energy surfaces defined by adjustable parameters. Resolved vibronic structure is often observed for the examples presented, providing key information for the application of straightforward theoretical models. Two types of tetragonal complexes are explored: first, d^2-configured metal-oxo complexes, and second, square-planar complexes of metal ions with the d^8 electron configuration.

ONE-DIMENSIONAL NORMAL COORDINATE MODEL

The potential energy surfaces for the initial and final states of a transition are a key aspect of any model used to calculate electronic spectra. All examples discussed in the following involve transition metal complexes with nondegenerate electronic ground states, leading to luminescence spectra arising from a transition to a single electronic state. The simplest quantitative model for a luminescence transition involves two harmonic potential energy curves along a single normal coordinate, as illustrated in Figure 6.1. The highest-energy transition in this model is the electronic origin of the luminescence spectrum, denoted as E_{00}. The luminescence band maximum is at E_{max}, also given in Figure 6.1. The potential energy minima are offset by

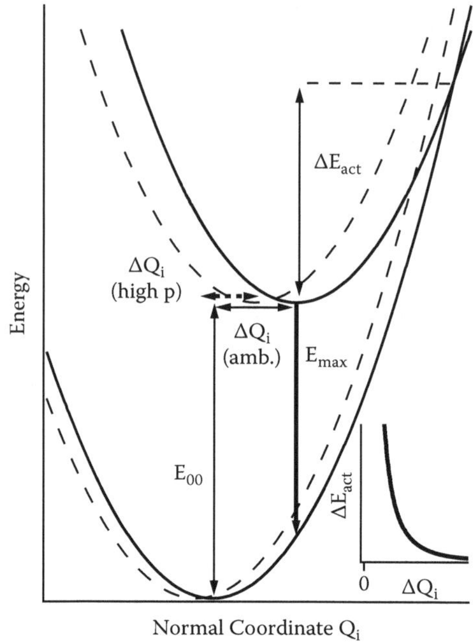

FIGURE 6.1 Potential energy curves at ambient pressure (solid) and at high pressure (dotted) for the ground and emitting states along a single normal coordinate Q_i. The spectroscopic parameters E_{00}, E_{max}, ΔQ_i, and ΔE_{act} are defined. The only change at high pressure is a decrease of ΔQ_i. The inset shows the variation of the activation energy ΔE_{act} with ΔQ_i.

an amount ΔQ_i along the normal coordinate Q_i. The final parameter of interest is the activation energy ΔE_{act}, which, in a simple classical view, determines the nonradiative relaxation rate constant: If it is high, the nonradiative relaxation processes are expected to be inefficient. The change of these four parameters with pressure will be analyzed from luminescence spectra for the transition metal complexes discussed in the following. The only other quantities needed to define the curves in Figure 6.1 are the vibrational frequencies of the mode associated with the normal coordinate Q_i in the ground and emitting states. The ground-state frequencies are found to change by very small amounts from pressure-dependent Raman spectra, and such frequency changes do not have a significant effect on the luminescence spectra.

The model in Figure 6.1 is chosen for a substantial offset ΔQ_i and for a situation where the emitting-state potential energy minimum is at a larger value of Q_i than the ground-state minimum, corresponding to weaker metal-ligand bonds in the emitting state. For this case, it is easy to qualitatively estimate the effect of external pressure on ΔQ_i: Its magnitude is expected to decrease because pressure affects the emitting-state minimum more strongly than the ground-state minimum, as illustrated by the dotted potential energy curves in Figure 6.1. This model therefore corresponds to a complex for which ΔQ_i decreases under high pressure, as indicated in Figure 6.1 by ΔQ_i (high pressure), which is set to a smaller value than ΔQ_i at ambient pressure. In order to obtain the dotted potential energy curves, the value of ΔQ_i was decreased by

9%, a value comparable to the 10% to 15% decreases of offsets ΔQ_i reported in the literature for halide complexes of chromium(III) between ambient pressure and 50 kbar.[32] Changes of the energies E_{00} and E_{max} are also expected with pressure. Their magnitudes and signs depend on the specific bonding situation and will be discussed in the following. The influence of pressure on the activation energy ΔE_{act} is shown in the inset to Figure 6.1: As ΔQ_i decreases, the activation energy increases strongly.

Calculated luminescence spectra for both sets of potential energy curves in Figure 6.1 are shown in Figure 6.2 at high and low resolution. Such calculations are easily carried out for harmonic and anharmonic potential energy surfaces.[26–30] The spectra shown as solid traces correspond to the ambient pressure potential energy curves denoted by solid curves in Figure 6.1. A long progression is observed as a consequence of the large offset ΔQ_i. The members of the progression are separated by the ground-state vibrational frequency of the mode with normal coordinate Q_i, and the intensity distribution within the progression depends strongly on the magnitude of ΔQ_i. Such progressions are observed for many transition metal complexes, most often at low temperature. The decrease of ΔQ_i in Figure 6.1 leads to the spectra shown as dotted traces in Figure 6.2. The energies of the maxima forming the resolved progression are independent of the magnitude of ΔQ_i, as illustrated by the dotted vertical line, but the intensity distribution within the progression changes significantly, showing an increase for the high-energy members of the progression and a decrease for the maxima at lower energy. The change of ΔQ_i with pressure can

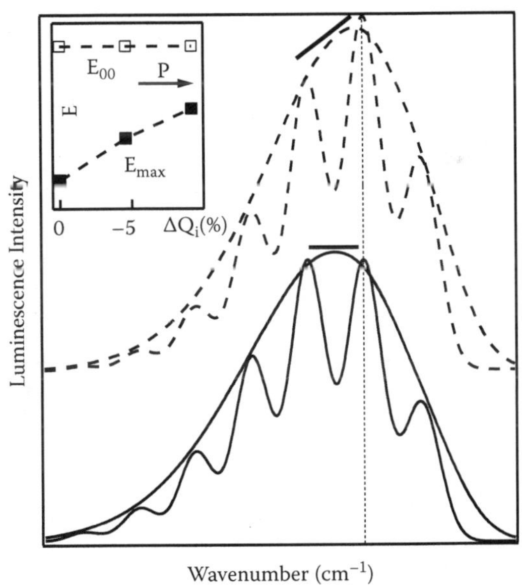

FIGURE 6.2 Luminescence spectra calculated from the potential energy curves in Figure 6.1. Spectra at ambient pressure are given by solid lines, and those at high pressure by dotted lines. The bars above the spectra illustrate the change in vibronic intensities resulting from a decrease of ΔQ_i in Figure 6.1. The inset shows the variation of E_{00} and E_{max} obtained from the calculated spectra.

be determined by fitting calculated spectra, as shown in Figure 6.2, to experimental spectra with resolved vibronic structure. Typically, ΔQ_i and E_{00} are treated as adjustable parameters, defining E_{max} and ΔE_{act}. At lower resolution, only an unresolved band with a single maximum is observed, as illustrated by the low-resolution envelopes in Figure 6.2. The maximum E_{max} of the spectrum at high pressure, shown as a dotted trace, is closer to the vertical dotted line than the maximum of the ambient pressure spectrum, denoted by the solid trace. The decrease of ΔQ_i with pressure is the only modification to any parameter value used to calculate the high-pressure spectra in Figure 6.2. It leads to a blue shift of the luminescence band maximum, as shown in the inset to Figure 6.2. The energy of the electronic origin E_{00} is constant in the high-pressure calculated spectra in the inset of Figure 6.2. The increase of E_{max} indicates that an interpretation of band maxima with pressure in terms of purely electronic models can be fallible: A pressure-induced shift of E_{max} occurs even if the electronic energy difference between the ground and emitting states, defining the electronic origin E_{00}, does not change. Vibronic effects, such as the decrease of the energy difference between E_{00} and E_{max} in the inset of Figure 6.2, need to be considered. The model in Figure 6.1 can be used to obtain quantitative parameter values from experimental spectra, as illustrated in the following.

PRESSURE EFFECTS ON VIBRONIC PROGRESSIONS AND LUMINESCENCE ENERGIES: METAL-OXO COMPLEXES

The luminescence spectra of many different metal-oxo complexes have been reported, and they often show distinct progressions in the metal-oxo stretching mode due to large offsets $\Delta Q_{metal\text{-}oxo}$ between the minima of the potential energy surfaces of the ground and emitting states along the metal-oxo normal coordinate.[39–46] In the following, we discuss the pressure effects on room temperature luminescence spectra of *trans*-dioxo rhenium(V) and mono-oxo molybdenum(IV) complexes, both containing metal centers with a d^2 electron configuration. All spectra were measured using a highly sensitive Raman microscope spectrometer with the 514.5 and 488.0 nm excitation lines of an argon ion laser. A diamond anvil cell was used to control the hydrostatic pressure on the sample crystals, and ruby luminescence was used for pressure calibration. The luminescence bands of the metal-oxo complexes discussed here are in the visible and near-infrared spectral regions and often show a dominant progression involving the metal-oxo stretching mode with a frequency of approximately 900 cm^{-1}.

The highest occupied and lowest unoccupied orbitals of these six-coordinate complexes arise from the t_{2g} orbitals in the O_h point group. The metal-oxo bonds are conventionally used to define the molecular z axis, leading to the occupied d_{xy} orbital (b_{2g} in D_{4h} point group symmetry), lower in energy than the empty $d_{xz,yz}$ (e_g in D_{4h} point group symmetry) orbitals.[39–45] Many complexes with different ancillary ligands in the xy plane show almost identical rhenium-oxo bond lengths of 1.765 Å, varying by only 0.001 Å for *trans*-dioxo complexes of rhenium(V) with ethylenediamine, pyridine, or 1-methylimidazole ligands in the xy plane.[47,48]

Figure 6.3 shows the pressure-dependent luminescence spectra of *trans*-$ReO_2(py)_4I$ and *trans*-$ReO_2(tmen)_2Cl$ at room temperature.[49–51] The abbreviations py

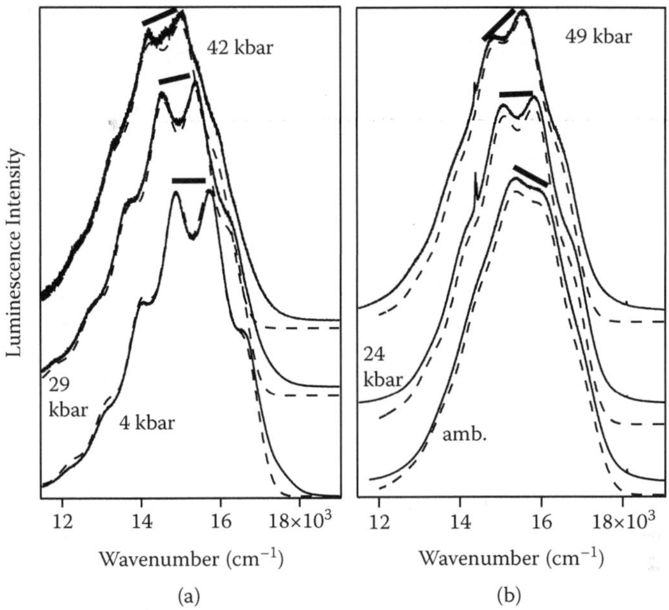

FIGURE 6.3 Experimental (solid lines) and calculated (dotted lines) pressure-dependent luminescence spectra of *trans*-$ReO_2(py)_4I$ (a) and *trans*-$ReO_2(tmen)_2Cl$ (b) at room temperature. Traces are offset along the ordinate for clarity, and all spectra are normalized to identical areas.

and tmen denote pyridine and tetramethylethylenediamine ligands, respectively. The progression in the metal-oxo mode is clearly visible in both spectra. As pressure increases, the intensity distribution within the progression changes toward higher intensities for the members of the progression at high energy. This is the change expected for a pressure-induced decrease in the offset $\Delta Q_{O=Re=O}$ along the metal-oxo coordinate, as discussed in the preceding section and shown in Figure 6.2. The overall change between ambient pressure and approximately 40 kbar is larger for *trans*-$ReO_2(tmen)_2Cl$ than for *trans*-$ReO_2(py)_4I$, as illustrated by the sloping lines above the spectra in Figure 6.3. In addition, a red shift of the entire luminescence band is observed as pressure increases, another consequence of the pressure-induced compression of metal-ligand bonds. The rhenium-nitrogen single bonds are weaker, and therefore more affected by pressure than the metal-oxo double bonds. These large bond length changes lead to stronger electrostatic crystal field changes for the filled d_{xy} orbital involving the metal-nitrogen bonds and to a higher destabilization than for the empty $d_{xz,yz}$ orbitals involved in the metal-oxo double bonds, whose energy is less affected by pressure. In the molecular orbital view, the metal-nitrogen π-antibonding character of the d_{xy} HOMO orbital leads to a strong increase of its energy with pressure, dominating the weaker energy increase of the metal-oxo π-antibonding $d_{xz,yz}$ LUMO orbitals and resulting in a red shift of the luminescence band.

Spectra calculated with one-dimensional potential energy curves along the rhenium-oxo coordinate are shown as dotted lines in Figure 6.3. They reproduce the

TABLE 6.1
Pressure Dependence of Room-Temperature Luminescence Parameters for Metal-Oxo Complexes, Obtained from Fits of Calculated Spectra to Experimental Spectra

Compound	E_{max} (cm^{-1}) ± ΔE_{max}/Dp (cm^{-1}/kbar)	$\Delta E_{00}/\Delta p$ (cm^{-1}/kbar)	$\Delta Q/\Delta p$ (Å/kbar^{-1})	$\hbar\omega_{metal\text{-}oxo}$ (cm^{-1}) + $\Delta\hbar\omega/\Delta p$ (cm^{-1}/kbar)
$ReO_2(py)_4I$[a]	15,360 – 15.7	–17.6	$-0.6 \cdot 10^{-4}$	905 + 0.53
$ReO_2(tmen)_2Cl$[b,c]	15,590 – 4.6	–8.4	$-1.9 \cdot 10^{-4}$	868 + 0.42
$ReO_2(en)_2Cl$[b,c]	13,780 – 6.8	–12.0	$-2.5 \cdot 10^{-4}$	898 + 0.37
MoOCl(CN-*t*-Bu)$_4$BPh$_4$[d]	11,950 + 12.0	n/a	≈0[e]	954 + 0.24
$MoOF(py)_4BPh_4$[d]	13,000 – 7.5	n/a	n/a	953 + 0.18

Note: Luminescence band maxima E_{max}, their pressure-induced changes and those of electronic origins E_{00}, offsets ΔQ_i along the metal-oxo stretching normal coordinate, and metal-oxo Raman frequencies are given.

[a] Grey et al.[51]
[b] Grey et al.[50]
[c] Grey et al.[53]
[d] Lanthier and Reber.[54]
[e] Estimated from the spectra in Figure 6.5.

experimental data precisely and lead to a quantitative determination of the parameters E_{00}, E_{max}, and $\Delta Q_{O=Re=O}$ for each pressure at which a spectrum was measured. The variations were found to be linear over the pressure range studied, and the slopes determined for the series of metal-oxo complexes compared here are summarized in Table 6.1. The band maxima E_{max} were determined from the calculated spectra by broadening each vibronic transition, as illustrated for the band envelopes in Figure 6.2. The pressure-induced variations of the parameters are illustrated in Figure 6.4 and show significant differences between these two compounds with similar ambient pressure luminescence properties. The band maxima for *trans*-$ReO_2(py)_4I$ show a much stronger red shift of –15.7 cm^{-1}/kbar than those of *trans*-$ReO_2(tmen)_2Cl$, where a shift of –4.6 cm^{-1}/kbar is obtained. The electronic origins for both compounds show a stronger red shift than the band maxima: –17.6 and –8.4 cm^{-1}/kbar for *trans*-$ReO_2(py)_4I$ and *trans*-$ReO_2(tmen)_2Cl$, respectively. This pressure-induced decrease of the energy difference between E_{00} and E_{max} again corresponds to the expectation for smaller $\Delta Q_{O=Re=O}$ values at high pressure, experimentally defined in the inset to Figure 6.2. This decrease, illustrated by the difference of the slopes for E_{00} and E_{max} in Figure 6.4, is approximately twice as large for *trans*-$ReO_2(tmen)_2Cl$ than for *trans*-$ReO_2(py)_4I$. Both the intensity distributions and the energies obtained from the calculated spectra indicate a decrease of $\Delta Q_{O=Re=O}$ with increasing pressure. The magnitude of this decrease for the two *trans*-dioxo complexes is compared in Figure 6.4b. The slopes differ by a factor of three, and $\Delta Q_{O=Re=O}$ values at 40 kbar are smaller by 7% and 2% than the ambient pressure values for *trans*-$ReO_2(tmen)_2Cl$ and *trans*-$ReO_2(py)_4I$,

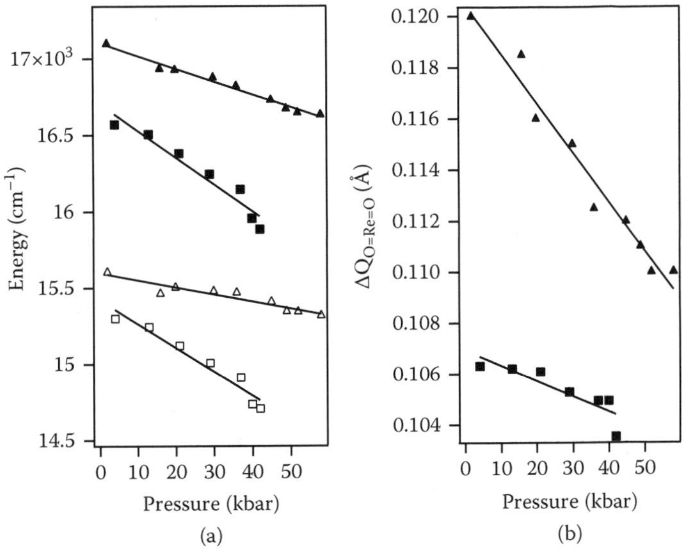

FIGURE 6.4 Pressure-induced variations of luminescence parameters for *trans*-$ReO_2(py)_4I$ (squares) and *trans*-$ReO_2(tmen)_2Cl$ (triangles). (a) Band maxima E_{max} (open symbols) and electronic origins E_{00} (solid symbols), (b) $\Delta Q_{O=Re=O}$. The slopes of the solid lines are given in Table 6.1.

respectively. The larger decrease for *trans*-$ReO_2(tmen)_2Cl$ is qualitatively obvious from the more prominent change in intensity distribution within the progression, illustrated in Figure 6.3 over an identical pressure range for both compounds. A more quantitative analysis of the spectra of *trans*-$ReO_2(tmen)_2Cl$ in Figure 6.3b reveals that the anharmonicity on the long-distance side of the ground state potential energy curve along the $Q_{O=Re=O}$ normal coordinate is trimmed off by high pressure.[49,50]

The different variations of $\Delta Q_{O=Re=O}$ with pressure for the two complexes in Figure 6.3 can be rationalized qualitatively from the effects of coupling between the ground state and excited states of identical symmetry. This effect is significant for *trans*-$ReO_2(tmen)_2^+$ but small for *trans*-$ReO_2(py)_4^+$.[45,50–52] It has been shown to more strongly influence lower-energy luminescence bands,[45,50] and therefore a larger pressure-induced variation of $\Delta Q_{O=Re=O}$ is expected for complexes such as *trans*-$ReO_2(en)_2^+$, where the pressure-induced decrease of the energy difference between E_{00} and E_{max} is more pronounced than for the *trans*-dioxo complexes with higher-energy luminescence bands in Figures 6.3 and 6.4, leading to a decrease of $\Delta Q_{O=Re=O}$ larger by 25% than for *trans*-$ReO_2(tmen)_2^+$, as summarized in Table 6.1. In addition to the luminescence energy, the size of the offset between ground- and emitting-state potential energy minima along the rhenium-ancillary ligand stretching coordinate also appears to influence the magnitude of the pressure-induced decrease of $\Delta Q_{O=Re=O}$. It is intuitively appealing to assume that large, monodentate ligands, such as pyridine, are more strongly affected by external pressure than compact, chelating ligands such as tetramethylethylenediamine, leading to the stronger red shift for *trans*-$ReO_2(py)_4I$, but this correlation is too simplistic, as a *trans*-dioxo complex with monodentate imidazole ligands shows a very small red shift of the luminescence band maximum E_{max} by only –2 cm^{-1}/kbar.[53] Pressure-dependent luminescence spectra reveal the important influence of the ancillary ligand, but it is obvious that a larger set of compounds needs to be studied in order to rationalize all observed effects. The comparison of the experimental and calculated luminescence spectra in Figure 6.3 leads to quantitative values for the parameters defining the ground- and emitting-state potential energy curves in Figure 6.3. The pressure-induced variations of $\Delta Q_{O=Re=O}$ and E_{00} or E_{max} are shown to be independent: A larger change of $\Delta Q_{O=Re=O}$ is observed for *trans*-$ReO_2(tmen)_2Cl$ than for *trans*-$ReO_2(py)_4I$, but the inverse order is obtained for the red shifts of E_{00} and E_{max}.

Mono-oxo complexes of d^2-configured metals provide an interesting comparison to *trans*-dioxo compounds. Molybdenum(IV) complexes are illustrative examples with easily discernible Mo-oxo progressions dominating the low-temperature luminescence spectra.[42,43,54] Vibronic progressions in the metal-oxo mode are shorter for mono-oxo compounds than for *trans*-dioxo complexes of both second- and third-row d-block metal ions.[42,43,54,55] Figure 6.5 shows the pressure-dependent luminescence spectra of $MoOCl(CN\text{-}t\text{-}Bu)_4BPh_4$. At ambient pressure and room temperature, the first and second members of the progression in the Mo-oxo modes are visible as a shoulder at approximately 12,700 cm^{-1} and as the overall maximum. Their relative intensities show no pressure-induced variation within experimental accuracy, as indicated by the sloped lines in Figure 6.5, in contrast to the spectra in Figure 6.3, where a change is easily observed. This indicates that the offset $\Delta Q_{Mo\text{-}oxo}$ changes very little over the pressure range in Figure 6.5, an important difference between the *trans*-dioxo and mono-oxo moieties.

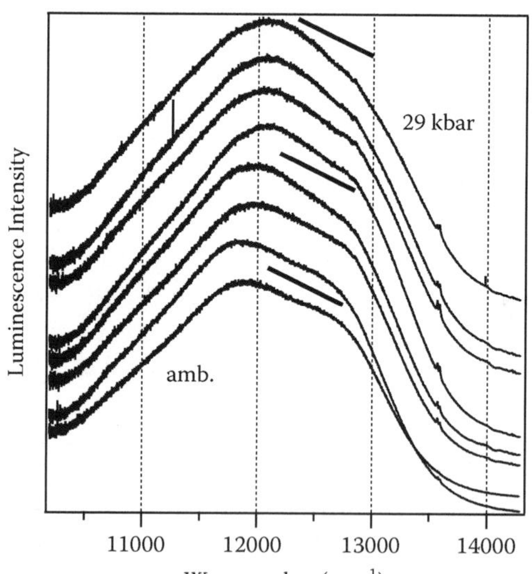

FIGURE 6.5 Pressure-dependent luminescence spectra of $MoOCl(CN\text{-}t\text{-}Bu)_4BPh_4$ at room temperature. The solid bars indicate the negligible variation of the intensity distribution within the vibronic progression in the Mo-oxo vibrational mode.

The band maximum in Figure 6.5 shows a blue shift of +12 cm^{-1}/kbar with pressure, in contrast to all *trans*-dioxo complexes in Figure 6.3 and Table 6.1, where a red shift is observed. This is a consequence of the strong π-acceptor character of the isocyanide ligands. Their π-bonding interactions with the metal d_{xy} orbital lead to a decrease in energy as the Mo-C bonds are compressed. A mono-oxo complex with pyridine ancillary ligands, $MoOF(py)_4BPh_4$, shows a pressure-induced red shift of −8 cm^{-1}/kbar for its band maximum, similar to the *trans*-dioxo complex of rhenium(V) with pyridine ligands.

The overview in this section is intended to illustrate that the predominantly metal-centered *d-d* transitions of metal-oxo complexes are well suited to a detailed exploration of pressure-induced luminescence effects caused by metal-ligand bonds with different characteristics, such as bond orders.

PRESSURE-INDUCED INCREASE OF LUMINESCENCE INTENSITIES AND PRESSURE-TUNED INTERMOLECULAR INTERACTIONS: SQUARE-PLANAR COMPLEXES

Square-planar complexes have long been of interest due to their open coordination sites along the fourfold rotation axis. Both intra- and intermolecular structural effects can be varied through external pressure, as reported recently in detailed variable-pressure structural studies of two-electron redox transformations in square-planar platinum(II) and palladium(II) halides[56] and intramolecular apical and equatorial metal-sulfur interactions in *cis*-[$PdCl_2$(1,4,7-trithiacyclononane)].[57] These studies show that metal-ligand bond lengths decrease by approximately 0.001 Å/kbar, and that changes of distances between a metal center and uncoordinated atoms are significantly larger, as illustrated by the observed decrease of the apical Pd···S distance by 0.005 Å/kbar in *cis*-[$PdCl_2$(1,4,7-trithiacyclononane)] between ambient pressure and 30 kbar.[57] The luminescence properties of square-planar complexes can be varied by slight changes of their environment, such as grinding crystals to a powder.[58] The pressure-dependent luminescence spectra of a variety of square-planar complexes of platinum(II) have been studied.[59–62] In the following, we focus on crystalline complexes with sulfur ligator atoms and *d-d* luminescence transitions. All structures show metal-metal distances longer than 8 Å, and no stacking of luminophores along the z axis, defined as the fourfold axis of the square-planar luminophore, occur.[60,62,63]

The series of complexes compared in the following are $M(SCN)_4^{2-}$ and $M(SeCN)_4^{2-}$, where M denotes palladium(II) and platinum(II). Low-temperature luminescence spectra show rich resolved vibronic structure with progressions involving the totally symmetric M-S stretching modes, as well as the nontotally symmetric stretching mode and the S-M-S bending mode.[60,61,64] The lowest-energy electronic transition from the singlet ground state to a triplet excited state involves the population of the σ-antibonding $d_{x^2-y^2}$ orbital and leads to a large change in metal-ligand bonding, giving rise to the vibronic structure of the luminescence spectra and to an expected blue shift of the luminescence band maximum due to shorter metal-ligand bonds at high pressure. Room-temperature luminescence spectra are shown for $Pd(SCN)_4^{2-}$ in Figure 6.6a. In contrast to the metal-oxo complexes discussed in the preceding section,

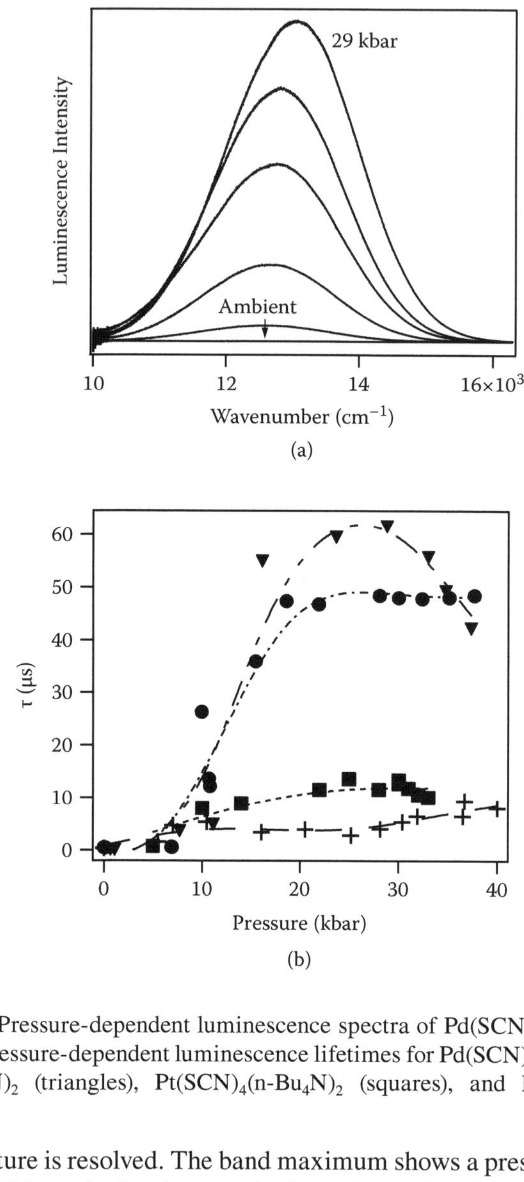

FIGURE 6.6 (a) Pressure-dependent luminescence spectra of $Pd(SCN)_4(n\text{-}Bu_4N)_2$ at room temperature. (b) Pressure-dependent luminescence lifetimes for $Pd(SCN)_4(n\text{-}Bu_4N)_2$ (circles), $Pd(SeCN)_4(n\text{-}Bu_4N)_2$ (triangles), $Pt(SCN)_4(n\text{-}Bu_4N)_2$ (squares), and $Pt(SeCN)_4(n\text{-}Bu_4N)_2$ (crosses).

no vibronic structure is resolved. The band maximum shows a pressure-induced blue shift of +29 cm^{-1}/kbar, similar in magnitude to the shifts observed for octahedral halide complexes of first-row transition metals.[9,32] The most obvious pressure effect is the dramatic increase of the luminescence intensity shown in Figure 6.6a.[61] The ambient pressure luminescence is very weak, and temperature-dependent spectra and lifetimes indicate that nonradiative relaxation processes dominate the excited-state deactivation at room temperature. External pressure leads to more competitive radiative rates, resulting in more intense luminescence. The pressure-dependent luminescence lifetimes in Figure 6.6b for all four compounds in this series show a distinct increase of the lifetime as pressure increases, indicating that nonradiative

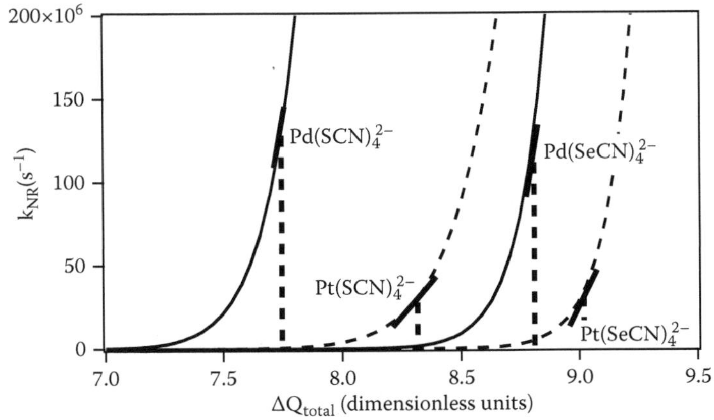

FIGURE 6.7 Variation of the nonradiative relaxation rate constant as a function of total offset between ground- and emitting-state potential energy minima for square-planar complexes.

relaxation rates decrease substantially with pressure.[60] At the highest pressures in Figure 6.6b, intensities and lifetimes decrease, likely due to efficient energy transfer among the closer-spaced complexes to quenching traps or pressure-induced structural imperfections. The increase of both luminescence intensities and lifetimes with pressure appears to be more pronounced for the palladium(II) complexes than for their platinum(II) analogs, as illustrated in Figure 6.6b, a difference that cannot be correlated with other phenomenological quantities, such as the pressure-induced shifts of the luminescence maxima. These shifts are +24 and +29 cm^{-1}/kbar for $Pt(SCN)_4^{2-}$ and $Pd(SCN)_4^{2-}$, respectively, but despite their similar magnitudes, very different enhancements of the luminescence intensities and lifetimes are observed for these two complexes in Figure 6.6b, indicating that the variation of other quantities, in particular the offsets ΔQ_i in Figure 6.1, are of importance.

The schematic view in Figure 6.1 is useful to qualitatively rationalize the pressure-induced decrease of the nonradiative relaxation rate. In the square-planar complexes, a decrease of ΔQ_i along several coordinates is expected as pressure increases, leading to a large increase of the activation energy ΔE_{act}, the classical barrier for nonradiative relaxation. This increase is strongly nonlinear, as illustrated in the inset to Figure 6.1. The ambient pressure offsets ΔQ_i can be determined for all compounds from low-temperature spectra with resolved vibronic structure. These are an important ingredient to models for nonradiative relaxation rates, in addition to vibrational frequencies of the modes associated with these normal coordinates. Temperature-dependent luminescence lifetimes lead to best-fit values for the adjustable parameters of established theoretical models for the nonradiative relaxation rate constants, qualitatively corresponding to the energy barrier ΔE_{act} in Figure 6.1 and the preexponential factor in the classical activation energy picture.[60] From this set of parameters, the variation of the nonradiative rate constant as the offsets ΔQ_i decrease is easily calculated without additional parameters. This variation is shown in Figure 6.7 for the four complexes as a function of the sum of all ΔQ_i values. External pressure

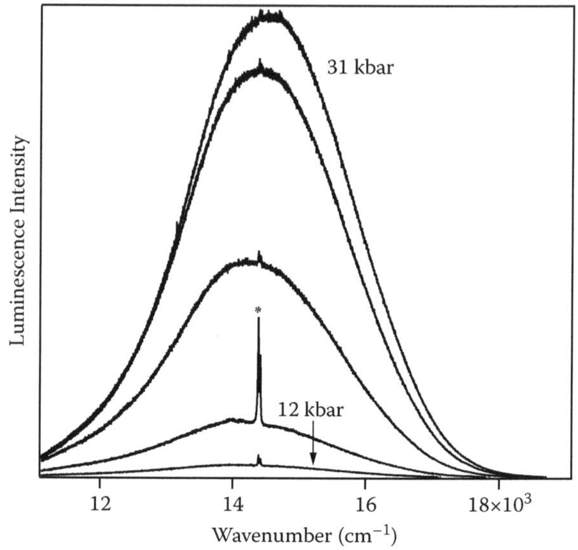

FIGURE 6.8 Pressure-dependent luminescence spectra of Pd(pyrrolidine-N-dithiocarbamate)$_2$ at room temperature. The asterisk denotes ruby luminescence used to calibrate the hydrostatic pressure in the diamond anvil cell.

causes a small decrease of ΔQ_{total}, and the slopes for each complex in Figure 6.7 indicate the magnitude of the variation of luminescence intensities and lifetimes. This simple approach immediately reveals larger slopes for palladium(II) complexes than for their platinum(II) analogs. The pressure-dependent luminescence intensities and lifetimes therefore provide a detailed view on excited-state relaxation, not accessible from ambient pressure data alone, and again emphasizing the importance of the offsets ΔQ_i, whose variation is neglected in purely electronic models.

The influence of individual offsets ΔQ_i on the increases of luminescence intensities and lifetimes still has to be explored in detail for these square-planar complexes. Complexes with chelating ligands are a first step in this direction. Figure 6.8 shows luminescence spectra of the Pd(pyrrolidine-N-dithiocarbamate)$_2$ complex.[65] Its luminescence energy and bandwidth are very similar to those of $Pd(SCN)_4^{2-}$, confirming that it originates from a *d-d* transition. The band maximum shows a blue shift of $+13\ cm^{-1}/kbar$. An obvious increase of the luminescence intensity is observed, but it is less pronounced than for the monodentate ligands in Figure 6.6a. It appears, therefore, that normal coordinates such as S-M-S bending that are blocked by the bidentate ligand play a significant role for the observed pressure effect. The changes in luminescence properties arising through external pressure from intramolecular ground- and emitting-state effects can again be rationalized in the context of the model defined in Figure 6.1.

Intermolecular interactions in square-planar platinum(II) complexes can have a significant influence on the *d-d* luminescence spectrum. An example based on the $[Pt(SCN)_4]^{2-}$ luminophore is shown in Figure 6.9. The trimetallic $\{Pt(SCN)_2[\mu\text{-SCN})Mn(NCS)(bipyridine)_2]_2\}$ complex has a luminescence maximum and bandwidth

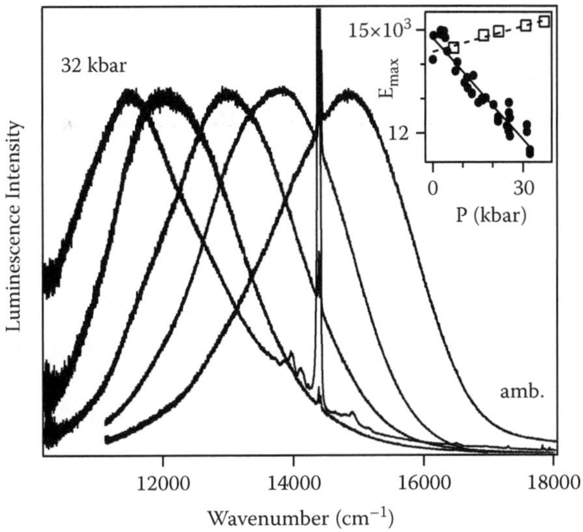

FIGURE 6.9 Pressure-dependent single-crystal luminescence spectra of the trimetallic complex {$Pt(SCN)_2[m$-$SCN)Mn(NCS)(bipyridine)_2]_2$}. The inset shows the strong red shift of the band maximum for the trimetallic complex (solid circles and line) compared to the blue shift observed for $Pt(SCN)_4(n\text{-}Bu_4N)_2$ (open squares and dotted line).

similar to those of $Pt(SCN)_4(n\text{-}Bu_4N)_2$ at ambient pressure, but its maximum shows a pressure-induced red shift of –99 cm^{-1}/kbar,[66] in contrast to the blue shift of +24 cm^{-1}/kbar observed for $Pt(SCN)_4(n\text{-}Bu_4N)_2$.[60] These different trends are shown in the inset of Figure 6.9. Neither the magnitude nor the sign of this red shift can be rationalized with the intramolecular effects discussed in the preceding paragraphs. It arises from intermolecular interactions between bipyridine ligands of neighboring complexes and the d_{z^2} orbital of the platinum(II) center. The intermolecular distances decrease strongly as pressure increases and influence the molecular electronic states involved in the metal-centered transition in specific structures, such as that of the trimetallic complex in Figure 6.9.[66] Pressure-induced red shifts of the *d-d* luminescence maxima have also been observed for platinum(II) complexes with 1,4,7-trithiacyclononane ligands,[67] illustrating the dominant influence of the decreasing apical Pt···S distance. Comparable red shifts of luminescence band maxima have been reported for other types of emission transitions and arise from metal-metal interactions in stacked structures with metal centers separated by distances on the order of 3 Å at ambient pressure,[12,59] indicative of the variety of interactions that can be probed by pressure-dependent spectroscopy.

CONCLUSIONS

This chapter summarizes how detailed insight on a variety of effects can be gained from pressure-dependent *d-d* luminescence bands. The exploration of such effects has become very accessible, due to sensitive detection with microscope spectrometers combined with established, versatile diamond anvil cells. An interesting direction of future research is focused on the variation of other relatively weak interactions using luminescence transitions other than the *d-d* bands discussed in this chapter and other spectroscopic techniques, such as resonance Raman measurements. Such studies will most likely require a combination of pressure-dependent spectroscopy and crystallography. Pioneering work in this area has been carried out for salts of $Pt(CN)_4^{2-}$[12] and $Au(CN)_2^-$,[7,14] but many intriguing pressure effects for compounds outside these two classes of late transition metal compounds remain to be discovered.

ACKNOWLEDGMENTS

Financial support from the Natural Science and Engineering Research Council (Canada) is gratefully acknowledged. We thank Professor Ian S. Butler (McGill University, Montreal, Canada) for encouraging us to measure pressure-dependent spectra, for many helpful discussions, and for the loan of a diamond anvil cell for early measurements.

REFERENCES

1. Grochala, W., R. Hoffmann, J. Feng, and N. W. Ashcroft. 2007. The chemical imagination at work in very tight places. *Angew. Chem. Int. Ed.*, 46, 3620.
2. Drickamer, H. G. 1990. Forty years of pressure tuning spectroscopy. *Ann. Rev. Mater. Sci.*, 20, 1.
3. Hemley, R. J. 2000. Effects of high pressure on molecules. *Ann. Rev. Phys. Chem.*, 51, 763.
4. Drickamer, H. G. 1986. Pressure tuning spectroscopy. *Acc. Chem. Res.*, 19, 329.
5. Drickamer, H. G. 1974. Electronic transitions in transition metal compounds at high pressure. *Angew. Chem. Int. Ed.*, 13, 39.
6. Kenney III, J. W. 1999. Pressure effects on emissive materials. In *Optoelectronic Properties of Inorganic Compounds*, ed. D. M. Roundhill and J. P. Fackler Jr., 231. New York: Plenum Press.
7. Bray, K. L. 2001. High pressure probes of electronic structure and luminescence properties of transition metal and lanthanide systems. *Top. Curr. Chem.*, 213, 1.
8. Grey, J. K., and I. S. Butler. 2001. Effects of high external pressure on the electronic spectra of coordination compounds. *Coord. Chem. Rev.*, 219–221, 713.
9. Wenger, O. S., and H. U. Güdel. 2002. Luminescence spectroscopy of V^{3+}-doped Cs_2NaYCl_6 under high pressure. *Chem. Phys. Lett.*, 354, 75.
10. Gütlich, P., V. Ksenofontov, and A. B. Gaspar. 2005. Pressure effect studies on spin-crossover compounds. *Coord. Chem. Rev.*, 249, 1811.
11. Jeftic, J., C. Ecolivet, and A. Hauser. 2003. External pressure and light influence on internal pressure in a spin-crossover solid $[Zn:Fe(ptz)_6](BF_4)_2$. *High Press. Res.*, 23, 359.
12. Gliemann, G., and H. Yersin. 1985. Spectroscopic properties of the quasi one-dimensional tetracyanoplatinate(II) compounds. *Struct. Bond.*, 62, 87.
13. Leyrer, E., F. Zimmermann, J. I. Zink, and G. Gliemann. 1985. Triboluminescence, photoluminescence, and high-pressure spectroscopy of tetracyanoplatinate salts. Determination of the pressure at triboluminescent sites. *Inorg. Chem.*, 24, 102.
14. Fischer, P., J. Mesot, B. Lucas, A. Ludi, H. H. Patterson, and A. Hewat. 1997. Pressure dependence investigation of the low-temperature structure of $TlAu(CN)_2$ by high-resolution neutron powder diffraction and optical studies. *Inorg. Chem.*, 36, 2791.
15. Riesen, H., and H. U. Güdel. 1987. Pressure tuning of exchange interactions in dinuclear chromium(III) complexes. *Inorg. Chem.*, 26, 2347.

16. Riesen, H., and H. U. Güdel. 1987. Effect of high pressure on the exchange interactions in binuclear chromium(III) complexes. *J. Chem. Phys.*, 87, 3166.
17. Coronado, E., M. C. Giménéz-Lopez, G. Levchenko, F. M. Romero, V. Garciá-Baonza, A. Milner, and M. Paz-Pasternak. 2005. Pressure-tuning of magnetism and linkage isomerism in iron(II) hexacyanochromate. *J. Am. Chem. Soc*, 127, 4580.
18. Takeda, K., I. Shirotani, and K. Yakushi. 2000. Pressure-induced insulator-to-metal-to-insulator transitions in one-dimensional bis(dimethylglyoximato)platinum(II), $Pt(dmg)_2$. *Chem. Mater.*, 12, 912.
19. Takagi, H. D., K. Noda, and S. Itoh. 2004. Piezochromism and related phenomena exhibited by palladium complexes. *Platinum Met. Rev.*, 48, 117.
20. Todres, Z. V. 2004. Recent advances in the study of mechanochromic transitions. *J. Chem. Res.*, 89.
21. Lee, Y.-A., and R. Eisenberg. 2003. Luminescence tribochromism and bright emission in gold(i) thiouracilate complexes. *J. Am. Chem. Soc.*, 125, 7778.
22. Grove, L. J., J. M. Rennekamp, H. Jude, and W. B. Connick. 2004. A new class of platinum(II) vapochromic salts. *J. Am. Chem. Soc*, 126, 1594.
23. Mansour, M. A., W. B. Connick, R. J. Lachicotte, H. J. Gysling, and R. Eisenberg. 1998. Linear chain Au(I) dimer compounds as environmental sensors: A luminescent switch for the detection of volatile organic compounds. *J. Am. Chem. Soc.*, 120, 1329.
24. Drickamer, H. G., and K. L. Bray. 1990. Pressure tuning spectroscopy as a diagnostic for pressure-induced rearrangements (piezochromism) of solid-state copper(II) complexes. *Acc. Chem. Res.*, 23, 55.
25. Moreno, M., J. A. Aramburu, and M. T. Barriuso. 2004. Electronic properties and bonding in transition metal complexes: Influence of pressure. *Struct. Bond.*, 106, 127.
26. Heller, E. J. 1981. The semiclassical way to molecular spectroscopy. *Acc. Chem. Res.*, 14, 368.
27. Zink, J. I., and K.-S. K. Shin. 1991. Molecular distortions in excited electronic states determined from electronic and resonance Raman spectroscopy. *Adv. Photochem.*, 16, 119.
28. Reber, C., and J. I. Zink. 1992. Unusual features in absorption spectra arising from coupled potential energy surfaces. *Comments Inorg. Chem.*, 13, 177.
29. Wexler, D., J. I. Zink, and C. Reber. 1994. Spectroscopic manifestations of potential surface coupling along normal coordinates in transition metal complexes. *Top. Curr. Chem.*, 171, 173.
30. Brunold, T., and H. U. Güdel. 1999. Luminescence spectroscopy. In *Inorganic Electronic Structure and Spectroscopy*, ed. E. I. Solomon and A. B. P. Lever, 259. Vol. I. New York: John Wiley & Sons.
31. Zink, J. I. 2001. Photo-induced metal-ligand bond weakening, potential surfaces, and spectra. *Coord. Chem. Rev.*, 211, 69.
32. Wenger, O. S., R. Valiente, and H. U. Güdel. 2001. Influence of hydrostatic pressure on the Jahn–Teller effect in the $^4T_{2g}$ excited state of $CrCl_6^{3-}$ doped $Cs_2NaScCl_6$. *J. Chem. Phys.*, 115, 3819.
33. Khodadoost, B., S. Lee, J. B. Page, and R. C. Hanson. 1988. Resonance Raman scattering and optical absorption studies of MnO_4^- in $KClO_4$ at high pressure. *Phys. Rev. B*, 38, 5288.
34. Drickamer, H. G., C. W. Frank, and C. P. Slichter. 1972. Optical versus thermal transitions in solids at high pressure. *Proc. Natl. Acad. Sci. USA*, 69, 933.
35. Okamoto, B. Y., W. D. Drotning, and H. G. Drickamer. 1974. The evaluation of configuration coordinate parameters from high pressure absorption and luminescence data. *Proc. Natl. Acad. Sci. USA*, 71, 2671.
36. Moreno, M. 2002. Effects of hydrostatic and chemical pressures on impurities determined through optical parameters. *High Press. Res.*, 22, 29.

37. Seijo, L., and Z. Barandarián. 2003. High pressure effects on the structure and spectroscopy of V^{3+} substitutional defects in Cs_2NaYCl_6. An *ab initio* embedded cluster study. *J. Chem. Phys.*, 118, 1921.
38. Cruz, S. A., and J. Soullard. 2004. Pressure effects on the electronic and structural properties of molecules. *Chem. Phys. Lett.*, 391, 138.
39. Winkler, J. R., and H. B. Gray. 1983. Emission spectroscopic properties of dioxorhenium(V) complexes in crystals and solutions. *J. Am. Chem. Soc.*, 105, 1373.
40. Winkler, J. R., and H. B. Gray. 1985. Electronic absorption and emission spectra of dioxorhenium(v) complexes. Characterization of the luminescent ^3E_g state. *Inorg. Chem.*, 24, 346.
41. Miskowski, V. M., H. B. Gray, and M. D. Hopkins. 1996. Electronic structure of metal-oxo complexes. In *Adv. in Trans. Met. Coord. Chem.*, ed. C.-M. Che and V. W.-W. Yam, 159. Vol. 1. Greenwich, CT: JAI Press.
42. Isovitsch, R. A., A. S. Beadle, F. R. Fronczek, and A. W. Maverick. 1998. Electronic absorption spectra and phosphorescence of oxygen-containing molybdenum(IV) complexes. *Inorg. Chem.*, 37, 4258.
43. Da Re, R. E., and M. D. Hopkins. 2002. Electronic spectra and structures of d^2 molybdenum-oxo complexes. Effects of structural distortions on orbital energies, two-electron terms, and the mixing of singlet and triplet states. *Inorg. Chem.*, 41, 6973.
44. Savoie, C., and C. Reber. 1998. Emitting state energies and vibronic structure in the luminescence spectra of *trans*-dioxorhenium(V) complexes. *Coord. Chem. Rev.*, 171, 387.
45. Savoie, C., and C. Reber. 2000. Coupled electronic states in trans-dioxo complexes of rhenium(V) and osmium(VI) probed by near-infrared and visible luminescence spectroscopy. *J. Am. Chem. Soc.*, 122, 844.
46. Kirgan, R. A., B. P. Sullivan, and D. P. Rillema. 2007. Photochemistry and photophysics of coordination compounds: Rhenium. *Top. Curr. Chem.*, 281, 45.
47. Lock, C. J. L., and G. Turner. 1978. A reinvestigation of dioxobis(ethylenediamine) rhenium(V) chloride and dioxotetrakis(pyridine)rhenium(V) chloride dihydrate. *Acta Cryst.*, B34, 923.
48. Bélanger, S., and A. L. Beauchamp. 1996. Preparation and protonation studies of *trans*-dioxorhenium(V) complexes with imidazoles. *Inorg. Chem.*, 35, 7836.
49. Grey, J. K., M. Triest, I. S. Butler, and C. Reber. 2001. Effect of pressure on the vibronic luminescence spectrum of a *trans*-dioxo rhenium(V) complex. *J. Phys. Chem. A*, 105, 6269.
50. Grey, J. K., I. S. Butler, and C. Reber. 2002. Effect of pressure on coupled electronic ground and excited states determined from luminescence spectra of *trans*-dioxorhenium(V) complexes. *J. Am. Chem. Soc.*, 124, 11699.
51. Grey, J. K., I. S. Butler, and C. Reber. 2004. Temperature- and pressure-dependent luminescence spectroscopy on the *trans*-$[ReO_2(pyridine)_4]^+$ complex—Analysis of vibronic structure, luminescence energies, and bonding characteristics. *Can. J. Chem.*, 82, 1083.
52. Newsham, M. D., E. P. Giannelis, T. J. Pinnavaia, and D. G. Nocera. 1988. The influence of guest-host interactions on the excited-state properties of dioxorhenium(V) ions in intracrystalline environments of complex-layered oxides. *J. Am. Chem. Soc.*, 110, 3885.
53. Grey, J. K., M. Marguerit, I. S. Butler, and C. Reber. 2002. Pressure-dependent Raman spectroscopy of metal-oxo multiple bonds in rhenium(V) and osmium(VI) complexes. *Chem. Phys. Lett.*, 366, 361.
54. Lanthier, E., J. Bendix, and C. Reber. 2010. Pressure-dependent luminescence spectroscopy of molybdenum(IV) oxo complexes. *Dalton Trans.*, 39, 3695.
55. Del Negro, A. S., Z. Wang, C. J. Seliskar, W. R. Heineman, B. P. Sullivan, S. E. Hightower, T. L. Hubler, and S. A. Bryan. 2005. Luminescence from the *trans*-dioxotechnetium(V) chromophore. *J. Am. Chem. Soc.*, 127, 14978.

56. Heines, P., H.-L. Keller, M. Armbrüster, U. Schwarz, and J. Tse. 2006. Pressure-induced internal redox reaction of $Cs_2[PdI_4]·I_2$, $Cs_2[PdBr_4]·I_2$, and $Cs_2[PdCl_4]·I_2$. *Inorg. Chem.*, 45, 9818.
57. Allan, D. R., A. J. Blake, D. Huang, T. J. Prior, and M. Schröder. 2006. High pressure co-ordination chemistry of a palladium thioether complex: Pressure versus electrons. *Chem. Commun.*, 4081.
58. Abe, T., T. Itakura, N. Ikeda, and K. Shinozaki. 2009. Luminescence color change of a platinum(II) complex solid upon mechanical grinding. *Dalton Trans.*, 711.
59. Wenger, O. S., S. García-Revilla, H. U. Güdel, H. B. Gray, and R. Valiente. 2004. Pressure dependence of Pt(2,2′-bipyridine)Cl_2 luminescence. The red complex converts to a yellow form at 17.5 kbar. *Chem. Phys. Lett.*, 384, 190.
60. Grey, J. K., I. S. Butler, and C. Reber. 2003. Pressure-induced enhancements of luminescence intensities and lifetimes correlated with emitting-state distortions for thiocyanate and selenocyanate complexes of platinum(II) and palladium(II). *Inorg. Chem.*, 42, 6503.
61. Grey, J. K., I. S. Butler, and C. Reber. 2002. Large pressure-induced increase in luminescence intensity for the $[Pd(SCN)_4]^{2-}$ complex. *J. Am. Chem. Soc.*, 224, 9384.
62. Hidvegi, I., W. Tuszynski, and G. Gliemannn. 1981. Luminescence of $K_2Pt(SCN)_4$ single crystals at high pressure. *Chem. Phys. Lett.*, 77, 517.
63. Rohde, J.-U., B. von Malottki, and W. Preetz. 2000. Kristallstrukturen, spektroskopische Charakterisierung und Normalkoordinatenanalyse von $(n\text{-}Bu_4N)_2M(ECN)_4$. *Z. Anorg. Allg. Chem.*, 626, 905.
64. Pelletier, Y., and C. Reber. 2000. Luminescence spectroscopy and emitting-state properties of $Pd(SCN)_4^{2-}$ in crystals. *Inorg. Chem.*, 39, 4535.
65. Genre, C., G. Levasseur-Thériault, and C. Reber. 2009. Emitting-state properties of square-planar dithiocarbamate complexes of palladium(II) and platinum(II) probed by pressure-dependent luminescence spectroscopy. *Can. J. Chem.*, 87, 1625.
66. Levasseur-Thériault, G., C. Reber, C. Aronica, and D. Luneau. 2006. Large pressure-induced red shift of the luminescence band originating from nonstacked square-planar $[Pt(SCN)_4]^{2-}$ in a novel trimetallic complex. *Inorg. Chem.*, 45, 2379.
67. Pierce, E., E. Lanthier, C. Genre, Y. Chumakov, D. Luneau, and C. Reber. 2010. The Interaction of Thioether Groups at the Open Coordination Sites of Palladium(II) and Platinum(II) Complexes Probed by Luminescence Spectroscopy at Variable Pressure. *Inorg. Chem.*, 49, 4901.

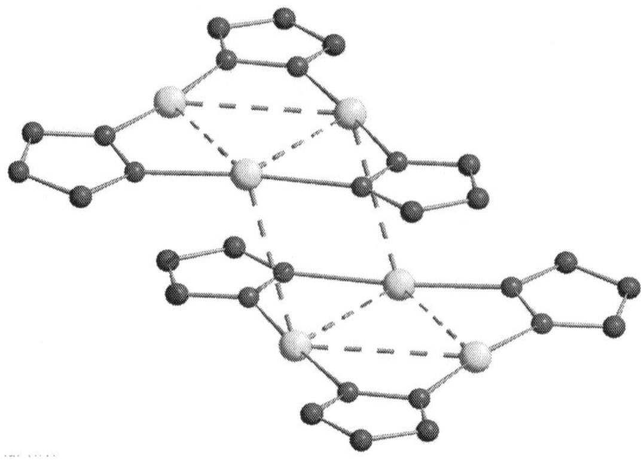

FIGURE 2.2 Schematic drawing of the $[Ag(pz)]_3$ trimers. Short intra- and intermolecular contacts are evidenced as fragmented lines.

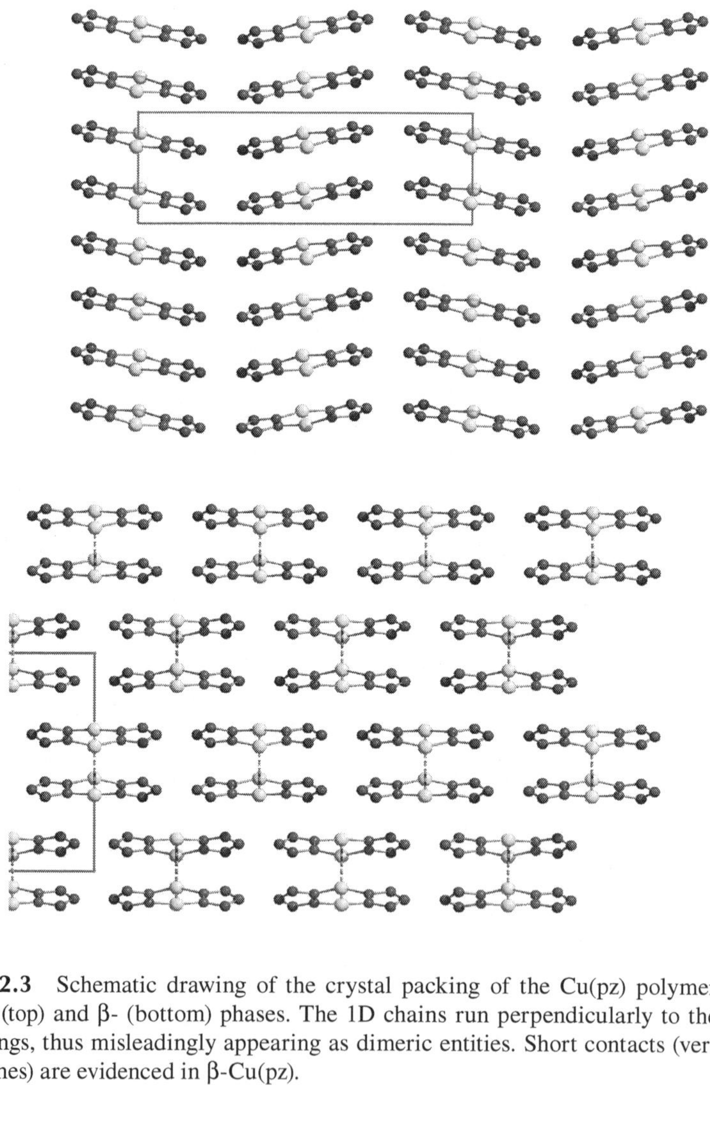

FIGURE 2.3 Schematic drawing of the crystal packing of the Cu(pz) polymeric chains in the α- (top) and β- (bottom) phases. The 1D chains run perpendicularly to the plane of the drawings, thus misleadingly appearing as dimeric entities. Short contacts (vertical fragmented lines) are evidenced in β-Cu(pz).

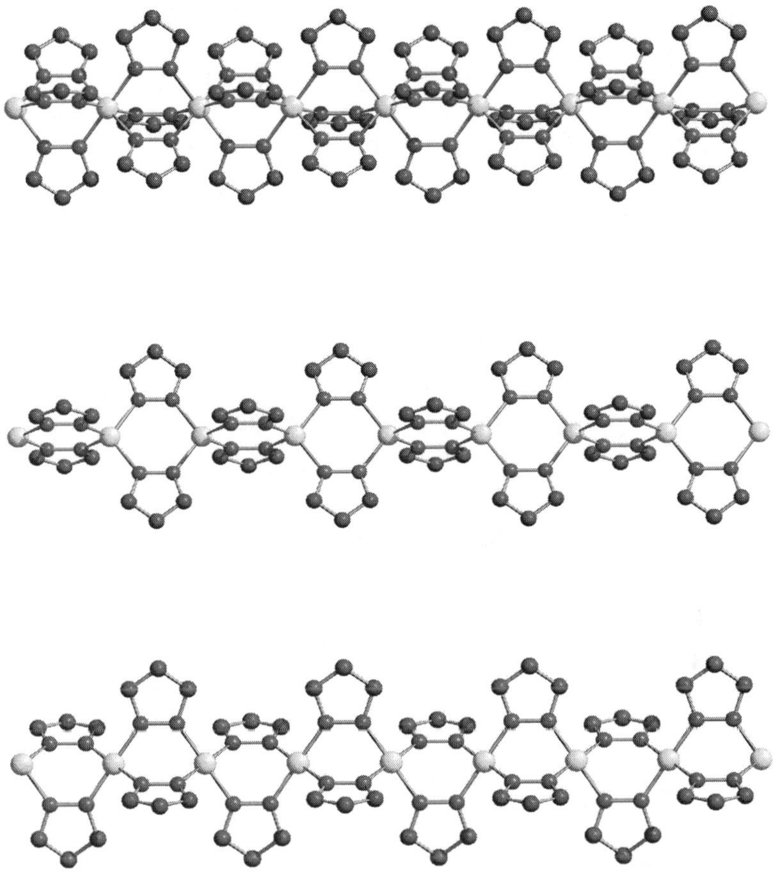

FIGURE 2.4 Schematic drawing of the $M(pz)_n$ polymeric chains. Top to bottom: $Fe(pz)_3$, $Co(pz)_2$, and $Ni(pz)_2$. In the two polymorphs of the latter, identical chains pack in pseudohexagonal (α) or pseudorectangular (β) fashion. In all cases, metal atoms are strictly collinear.

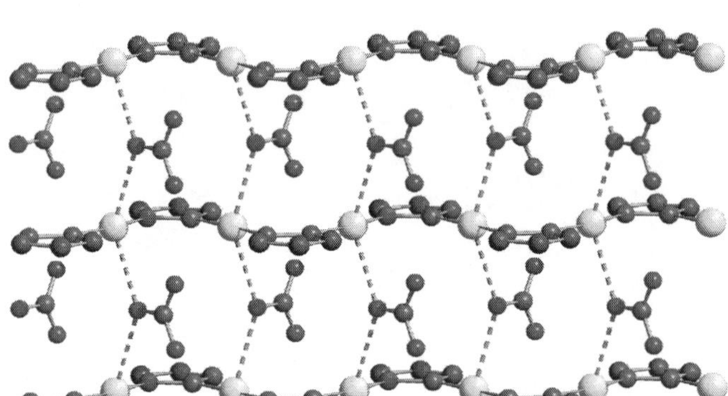

FIGURE 2.5 Schematic drawing of a portion of the $[Hg(pz)]_n^{n+}$ chain, surrounded by loosely interacting (dashed lines) nitrate ions.

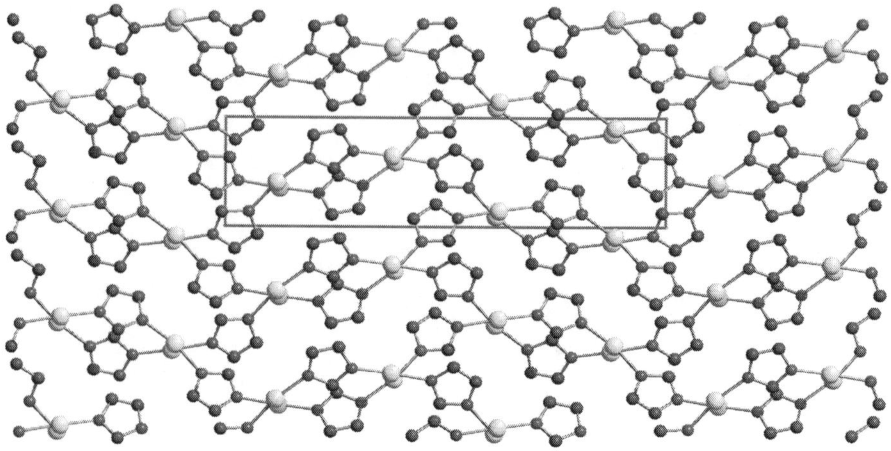

FIGURE 2.7 Schematic drawing of the crystal packing of Ag(im), viewed down [010]. The one-dimensional polymeric chains run nearly parallel to the horizontal cell axis, **c**.

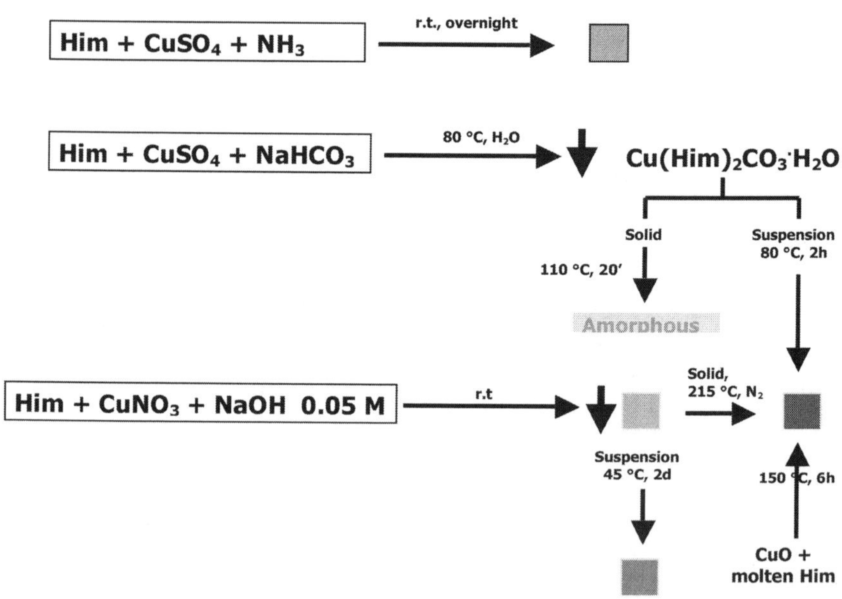

FIGURE 2.8 Reaction conditions employed for the preparation of monophasic samples of the different $Cu(im)_2$ polymorphs, each one indicated by a squared color label.

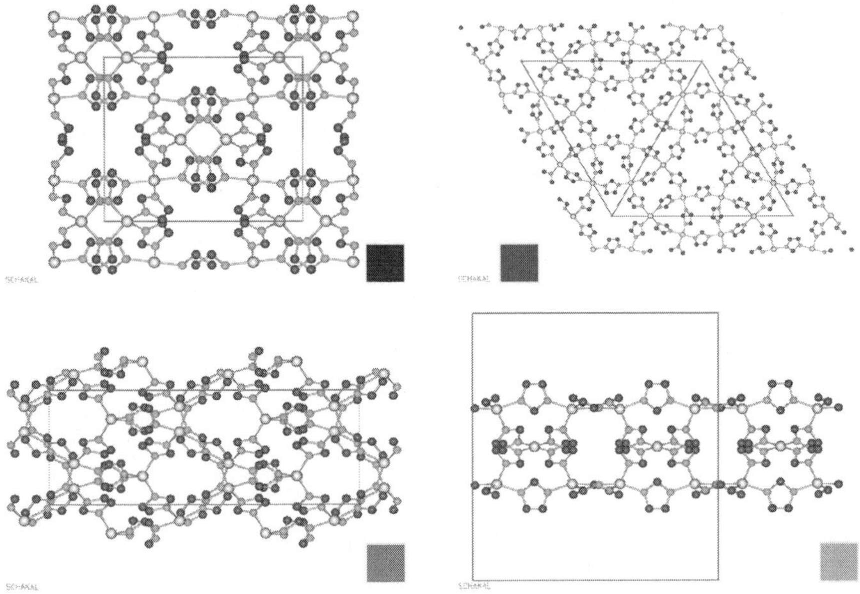

? XRPD pattern similar to $Ni(im)_2$ yellow (square planar d^8)

FIGURE 2.9 Schematic drawing of the crystal packing of $Cu(im)_2$: left to right, top to bottom: J, B, O, and G polymorphs. The structure of the pink (P) phase is still unknown.

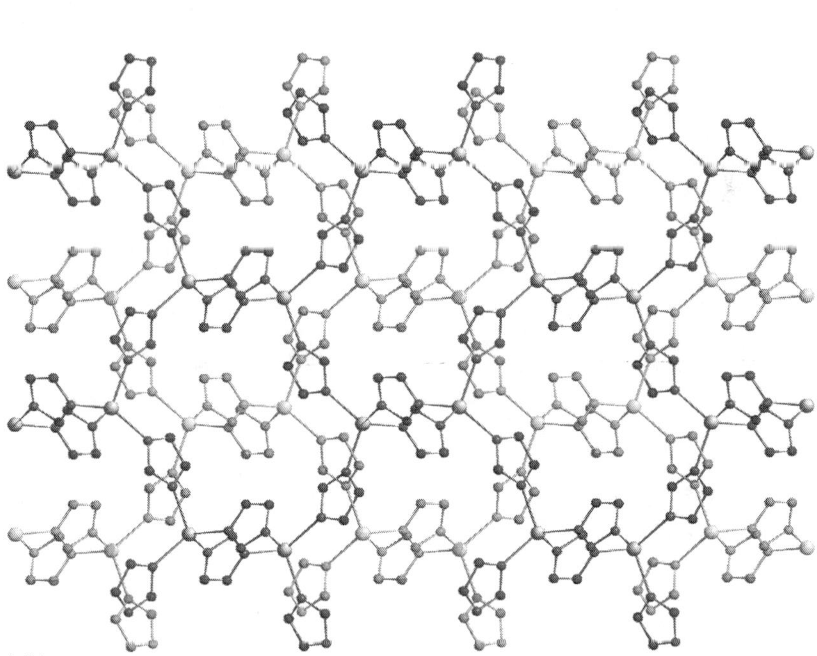

FIGURE 2.10 Schematic drawing of the $M(im)_2$ (M = Cd, Hg) species. The two interpenetrating diamondoid frameworks are shown in red and green.

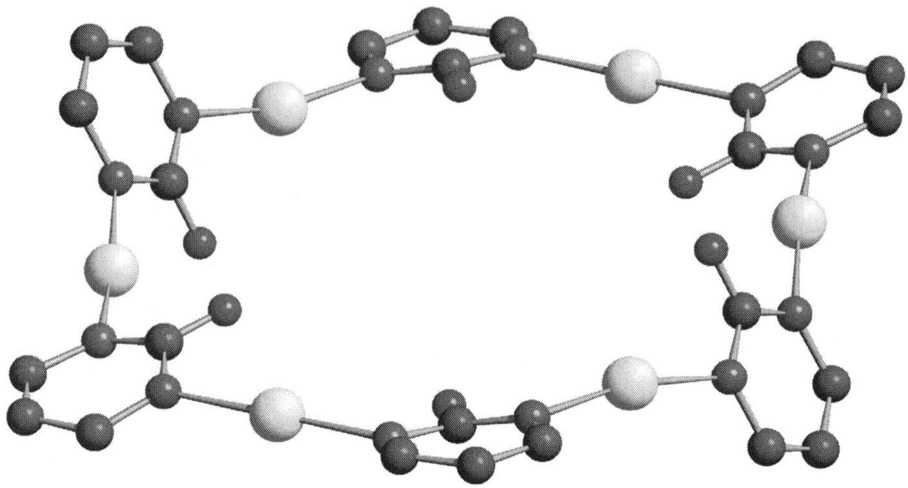

FIGURE 2.12 Schematic drawing of the $[Ag(2\text{-pymo})]_6$ molecule, possessing crystallographic C_{2h} symmetry. Note the folded character of the whole ring.

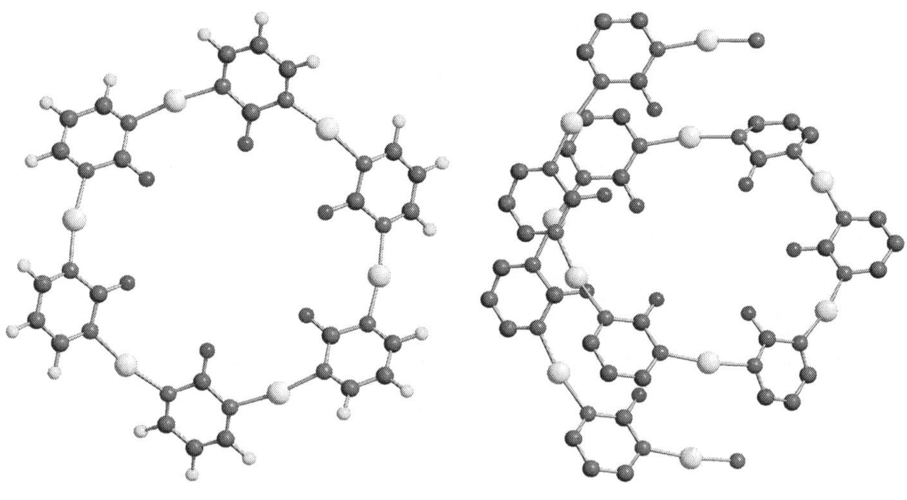

FIGURE 2.13 Molecular drawing of (a) the $[Cu(2\text{-pymo})]_6$ hexamer (similar to that found in the silver analogue) and (b) a portion of the infinite helix of $[Cu_2(2\text{-pymo})_2]_n$.

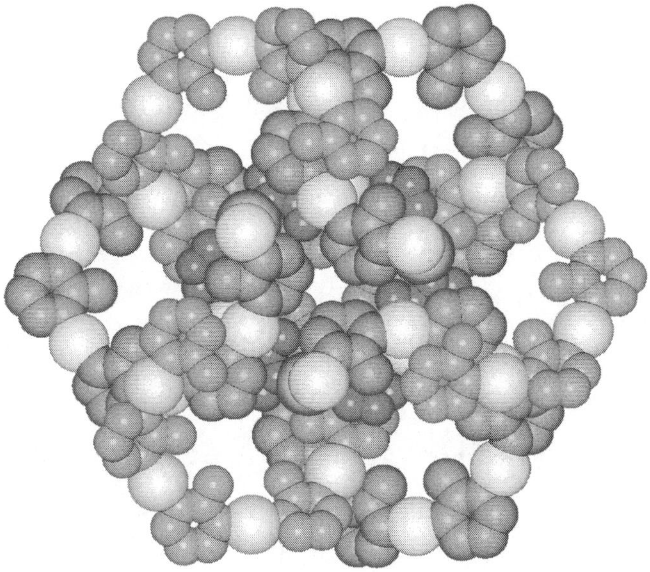

FIGURE 2.14 A packing diagram of the $1/6[Cu(2\text{-pymo})]_6 \cdot 1/n[Cu_2(2\text{-pymo})_2]_n$ species, viewed down [001]; helices of different polarity are depicted by different colors (yellow and green); within this trigonal packing of helices, *closed* cavities about the origin host the $[Cu(2\text{-pymo})]_6$ hexamer (red).

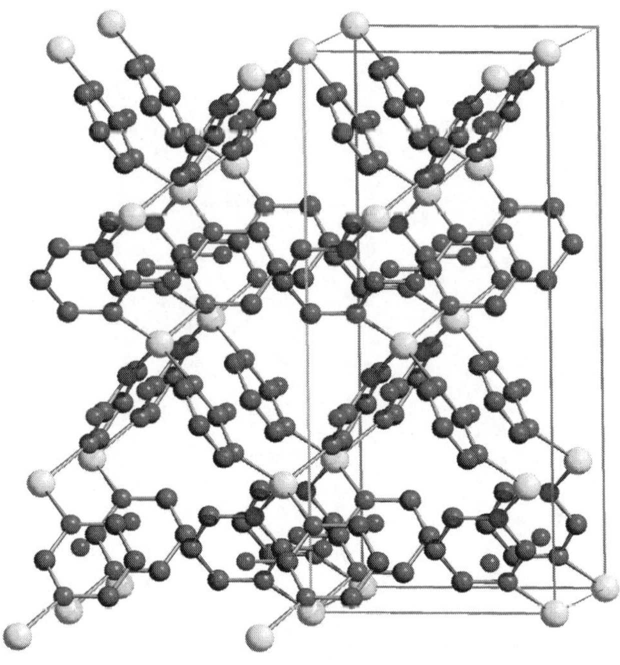

FIGURE 2.15 Partial drawing of the 3D diamondoid network in $Co(2\text{-pymo})_2$.

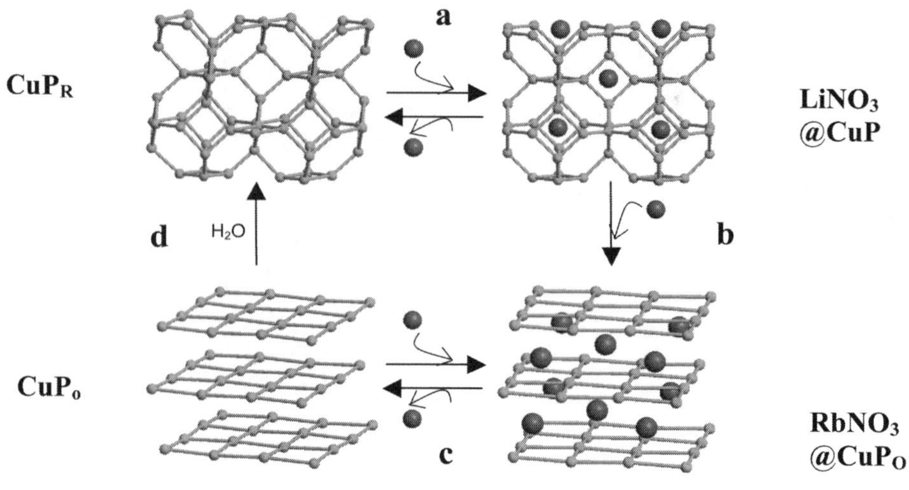

FIGURE 2.18 Guest-induced transformations in the Cu(2-pymo)$_2$ (**CuPR**) framework. (a) Incorporation of n/3 MNO_3. (b) Additional incorporation of 1/6 MNO_3. (c) Removal of 1/2 MNO_3. (d) Water addition. For the M cations, see text. The balls and sticks denote Cu and pyrimidin-2-olate-*N,N′*-bridges, respectively. Coordinates from the crystal structures of **CuPR**, **LiNO$_3$@CuPR**, and **RbNO$_3$@CuPO**.

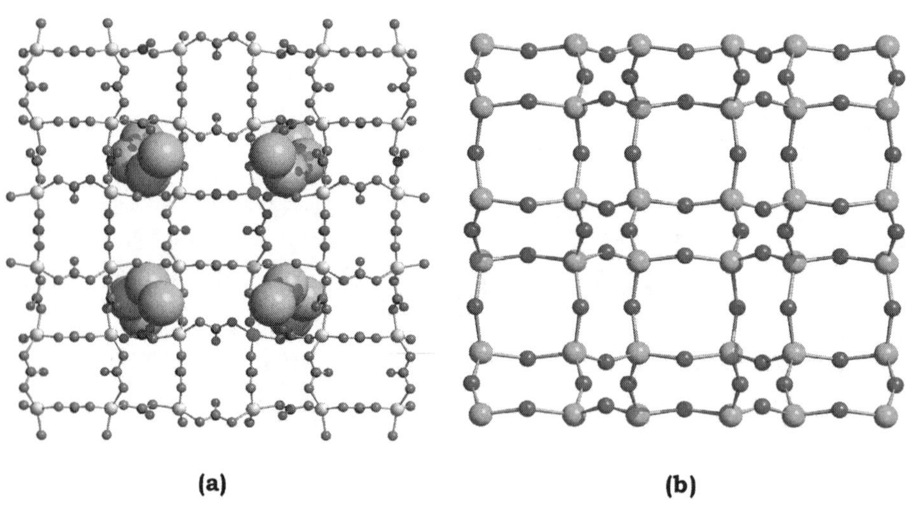

FIGURE 2.22 View, down [001], of (a) the crystal structure of Cu(F-pymo)$_2$, as compared to (b) the inorganic gismondine. Highlighted with PCK style, the four helical channels comprised within one unit cell.

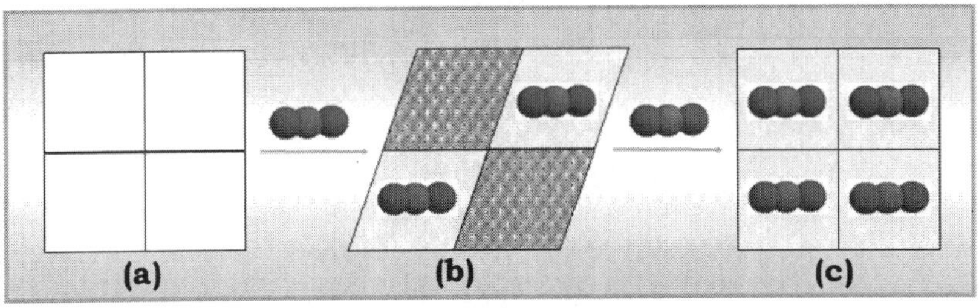

FIGURE 2.23 Structural modifications experienced by the Cu(F-pymo)$_2$ MOF upon CO_2 inclusion: (a) activated, void framework; (b) start of the CO_2 settling within two opposite helical channels out of the four of the unit cell, with concomitant rhombic distortion; (c) settling of the CO_2 molecules in the remaining two helical channels, with restoring of the pristine tetragonal symmetry. The orientation of the CO_2 molecules within the framework has been arbitrarily assigned.

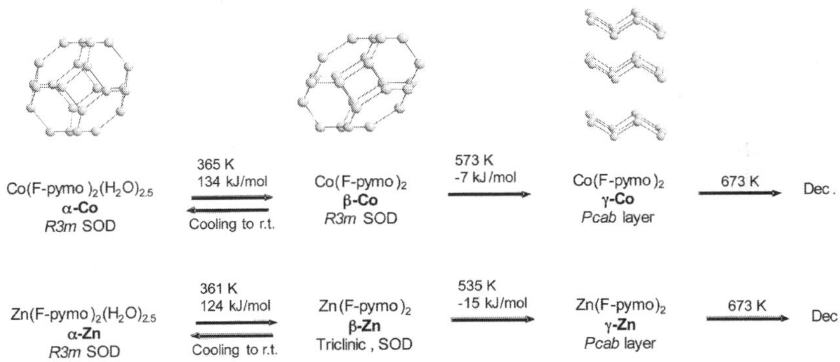

FIGURE 2.24 Schematic representation of the thermal behavior of the M(F-pymo)$_2$(H_2O)$_{2.5}$ species (**β-M**, M = Co, Zn), as retrieved from the concomitant usage thermal analysis and thermodiffractometry.

FIGURE 2.25 Custom-made temperature-controlled sample holder mounted on our x-ray powder diffractometer.

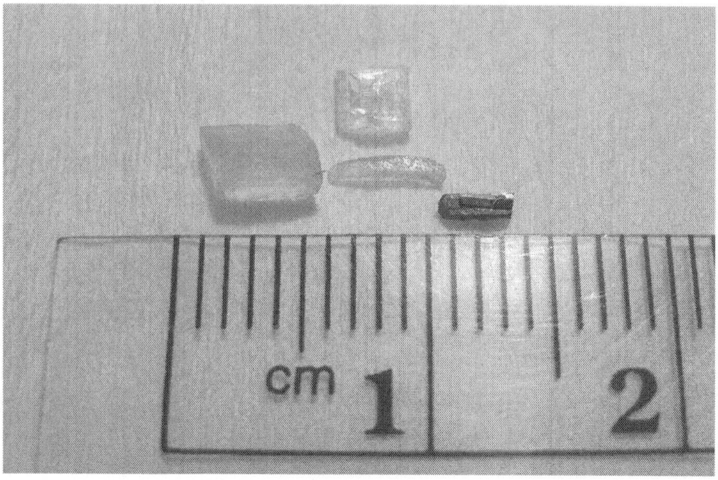

FIGURE 3.3 Some representative samples from the IPNS. Although a minimum sample size of approximately 1 mm^3 in volume is required for the typical single crystal experiment, in practice larger samples are often desirable.

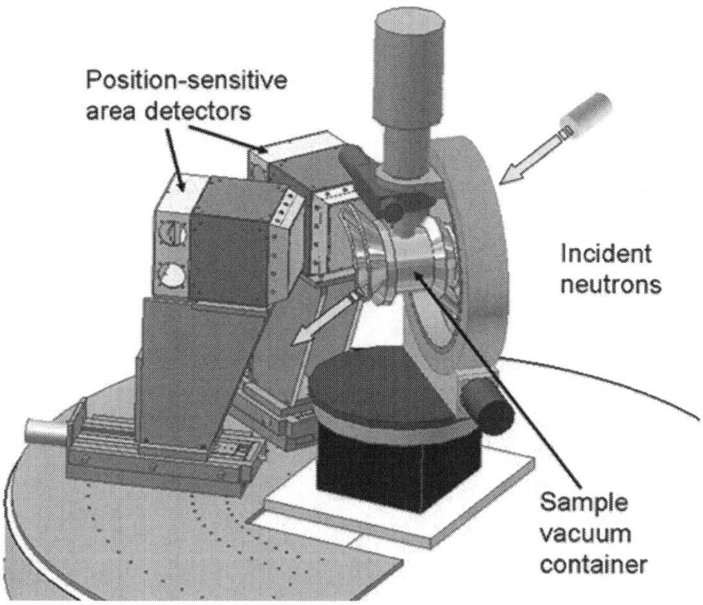

FIGURE 3.6 Diagram of the IPNS SCD instrument now at Los Alamos. The sample is mounted in the center of the vacuum chamber and can be rotated 90° about the χ circle and 360° about φ. The closed-cycle refrigerator is mounted vertically on the φ axis. Two position-sensitive area detectors are centered at 75° and 120° scattering angles from the sample, and can cover a large volume of reciprocal space. The crystal is stationary during the collection of each data frame; approximately 22 settings of the diffractometer are required to cover one hemisphere of reciprocal space.

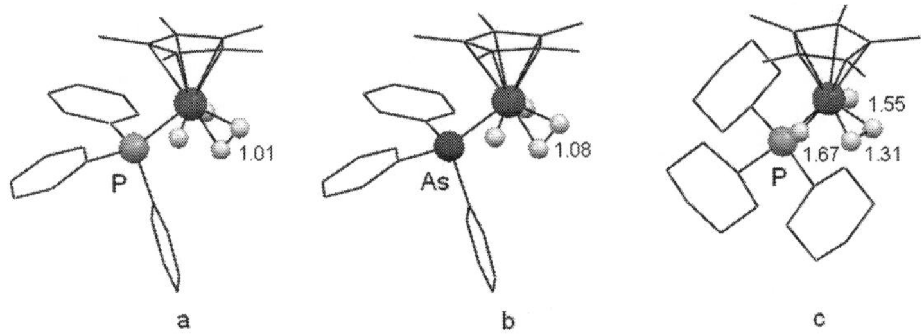

FIGURE 3.10 Neutron structures of $[Cp^*Os(H)_2(\mu\text{-}H_2)L]^+$ complexes (L = PPh_3, $AsPh_3$, PCy_3), showing variation in the H-H bond as a function of sterics and electronics. Bond distances in Å.

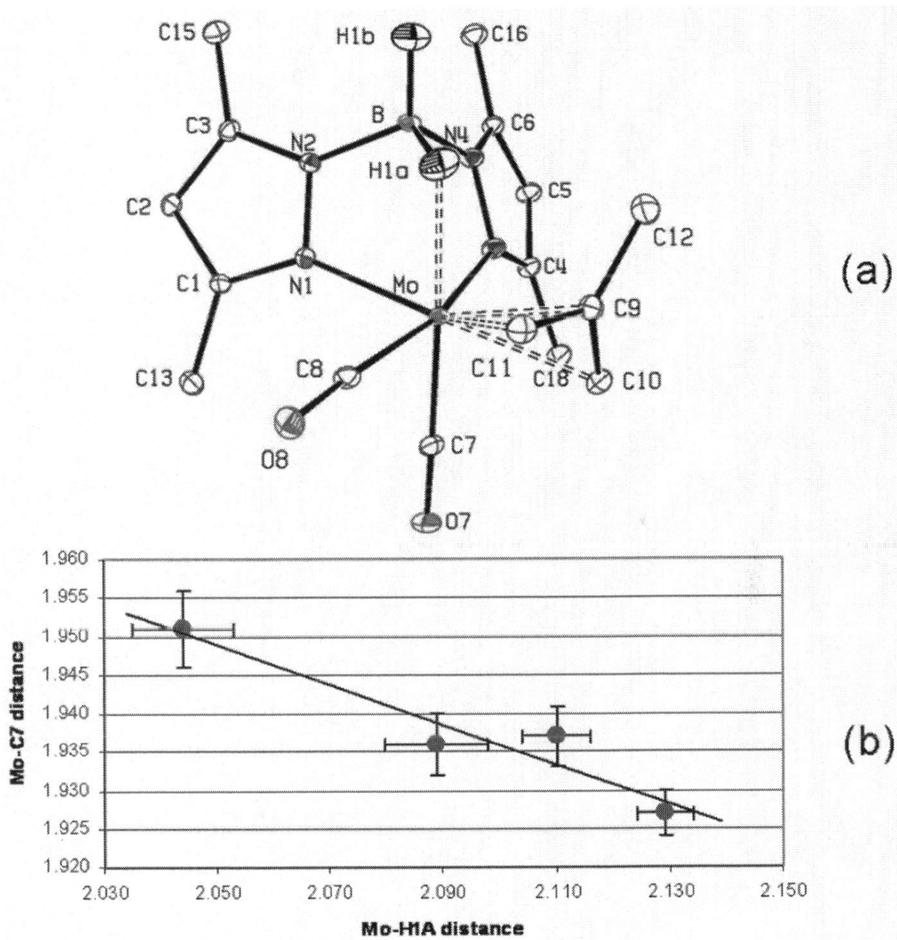

FIGURE 3.12 (a) The neutron structure of $Bp^*[Mo(CO)_2(\eta^3\text{-}C_3H_4Me)]$. (b) Plot of Mo-C7 distance versus Mo-H1A distance for a series of substituted scorpionates. As the substituents on the pyrazolylborate ligands become more electron withdrawing, the agostic Mo-H interaction becomes stronger (shorter Mo-H distance). This is reflected in the *trans* influence, where Mo-CO distances increase with decreasing Mo-H distance.

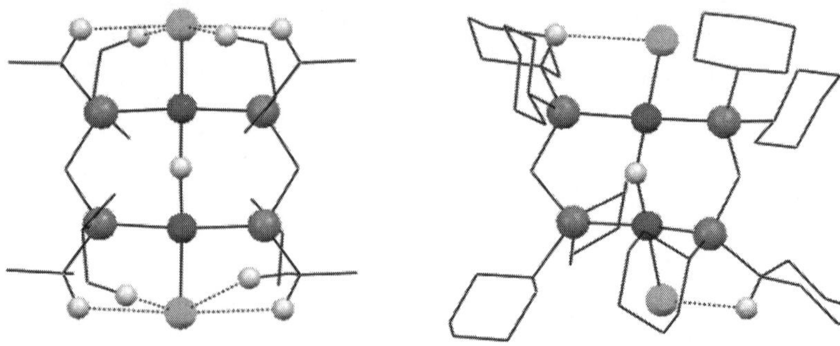

FIGURE 3.17 Comparison of the unsupported bridging hydride complexes [(dippm)$_2$Ni$_2$Cl$_2$](μ-H) (neutron structure, left) and [(dcpm)$_2$Ni$_2$Cl$_2$](μ-H) (x-ray structure, right). C-H···Cl contacts shorter than the sum of the van der Waals radii (2.95 Å) are shown as dashed lines. The sterics of the phosphine ligand are suggested to be the influencing factor on the geometry of the bridging hydride. In the linear example, the close approach of the isopropyl groups "locks" the chlorides into place, resulting in a linear hydride complex. Hydrogen atoms not involved in the hydrogen bridges have been omitted for clarity.

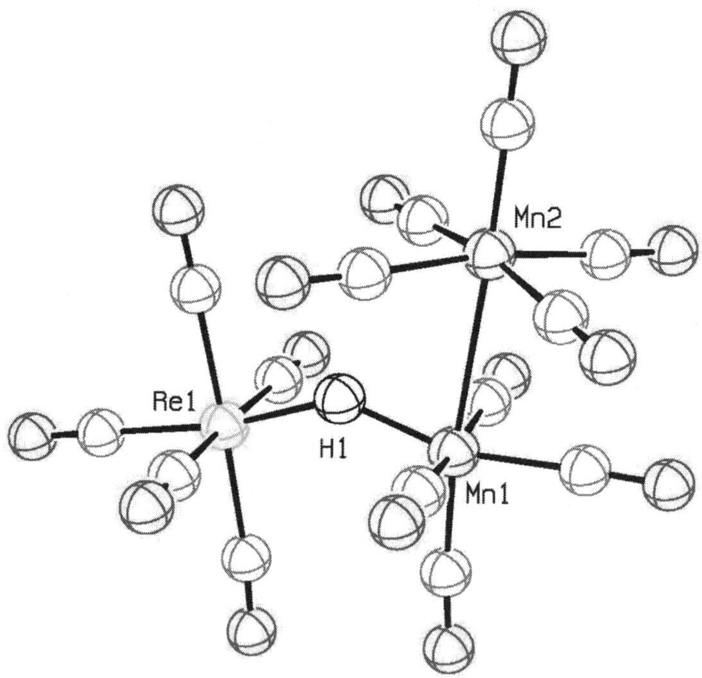

FIGURE 3.18 Neutron diffraction structure of (CO)$_5$Re(μ-H)Mn(CO)$_4$Mn(CO)$_5$. The Mn2 site was found to be 9.2% occupied by Re due to a co-crystallization of the isomorphous (CO)$_5$Re(μ-H)Mn(CO)$_4$Re(CO)$_5$.

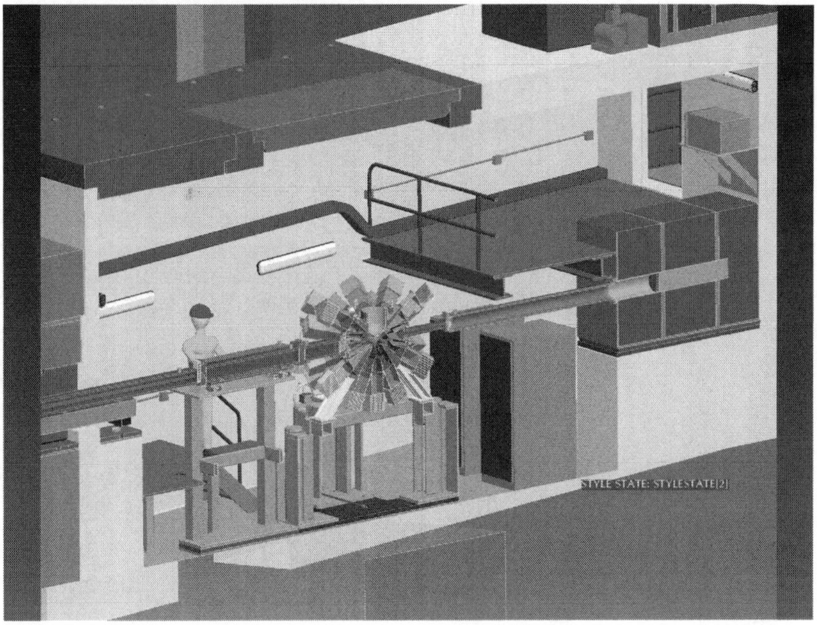

FIGURE 3.21 A cutaway diagram of the TOPAZ single-crystal diffractometer currently under construction at the SNS. This instrument will be ideal for smaller sample sizes and larger unit cells, both of which limit the number of currently feasible problems in single-crystal neutron diffraction.

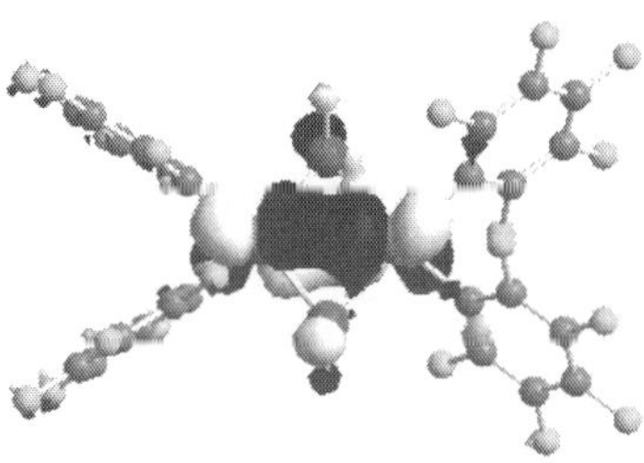

HOMO-15

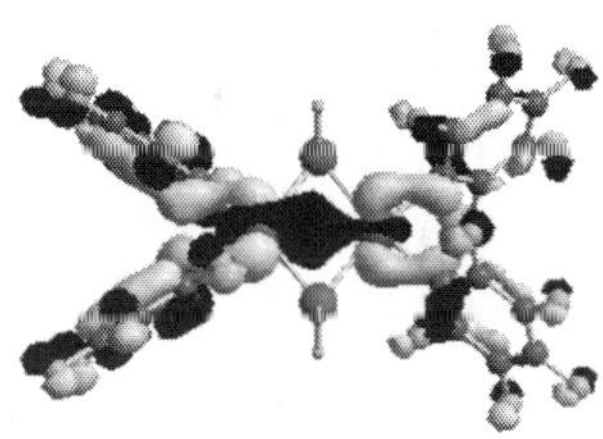

HOMO-18

Pt-Pt (σ-MO)

SCHEME 4.1 The MOs contributing to the formation of the Pt(III)-Pt(III) bonds in $[(C_6F_5)_2Pt(\mu\text{-}PH_2)_2Pt[(C_6F_5)_2]$ and $[(CF_3)_2Pt(\mu\text{-}PH_2)_2Pt(\mu\text{-}PH_2)_2Pt(CF_3)_2]$ complexes.

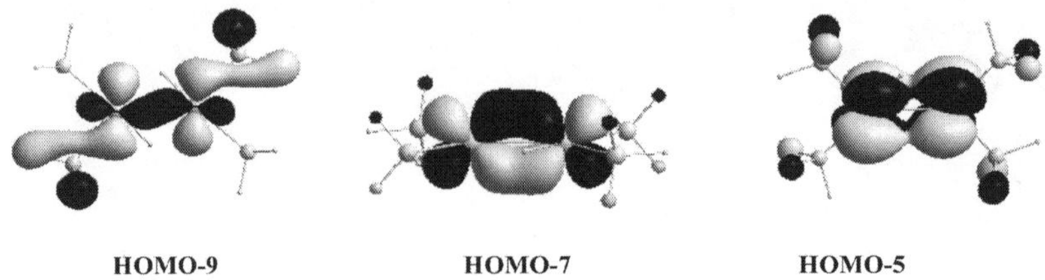

SCHEME 4.2 The most relevant molecular orbital interactions describing the Pt-Pt interactions in the model complex $[(H_3Si)(PH_3)Pt(\mu\text{-}H)_2Pt(H_3Si)(PH_3)]$.

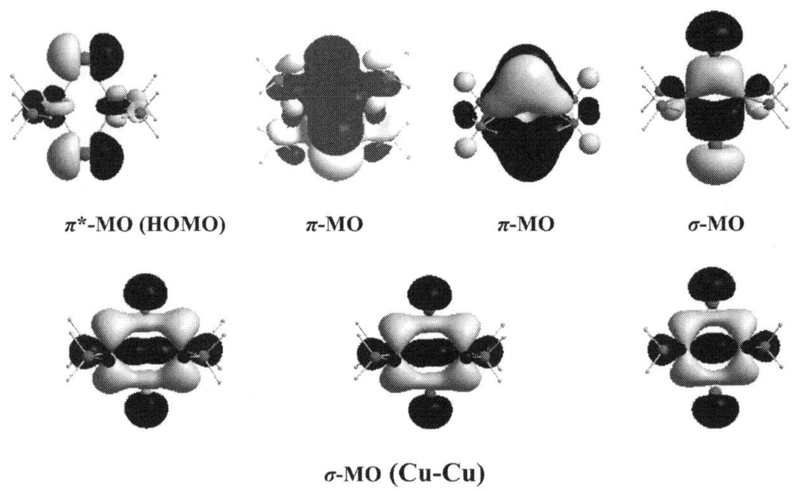

SCHEME 4.3 The most important molecular orbitals of the halo-bridged copper(I) dimers.

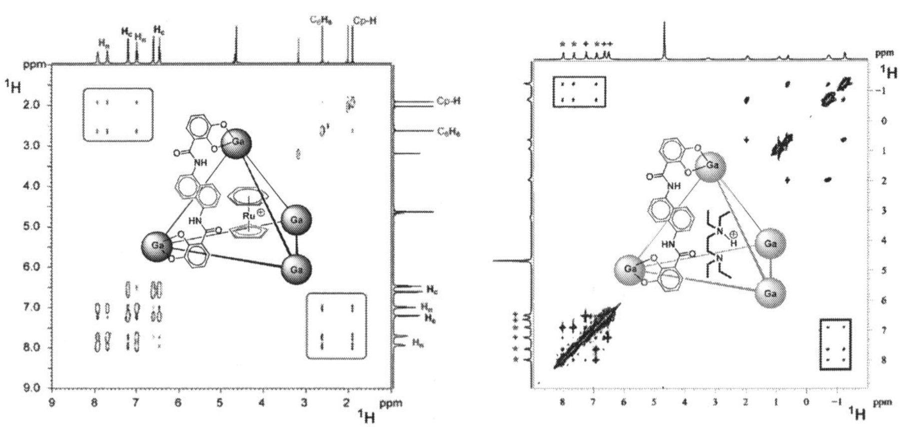

FIGURE 5.6 ¹H-NOESY spectra of the supramolecular adducts formed by the multianionic Ga_4L_6 host and organometallic (left) and organic (right) guests. HDO cross-peaks are deleted for clarity in both spectra. (Adapted from Pluth et al., *J. Am. Chem. Soc.*, 129, 11459–11467, 2007; Fiedler et al., *Inorg. Chem.*, 43, 846–848, 2004.)

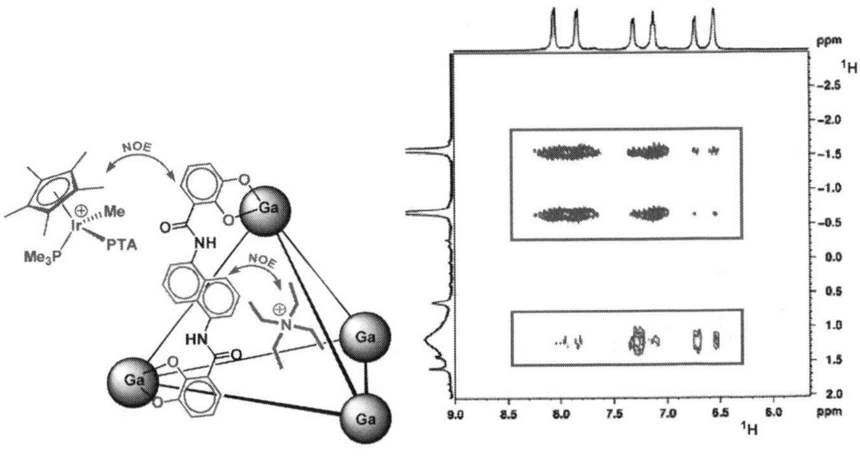

FIGURE 5.7 A section of the ^{1}H-NOESY spectrum of the anionic inclusion complex ($Ga_4L_6 \subset NEt_4^+$) ion paired with an external Cp*(PMe_3)Ir(Me)(PTA)$^+$ organometallic cation. The red box highlights correlations between encapsulated NEt_4^+ and the three symmetry equivalent Ga_4L_6 host naphthyl protons. The blue box highlights correlations between broad exterior Cp* peaks of the organometallic cation at 1 ppm and the three symmetry equivalent Ga_4L_6 host catecholate protons. (Adapted from Leung et al., *J. Am. Chem. Soc.*, 128, 9781–9797, 2006.)

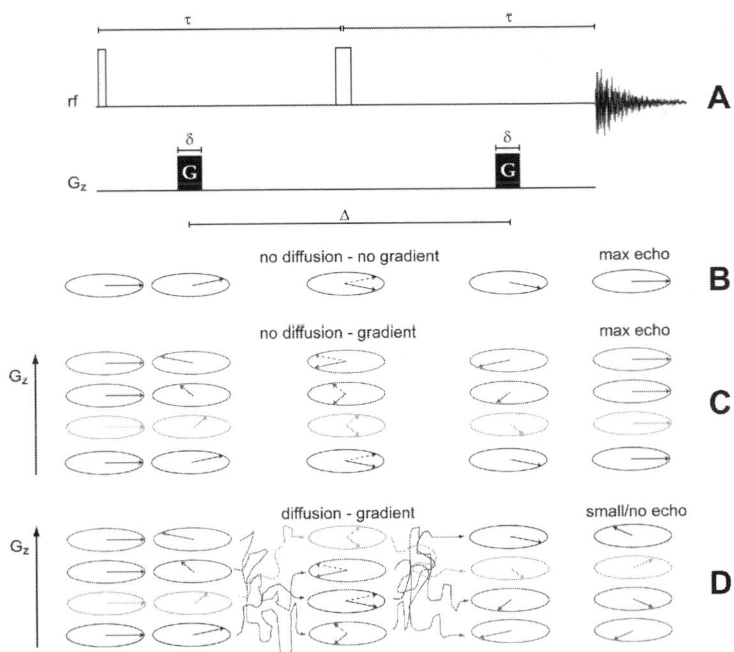

FIGURE 5.18 (a) Pulse sequence for the PGSE experiment. (b–d) Vectorial representation of the time evolution of the magnetization in the *xy* plane in the reference rotating frame in the absence of diffusion and gradients (b), absence of diffusion and presence of gradients (c), and presence of diffusion and gradients (d). Colors (black, green, red, and blue) represent hypothetical sample slices subjected to different values of the gradient strength.

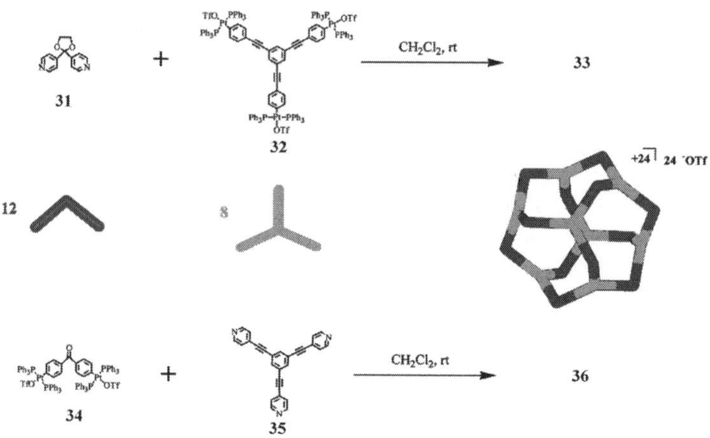

SCHEME 5.4

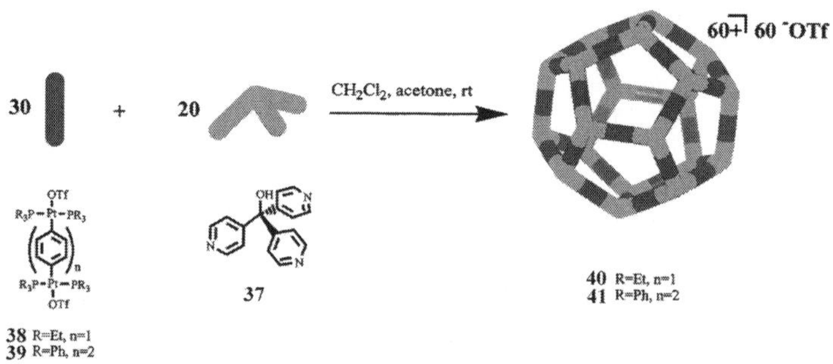

SCHEME 5.5

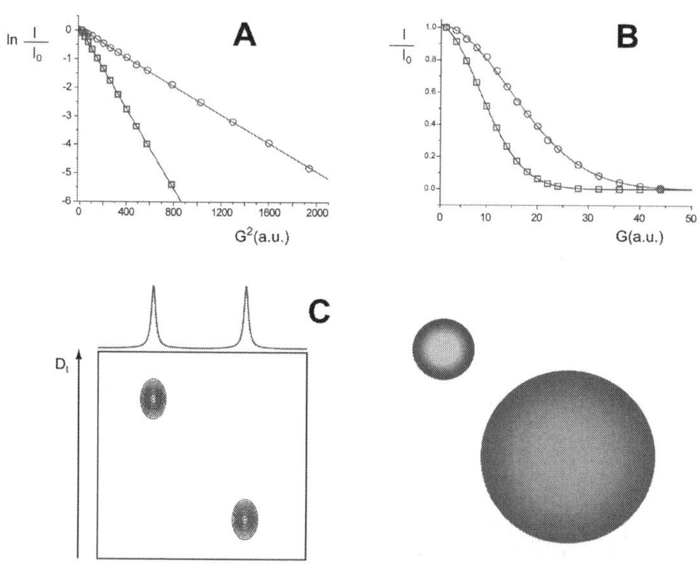

FIGURE 5.20 Schematization of PGSE data editing for a hypothetical mixture composed of a large (red), slow-diffusing species and a small (blue), fast-diffusing species.

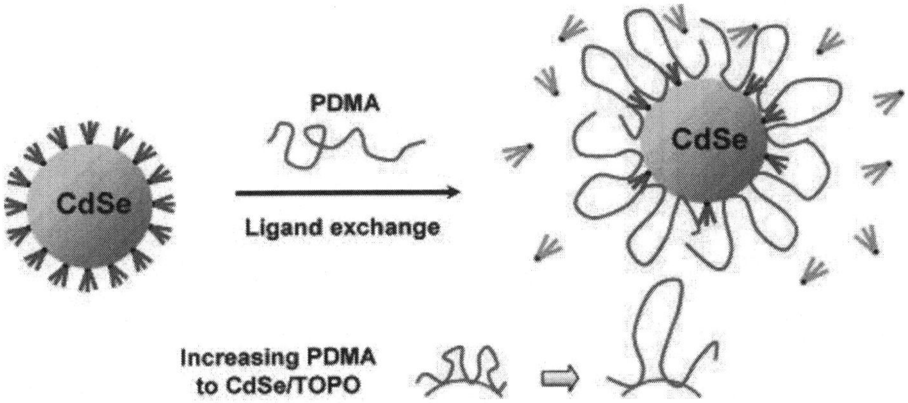

FIGURE 5.27 Sketch of the replacement of TOPO by PDMA on the surface of the QD and different binding conformation of PDMA. (Reproduced from Shen et al., *J. Phys. Chem. B*, 112, 1626–1633, 2008. With permission.)

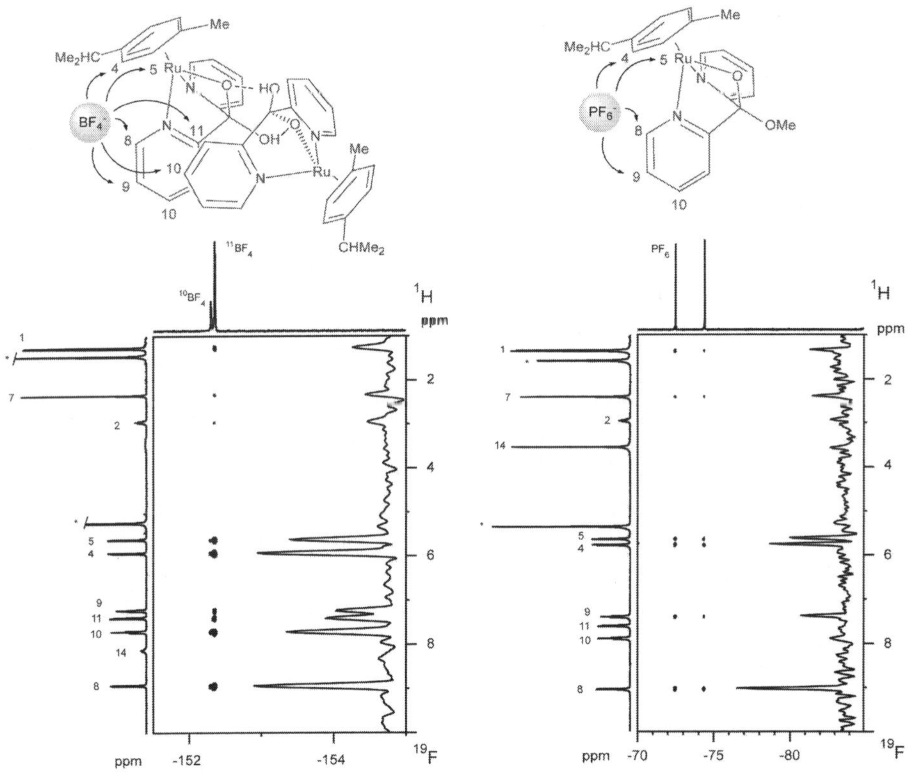

FIGURE 5.31 $^{49}F,^{1}H$-HOESY NMR spectra (376.65 MHz, 296 K) of complexes **48** (left) and **49** (right) in CD_2Cl_2. The asterisk denotes the residues of nondeuterated solvent and water. (Reproduced from Zuccaccia et al., *Organometallics*, 26, 6099–6105, 2007. With permission.)

Index

E

F

G

H

I

J

K

L

M

O

P

Q

R

S

An environmentally friendly book printed and bound in England by www.printondemand-worldwide.com

PEFC Certified
This product is from sustainably managed forests and controlled sources
www.pefc.org
PEFC
PEFC/16-33-415

FSC
www.fsc.org
MIX
Paper from responsible sources
FSC® C004959

This book is made entirely of sustainable materials; FSC paper for the cover and PEFC paper for the text pages.

#0091 - 310713 - C0 - 234/156/13 [15] - CB